The Biotol Project

The BIOTOL team

OPEN UNIVERSITEIT, THE NETHERLANDS
Prof M. C. E. van Dam-Mieras
Prof W. H. de Jeu
Prof J. de Vries

UNIVERSITY OF GREENWICH (FORMERLY THAMES POLYTECHNIC), UK
Prof B. R. Currell
Dr J. W. James
Dr C. K. Leach
Mr R. A. Patmore

This series of books has been developed through a collaboration between the Open universiteit of the Netherlands and University of Greenwich (formerly Thames Polytechnic) to provide a whole library of advanced level flexible learning materials including books, computer and video programmes. The series will be of particular value to those working in the chemical, pharmaceutical, health care, food and drinks, agriculture, and environmental, manufacturing and service industries. These industries will be increasingly faced with training problems as the use of biologically based techniques replaces or enhances chemical ones or indeed allows the development of products previously impossible.

The BIOTOL books may be studied privately, but specifically they provide a cost-effective major resource for in-house company training and are the basis for a wider range of courses (open, distance or traditional) from universities which, with practical and tutorial support, lead to recognised qualifications. There is a developing network of institutions throughout Europe to offer tutorial and practical support and courses based on BIOTOL both for those newly entering the field of biotechnology and for graduates looking for more advanced training. BIOTOL is for any one wishing to know about and use the principles and techniques of modern biotechnology whether they are technicians needing further education, new graduates wishing to extend their knowledge, mature staff faced with changing work or a new career, managers unfamiliar with the new technology or those returning to work after a career break.

Our learning texts, written in an informal and friendly style, embody the best characteristics of both open and distance learning to provide a flexible resource for individuals, training organisations, polytechnics and universities, and professional bodies. The content of each book has been carefully worked out between teachers and industry to lead students through a programme of work so that they may achieve clearly stated learning objectives. There are activities and exercises throughout the books, and self assessment questions that allow students to check their own progress and receive any necessary remedial help.

The books, within the series, are modular allowing students to select their own entry point depending on their knowledge and previous experience. These texts therefore remove the necessity for students to attend institution based lectures at specific times and places, bringing a new freedom to study their chosen subject at the time they need and a pace and place to suit them. This same freedom is highly beneficial to industry since staff can receive training without spending significant periods away from the workplace attending lectures and courses, and without altering work patterns.

SOFTWARE IN THE BIOTOL SERIES

BIOcalm interactive computer programmes provide experience in decision making in many of the techniques used in Biotechnology. They simulate the practical problems and decisions that need to be addressed in planning, setting up and carrying out research or development experiments and production processes. Each programme has an extensive library including basic concepts, experimental techniques, data and units. Also included with each programme are the relevant BIOTOL books which cover the necessary theoretical background.

The programmes and supporting BIOTOL books are listed below.

Isolation and Growth of Micro-organisms
Book: *In vitro* Cultivation of Micro-organisms
Energy Sources for Cells

Elucidation and Manipulation of Metabolic Pathways
Books: *In vitro* Cultivation of Micro-organisms
Energy Sources for Cells

Gene Isolation and Characterisation
Books: Techniques for Engineering Genes
Strategies for Engineering Organisms

Applications of Genetic Manipulation
Books: Techniques for Engineering Genes
Strategies for Engineering Organisms

Extraction, Purification and Characterisation of an Enzyme
Books: Analysis of Amino Acids, Proteins and Nucleic Acids
Techniques used in Bioproduct Analysis

Enzyme Engineering
Books: Principles of Enzymology for Technological Applications
Molecular Fabric of Cells

Bioprocess Technology
Books: Bioreactor Design and Product Yield
Product Recovery in Bioprocess Technology
Bioprocess Technology: Modelling and Transport Phenomena
Operational Modes of Bioreactors

Technological Applications of Immunochemicals

BOOKS IN THE BIOTOL SERIES

The Molecular Fabric of Cells
Infrastructure and Activities of Cells

Techniques used in Bioproduct Analysis
Analysis of Amino Acids, Proteins and Nucleic Acids

Principles of Cell Energetics
Energy Sources for Cells
Biosynthesis and the Integration of Cell Metabolism

Genome Management in Prokaryotes
Genome Management in Eukaryotes

Crop Physiology
Crop Productivity

Functional Physiology
Cellular Interactions and Immunobiology
Defence Mechanisms

Bioprocess Technology: Modelling and Transport Phenomena
Operational Modes of Bioreactors

In vitro Cultivation of Micro-organisms
In vitro Cultivation of Plant Cells
In vitro Cultivation of Animal Cells

Bioreactor Design and Product Yield
Product Recovery in Bioprocess Technology

Techniques for Engineering Genes
Strategies for Engineering Organisms

Principles of Enzymology for Technological Applications
Technological Applications of Biocatalysts
Technological Applications of Immunochemicals

Biotechnological Innovations in Health Care

Biotechnological Innovations in Crop Improvement
Biotechnological Innovations in Animal Productivity

Biotechnological Innovations in Energy and Environmental Management

Biotechnological Innovations in Chemical Synthesis

Biotechnological Innovations in Food Processing

A Compendium of Good Practices in Biotechnology

BIOTOL

BIOTECHNOLOGY BY OPEN LEARNING

Technological Applications of Immunochemicals

PUBLISHED ON BEHALF OF :

Open universiteit and **University of Greenwich (formerly Thames Polytechnic)**

Valkenburgerweg 167
6401 DL Heerlen
Nederland

Avery Hill Road
Eltham, London SE9 2HB
United Kingdom

Butterworth-Heinemann Ltd
Linacre House, Jordan Hill, Oxford OX2 8DP

A member of the Reed Elsevier group

OXFORD LONDON BOSTON
MUNICH NEW DELHI SINGAPORE SYDNEY
TOKYO TORONTO WELLINGTON

First published 1994

British Library Cataloguing in Publication Data
A catalogue record for this book is
available from the British Library

Library of Congress Cataloguing in Publication Data
A catalogue record for this book is
available from the Library of Congress

ISBN 0 7506 0508 1

Composition by University of Greenwich
(formerly Thames Polytechnic)
Printed and Bound in Great Britain by
Martins the Printers, Berwick-upon-Tweed

Contributors

AUTHOR

Dr L. S. English, University of the West of England, Bristol, UK

EDITORS

Prof M. C. E. van Dam-Mieras, Open universiteit, Heerlen, The Netherlands

Dr C. K. Leach, De Montfort University, Leicester, UK

SCIENTIFIC AND COURSE ADVISORS

Prof M. C. E. van Dam-Mieras, Open universiteit, Heerlen, The Netherlands

Dr C. K. Leach, De Montfort University, Leicester, UK

ACKNOWLEDGEMENTS

Grateful thanks are extended, not only to the authors, editors and course advisors, but to all those who have contributed to the development and production of this book. They include Mrs A. Allwright and Miss J. Skelton.

The development of this BIOTOL text has been funded by **COMETT, The European Community Action Programme for Education and Training for Technology**. Additional support was received from the Open universiteit of The Netherlands and by University of Greenwich (formerly Thames Polytechnic).

Contents

How to use an open learning text

An open learning text presents to you a very carefully thought out programme of study to achieve stated learning objectives, just as a lecturer does. Rather than just listening to a lecture once, and trying to make notes at the same time, you can with a BIOTOL text study it at your own pace, go back over bits you are unsure about and study wherever you choose. Of great importance are the self assessment questions (SAQs) which challenge your understanding and progress and the responses which provide some help if you have had difficulty. These SAQs are carefully thought out to check that you are indeed achieving the set objectives and therefore are a very important part of your study. Every so often in the text you will find the symbol Π, our open door to learning, which indicates an activity for you to do. You will probably find that this participation is a great help to learning so it is important not to skip it.

Whilst you can, as an open learner, study where and when you want, do try to find a place where you can work without disturbance. Most students aim to study a certain number of hours each day or each weekend. If you decide to study for several hours at once, take short breaks of five to ten minutes regularly as it helps to maintain a higher level of overall concentration.

Before you begin a detailed reading of the text, familiarise yourself with the general layout of the material. Have a look at the contents of the various chapters and flip through the pages to get a general impression of the way the subject is dealt with. Forget the old taboo of not writing in books. There is room for your comments, notes and answers; use it and make the book your own personal study record for future revision and reference.

At intervals you will find a summary and list of objectives. The summary will emphasise the important points covered by the material that you have read and the objectives will give you a check list of the things you should then be able to achieve. There are notes in the left hand margin, to help orientate you and emphasise new and important messages.

BIOTOL will be used by universities, polytechnics and colleges as well as industrial training organisations and professional bodies. The texts will form a basis for flexible courses of all types leading to certificates, diplomas and degrees often through credit accumulation and transfer arrangements. In future there will be additional resources available including videos and computer based training programmes.

Preface

The ability of the immune system to distinguish between self and non-self and, in many instances, between normal and changed self is of fundamental importance to the well-being of individuals. These abilities are of fundamental importance in protecting animals against biological and chemical threats that may be present in the environment and from the consequences of molecular and cellular dysfunction arising in their tissues. The system pocesses these remarkable properties through a wide range of sophisticated physiological, cellular and molecular mechanisms. These mechanisms and their role in defence are the topics of a related BIOTOL text 'Cellular Interactions and Immunobiology'. The importance of the immune system to human society is not, however, confined to its natural roles in defence. It has both commercial and social importance way beyond these natural roles. Increasingly, the cells and molecular products of the immune system are being employed in a wide variety of activities. This text describes the practical applications of the products and components of the immune system.

This text has been written on the assumption that the reader is familiar with the cells and molecules of the immune system and the processes by which the immune system protects individuals. The ideal background is provided by the BIOTOL text 'Cellular Interactions and Immunobiology' but there are a number of other texts on immunology on market which would provide a suitable background. Despite this assumption, the first chapter gives a general overview of this system ensuring that readers have the core information that is necessary for comprehension of material covered in later chapters.

The detection and measurement of antigen-antibody reactions and their application in a wide variety of situations are described in Chapter 2. Coverage includes descriptions of precipitation and agglutination techniques, enzyme - and radiolabelled - immunoassays, immunohistochemical techniques and the use of antibodies in Western blotting.

In Chapter 3, we discuss the use of conventional immunisation procedures for producing antibodies. We extend this discussion in Chapter 4 to hybridoma technology and the production of monoclonal antibodies. The potential benefits of modifying antibody structure through molecular and genetic techniques are described in Chapter 5. We complete our discussion of monoclonal and engineered antibodies in Chapter 6 in which the strategies and techniques for purifying antibodies from hybridoma and fermentation broths are described.

In Chapter 7 we describe, in outline, the discovery and diversity of cytokines and their biological properties. Included in this chapter are the methods used to measure cytokines and strategies used for their production *in vitro*. We also describe the potential of using cytokines to achieve clinical objectives. A recurrent theme throughout the text is the application of immunochemicals in both clinical and non-clinical circumstances. In Chapter 8 we bring together these applications and describe some prospective applications of immunochemicals.

The development of new techniques to produce immunochemicals and the introduction of procedures which enable us to modify the structures and activities of antibodies have important consequences in the applications of immunochemicals in vaccination, diagnostics and therapeutics. These developments have, however, greatly extended the importance of immunochemicals way beyond their conventional applications in medicine and health care. This text will enable readers to gain the knowledge necessary for them to contribute to these economically and socially important developments.

Scientific and Course Advisors: Professor M.C.E. van Dam-Mieras
Dr C.K. Leach

Overview of the immune system

Overview of the immune system

1.1 General introduction

This text is the second of two BIOTOL texts on aspects of immunity. In the first, 'Cellular Interactions and Immunobiology', the mammalian immune system is examined in depth covering the physiological properties and the cellular and molecular mechanisms which operate within the system. This second text has been produced on the assumption that the reader has either studied 'Cellular Interactions and Immunobiology' or has gained an understanding of the immune system from studies elsewhere. The emphasis here has been placed on the use of immunochemicals in a wide variety of applications. Nevertheless, the text begins with a review of mammalian immunity to provide a context in which the remainder of the text may be studied.

The information density in this first chapter is high and those with no previous experience of the immune system will perhaps find it difficult. If this is the case, we would recommend that the reader first study the text 'Cellular Interactions and Immunobiology' or, as an alternative, Chapters 5 and 6 from the BIOTOL text 'Defence Mechanisms'. Either will provide a sound appreciation of immune mechanisms.

For those who feel that they are already familiar with the immune system, this first chapter will provide opportunities for revision and updating.

The chapter begins with a consideration of T and B cells and the nature of antigens and immunogens and moves on to consider the roles of B cells in humoral responses and T cells in cell mediated responses. It also examines the relationship between innate and acquired (adaptive) immunity. The interactions between cells involved in the immune response including the nature and importance of Major Histocompatibility Complex (MHC) are also described. This naturally leads to a discussion of the processing and presentation of antigens by which intercellular antigens associate with MHC Class II components and stimulate B cell activation whilst intracellular antigens associate with MHC Class I components and stimulate T cell activation.

The chapter also gives an account of the structures of antibodies and describe their biological functions. The molecular mechanisms by which a very large repertoire of antibody specificities are derived are described. In the final part of the chapter, we briefly outline the physiology of the immune response.

This is a long chapter reflecting both the complexity of the immune response and also the extent of our knowledge of this system. We recommend that readers do not attempt to complete this chapter in just one sitting. It is better to tackle it in stages and ensure that the concepts have been fully understood.

1.2 General features of immunity

1.2.1 T and B cell receptors and nonself

During development, animals learn to recognise their own tissue components which are exposed to the immune system as being self. Consequently any organic molecule which gains entry into the animal (we will call the animal or Man, the host) and is not structurally identical, is recognised as being nonself or foreign. Such a compound is called an antigen. Antigens are specifically recognised by T and B lymphocytes carrying

host

foreign

cell surface receptors which bind to antigens. When an antigen invades the host, the T and B lymphocytes which bear receptors specific for some part of the antigen initiate an immune response in an attempt to eliminate the invader.

1.2.2 Antigens and immunogens

immunogens

Antigens which induce an immune response are called immunogens. These antigens are foreign to the host, generally of high molecular weight and chemically complex. The majority are proteins, the next major group are polysaccharides and a few are lipoproteins. Many antigens of low molecular weight fail to induce a response on their own. However, if they are covalently linked to an immunogenic molecule such as a protein they can then induce the production of specific antibodies or activate T cells. Such antigens are called haptens and the immunogenic molecule to which they are bound, the carrier. We can therefore conclude that all immunogens are antigens but not all antigens are immunogens. The true definition of an antigen is that an antigen is a component which is bound by antibody or T cell receptor whereas an immunogen can be defined as a component or molecule which induces an immune response.

haptens

carrier

The majority of immune responses in the lifetime of an individual are to micro-organisms such as bacteria, viruses, yeasts and protozoa which are very complex carrying many different antigenic molecules. These include bacterial capsular antigens, O and H antigens, teichoic acids and other polysaccharides, viral haemagglutinins and coat proteins and many others. These are obviously structurally very different from the structural components of Man and generally induce very strong immune responses. Non-microbial antigens include many plant products, insect and snake venoms and antigens in food to name just a few.

Very minor structural differences are also sufficient to induce an immune response. For instance, mouse albumin is immunogenic in Man reflecting minor differences in the amino acid sequences of mouse and human albumin. The A and B blood group antigens only differ by a single sugar residue but induce anti-A and anti-B specific antibodies in B and A individuals respectively.

SAQ 1.1

Which of the following items would you consider to be immunogenic in Man ie would induce antibodies?

1) Human albumin.

2) An ABO-compatible blood transfusion.

3) *Streptococcus pneumoniae.*

4) Killed virus.

5) Host lymphocytes.

6) Donor cells from another individual.

1.2.3 B cells and the humoral response

antibodies

epitope

affinity

cross-reactive

B cells respond to antigens by producing copious amounts of antibodies which are basically soluble forms of the B cell receptors. Each receptor and the corresponding antibodies only bind to a small area of the whole antigenic molecule called an epitope which is about the size of a hexasaccharide or hexapeptide. Most antigens have many epitopes and thus induce many B cells to respond since each one is specific for one particular epitope. In some instances, two epitopes may be structurally related and antibodies specific for one epitope may also bind the other albeit with a different binding strength or affinity. We call these antibodies cross-reactive antibodies. We have summarised these ideas in Figure 1.1.

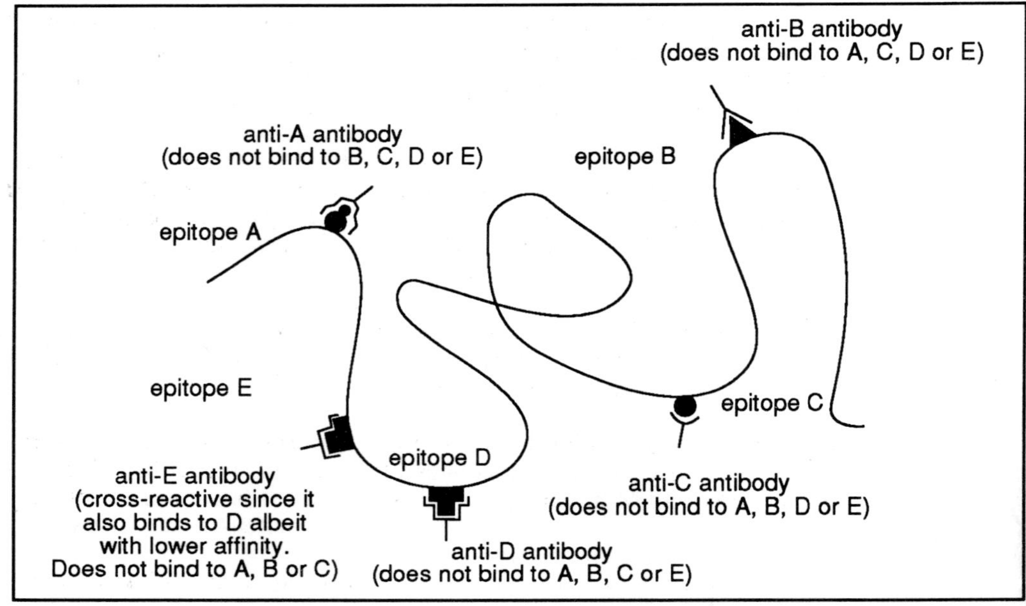

Figure 1.1 Antibodies are specific for epitopes on antigens. Note that, although anti-E will bind to the closely related epitope D, anti-D does not bind to E. Also note that antibodies are not usually named as epitope specific but antigen specific. For instance, if the antigenic molecule carrying all these epitopes was albumin, then all the antibodies would be anti-albumin antibodies.

dominant epitopes

The dominant epitopes, which can be of linear or conformational form, are generally exposed on the antigen surface where they can be bound by specific B cell receptors inducing antibody production. These antibodies are pivotal in causing destruction of whole antigen, be it a protein or microbe, because they can bind to the antigen immediately it gains entry to the host. Once antigens are broken down in the host, antibodies may also be generated to internal or hidden epitopes but these are not protective antibodies since they do not bind to the intact antigens.

1.2.4 T cells and the cell mediated response

T cell receptors bind MHC + peptide

T cells, in direct contrast, do not respond to extracellular antigens. T cell receptors, although antigen specific like antibodies, are not antibodies. Unlike B cells, T cells cannot 'see' native antigens such as whole bacteria, viruses or foreign proteins but recognise short peptides derived from these antigens which are expressed on cells such as those infected with micro-organisms, tumour cells and foreign cells (transplants). These peptides are always found attached to specialised molecules on the cell surface which belong to the Major Histocompatibility Complex (MHC) and which are divided into MHC Class I and MHC Class II.

cell mediated responses, cytotoxic T cells (T_C), delayed type hypersensitivity T cells (T_{DTH}), cytokines

In cell mediated responses, recognition by T cells of MHC I-peptides results in killing of the cell expressing the complex. Examples of cells that are killed by this mechanism are virally infected cells, tumour cells and foreign cells. The T cells responsible are called cytotoxic T cells (T_c). Cells infected with non-viral pathogens express MHC II-peptides which are recognised by different T cells called delayed type hypersensitivity T cells (T_{DTH}). These cells secrete mediators called cytokines which activate the infected cell resulting in destruction of the intracellular pathogens.

helper T cells T_H

T suppressor

Other T cells (helper T cells T_H) control the responses of B cells to antigen, again by cytokines. In fact, the majority of B cells cannot respond in their absence. A final subpopulation, T suppressor (T_S) cells negatively regulate responses of B cells and those of all the other T cells ie they generally prevent responses getting out of control. The basic mechanisms for these T cell responses are shown in Figure 1.2.

The picture given in Figure 1.2 is somewhat simplified. We will learn later that, for example, the responses of B cells to antigen are not only regulated by T_H and T_S cells but also depend upon the processing of antigens/immunogens by so called antigen presenting cells (APCs). We will describe these processes in a later section.

Π One way of trying to remember the information given in Figure 1.2 is to redraw it for yourself onto a fresh sheet of paper. You will find this useful to refer to when you are reading later sections.

Many responses involve both humoral and cell mediated elements. For example, responses to many viruses involve cytotoxic T cells and anti-viral antibodies. Figure 1.3 should help you to distinguish between B and T cell recognition of antigens.

Π In Figure 1.3 there is a cell which might be described as an 'antigen-presenting' cell. Which one is it?

You should have identified cell B. It takes in antigens and process them to produce low molecular weight products which become associated with MHC molecules on the surface of the cell. These peptides, associated with MHC molecules on the surface of the cell bind to receptors on T cells. In other words, cell B represents a cell which 'presents' the antigen to the T cell.

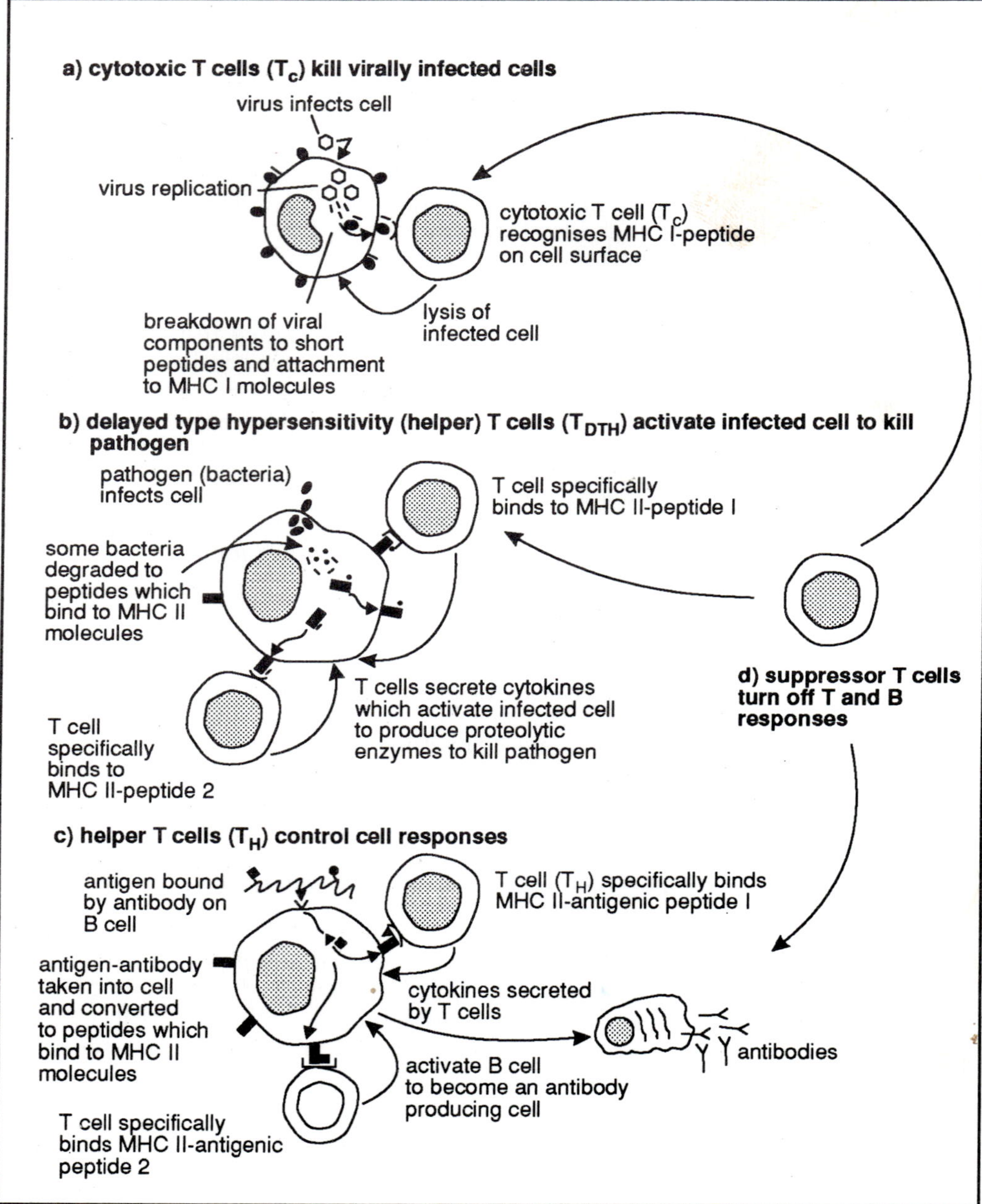

Figure 1.2 Major activities of T cells. Note that T cells only recognise antigen in the form of short peptides attached to either MHC I (cytotoxic T cells) or MHC II (helper T cells). Also notice that T cells are antigen specific - they only bind to one peptide attached to MHC. This is indicated in b) and c) but is also true for a) and d).

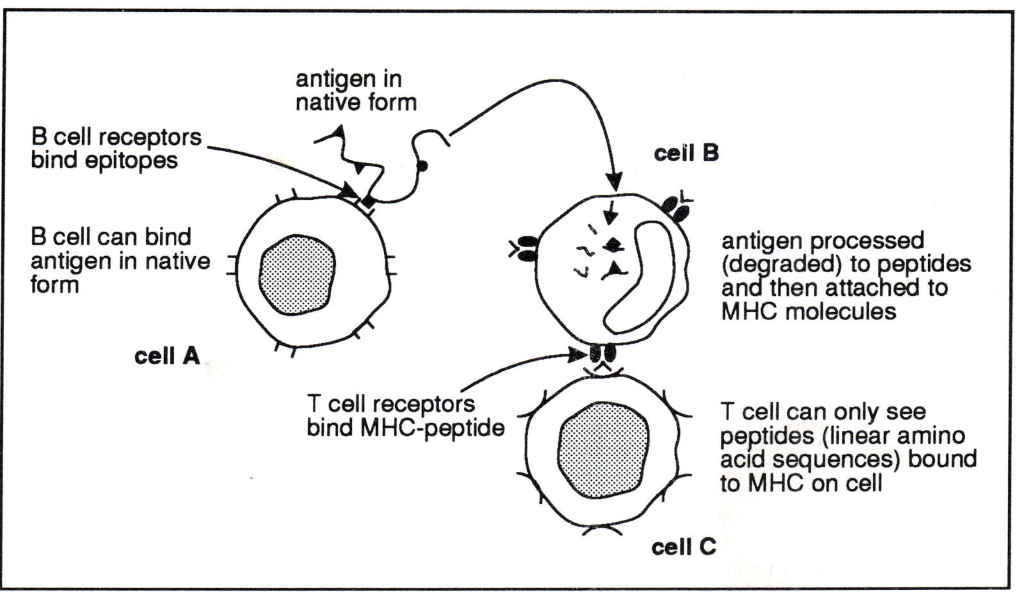

Figure 1.3 Binding characteristics of T and B cell receptors. For A, B and C, see intext activity.

Π Make a list of all the different types of lymphocytes, their receptors and the main functions of these lymphocytes and receptors.

Type of lymphocyte	Major function(s)
B cell	Produce antigen specific antibodies which bind to antigen and promote its destruction.
B cell receptor	Modified form of antibody which can bind to native (whole) antigen.
Helper T cell (T_H)	Activates B cells to respond to antigen. Cytokines, secreted by T cells, promote development of memory B cells and antibody producing plasma cells from progeny of activated B cell. Also activates other T cells.
T_H receptor	Recognises MHC II - peptide on B cell.
Cytotoxic T cell (T_C)	Kill virally infected cells, tumour cells and foreign cells (eg transplants).
T_C receptor	Recognises MHC I-peptide on cell surface.
Delayed hypersensitivity T cell (T_{DTH})	Possibly the same population as T_H. Activates cells infected with micro-organisms other than viruses to a high metabolic state (synthesis of enzymes and other agents) resulting in killing of pathogens inside the cell.
T_{DTH} receptor	Recognises MHC II-peptide on surface of infected cell.
Suppressor T cell (T_S)	Regulates the magnitude of T and B responses principally by suppressing T_H activities.
T_S receptor	Not well defined, maybe MHC I-peptide.

Table 1.1 Classification of antigen specific cells.

SAQ 1.2

Match the items in the left column with those in the right column using each item only once.

1) Conformational epitope	A) Amino acid sequence
2) MHC I-peptide	B) Helper T cell
3) Linear determinant	C) Antigen specific
4) MHC II-peptide	D) Cytotoxic T cell
5) Antibody	E) Molecular folding

1.2.5 T and B cell memory

T and B cells are continually being produced in the bone marrow during a major part of the life of an individual and each one has to encounter the antigen for which it is specific very early in its life or it does not survive.

primary response

clonal expansion

clone

When T and B cells first respond to antigen (the primary response) in either a humoral or cell mediated response, they undergo clonal expansion to produce various effector T cells or B plasma cells, the progeny of each cell being called a clone. This process, however, is rather slow taking a few days or even weeks. If a pathogen is involved, the individual may succumb to disease before sufficient levels of antibodies and/or T cells are produced which can successfully combat the effects of the pathogen leading to final recovery.

memory cells, secondary or anamnastic response, immunological memory, memory is antigen specific

However, when the individual encounters the antigen a second time, the situation is different since during the primary response some of the T and B cell progeny develop into memory cells which are able to react much faster and with more vigour to a subsequent incursion by antigen (secondary or anamnastic response). This usually results in the elimination of the antigen before it causes disease. Thus it seems that the immune system remembers the previous encounter with antigen and we say it has immunological memory. This T and B memory is antigen specific only affording protection against the antigen which induced the response.

Π Write down what you consider to be the two major characteristics of responses mediated by T and B cells.

The two major characteristics of the acquired immunity mediated by T and B cells are that these type of immunity are highly specific and that, once activated, these type of immunity involves memory.

1.2.6 Adaptive and innate immunity

acquired or adaptive immunity

Since immunity involving T and B cells is acquired by each individual by encountering antigens, we call these responses acquired or adaptive immunity. As we have seen, the major characteristics of this type of immunity are specificity and memory.

innate or natural immunity

There is another more primitive type of immunity which possesses neither specificity nor memory. This is called innate or natural immunity which means that an individual is born with this protection. It includes three major components, physical barriers, cells and soluble factors.

Because antibodies, in most instances, promote the destruction of antigens via activation of innate immunity, it is necessary for you to know the most important mechanisms and to realise that, in the absence of antibodies, this system may be totally ineffective in the fight against infection.

1.2.7 Innate immune mechanisms

Physical barriers

Physical barriers prevent access of some pathogens and other foreign matter to the tissues of the host. These include the skin and the epithelial/mucous layer lining the gut, respiratory tract and urinogenitary tract. Some pathogens, such as viruses have to gain entry to the epithelial cells to initiate infection whereas others, such as some of the bacterial genus *Streptococcus*, only need to colonise the external surface of the cells resulting in symptoms such as a sore throat. We show you some of the mechanisms in Figure 1.4.

Once the antigen, in most cases a pathogen, breaches the physical barriers it encounters other innate immune defence mechanisms which can be classified as either cellular or soluble in nature.

Cellular mechanisms

neutrophils and macrophages

reticulo-endothelial system

These include phagocytic cells such as neutrophils and macrophages which are attracted to the inflammatory site (where antigen is deposited) and which engulf and destroy some pathogens. Other cells, closely related to macrophages and forming the reticuloendothelial system are strategically placed in many areas of the body to intercept and destroy pathogens and other foreign matter in a similar manner. These include Kupffer cells lining the liver sinusoids through which blood flows, endothelial cells lining blood and lymph vessels and alveolar macrophages in the lungs.

phagocytes

bacterial adhesins

Phagocytes do not possess antigen specific receptors like those on T and B cells. There are, however, some nonspecific mechanisms which promote adhesion of some pathogens to these cells. Some bacteria possess bacterial adhesins on their surfaces which attach to phagocytes thus promoting engulfment of the pathogens. Some of the soluble factors discussed below also facilitate attachment of pathogens to the phagocytes.

Soluble factors

These include complement and acute phase proteins found in body fluids, lysozyme found in external secretions and interferons secreted by virally infected cells.

complement

chemotactic factors

opsonisation

Some pathogens activate complement resulting in the production of chemotactic factors which attract phagocytes to the site of infection. These cells possess receptors for C3b molecules which are products of complement activation and which attach to a wide range of pathogens by a process known as opsonisation. The phagocytes can then engulf and destroy the pathogens. Complement itself can also kill many pathogens without the help of phagocytic cells.

acute phase proteins

lysozyme

interferons

Acute phase proteins, produced in large quantities during infection, may bind to pathogens and promote killing sometimes by activation of complement. Lysozyme, found in external secretions such as saliva and tears breaks down bacterial cell walls. Finally, interferons, produced by virally infected cells, promote the synthesis of anti-viral proteins by neighbouring cells which may, in some cases, prevent the spread of the virus.

We have summarised the major features of innate immunity in Figure 1.4. We suggest that you spend some time on this figure and then attempt SAQ 1.3.

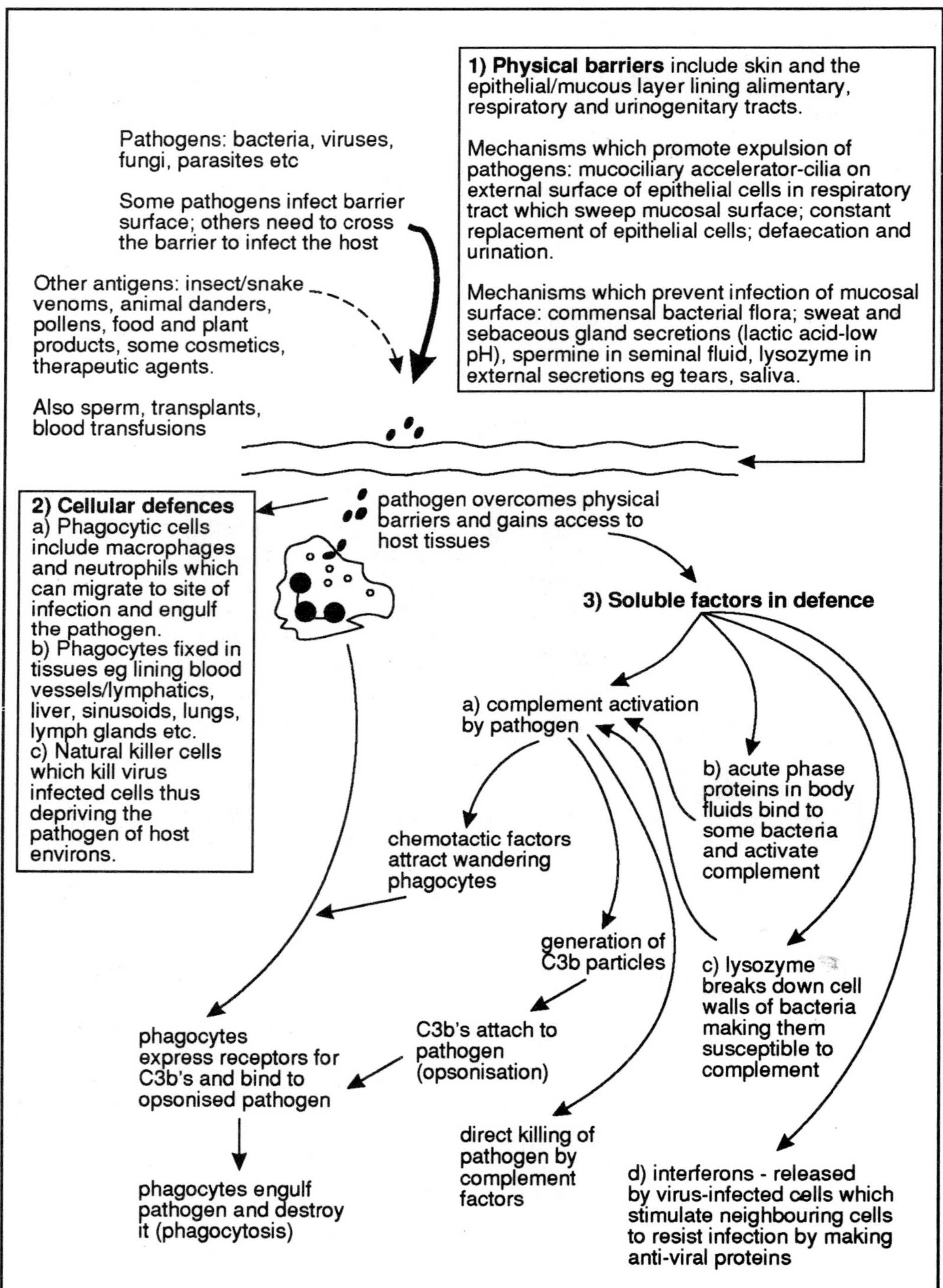

Figure 1.4 The physical barriers, cells and soluble components of innate immunity.

SAQ 1.3

Which one of the following statements is incorrect?

1) Mechanisms which may prevent infection of the host include defaecation, urination, presence of commensal bacterial species on the mucosa and the low pH of sebaceous secretions in skin.

2) Some acute phase proteins can promote killing of some pathogens by activating complement.

3) Opsonisation is the uptake of C3b coated pathogens by phagocytes and subsequent killing.

4) Antigens include low molecular weight agents such as some drugs which are not immunogenic.

5) Interferons do not kill viruses.

Innate immunity may be ineffective

In spite of its effectiveness against some pathogens, innate immunity provides little protection against many others for two reasons. Firstly, there is no antigen specificity and many antigens escape detection. Secondly, only a restricted number of pathogens can activate nonspecific innate immune mechanisms. For example, not all bacteria express bacterial adhesins recognised by phagocytes and only a few directly activate complement.

Π See if you can think of a way in which innate immunity can be enhanced so that phagocytic cells and complement can act on a much wider range of antigens?

One way would be as follows. If antibodies are produced which, when they bind to their target antigens, they might be able to activate innate immune mechanisms such as complement activation and phagocytosis. Thus the antibodies identify most antigens which enter the host and the innate immune mechanisms then destroy them - a perfect partnership!

1.2.8 The major components of immunity summarised

Π You have now been introduced to all the major participants in immune responses. Construct a diagram making suitable connections between the following: immunity subdivided into innate and adaptive immunity, then show components of innate immunity (cells and soluble factors only) and adaptive immunity (B cells, antibodies, T cell subpopulations and functions) finally show the connection between antibodies and innate immunity. Check your diagram with Figure 1.5, which also gives some additional information.

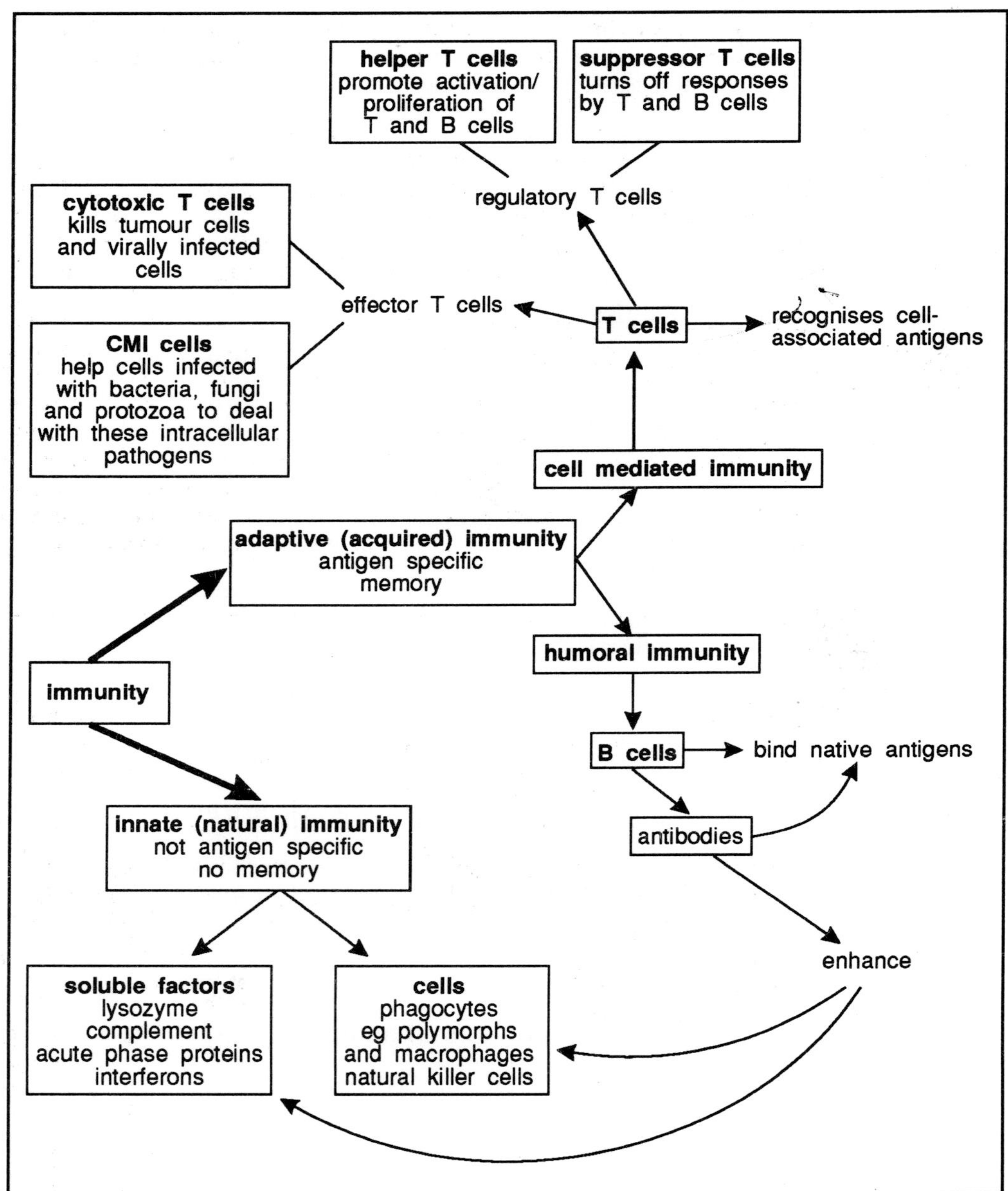

Figure 1.5 Innate and acquired immunity. CMI = cell mediated immune cells.

1.3 Cell interactions in B cell responses

We earlier described the humoral response as the activation of B cells followed by clonal expansion, antibody production and the generation of memory cells. All of these events require T cell help in a majority of B cell responses. Let us now examine the molecular basis for these T-B interactions. We need to discuss a rather important group of surface components called the Major Histocompatibility components. We discuss the organisation of the genes coding for these components first, then deal with their structures before examining their roles in the immune response.

1.3.1 MHC molecules

HLA (human leukocyte antigens)

The Major Histocompatibility Complex genes are situated on the short arm of chromosome 6 and are called HLA (human leukocyte antigens) in Man and on Chromosome 17 (called H-2) in mouse. We shall concern ourselves with HLA only. Figure 1.6 depicts a simplified version of this complex and shows that there are 3 classes of MHC genes together with other minor genes within this region. MHC Class III genes encode some complement products and are not involved with cell interactions.

HLA Class I loci

allogeneic

The HLA system is extremely polymorphic there being many alleles at each locus. There are 3 major HLA Class I loci, HLA-A, HLA-B and HLA-C of which there are about 25, 50 and 10 alleles respectively. Each individual inherits one allele from each locus from each parent making a total of 6 different MHC I molecules. You can conclude that there is little chance that any one individual inherits precisely the same set as another individual which means that we all express a different phenotype ie we are all allogeneic. Each of these MHC Class I genes encodes a single polypeptide α-chain of about 44kD which is expressed along with an invariant chain, β_2-microglobulin, on all nucleated cells and platelets. These products are involved with the activation of cytotoxic T cells.

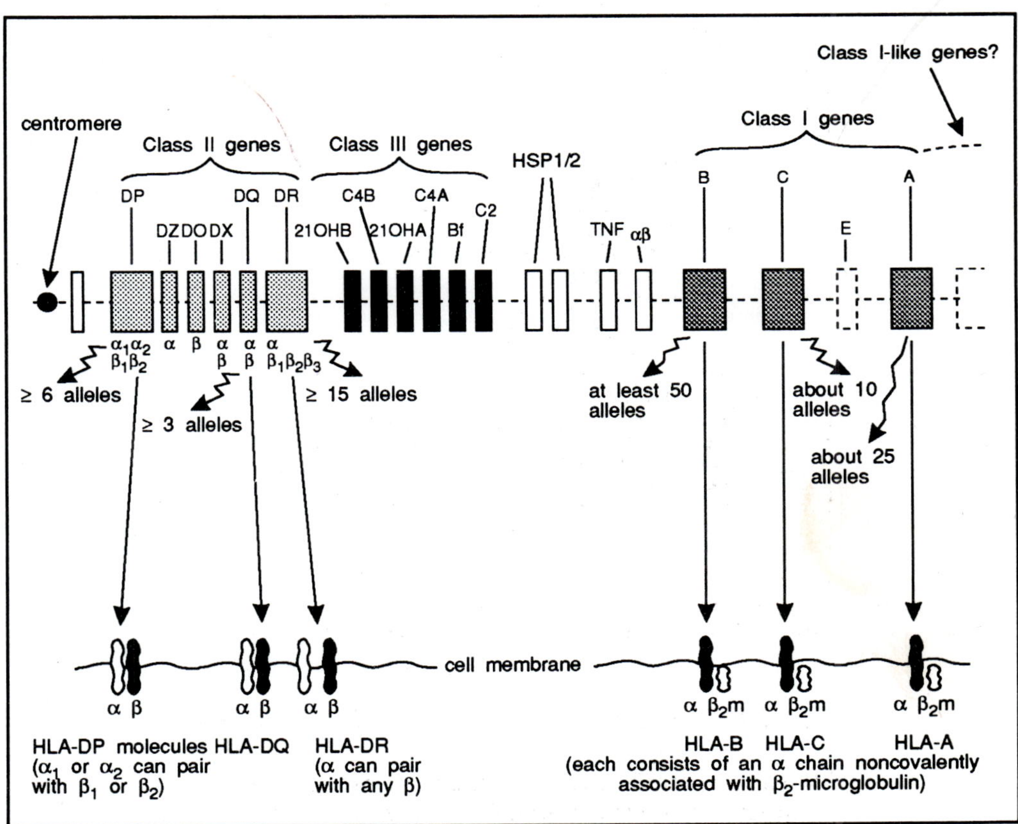

Figure 1.6 The HLA region and the gene products (not to scale). Just concentrate on the Class I and II genes. Notice that, as well as the DP, DQ and DR loci there are other ill defined loci (DZ, DO and DX) in MHC Class II.

MHC Class II, antigen presenting cells

All B cells constitutively (always) express many thousands of MHC Class II molecules on their surfaces. They are members of a group of select cells called antigen presenting cells (APCs) which, by expressing MHC Class II-peptide complexes, activate T cells.

The major MHC Class II gene loci are DP, DQ and DR which encode α and β polypeptides expressed as noncovalently associated $\alpha\beta$ pairs on the cell surface. Since there appears to be random association between α and β chains there may be 20 or more different MHC Class II products expressed on the surface of each antigen presenting cell. MHC Class II genes are also polymorphic like MHC Class I genes and individuals are likely to inherit a unique set. Each α and β chain possess 2 extracellular domains and the polymorphic regions of the α_1 and β_1 domains come together to form a binding cleft for the peptide (see Figure 1.7).

SAQ 1.4

Figure 1.7 shows the major structural features of the MHC Class I and II molecules. Study this figure and then complete the comparative table below indicating your choices with a + or -.

Characteristic	MHC I	MHC II
1) Associates with β_2-microglobulin		
2) 3 domains/chain		
3) Polymorphism		
4) Activates cytotoxic T cells		
5) Binds peptides		
6) Activates helper T cells		
7) Humoral immunity connection		
8) $\alpha\beta$ heterodimer		
9) Expressed on APCs		
10) Related to immunoglobulins		

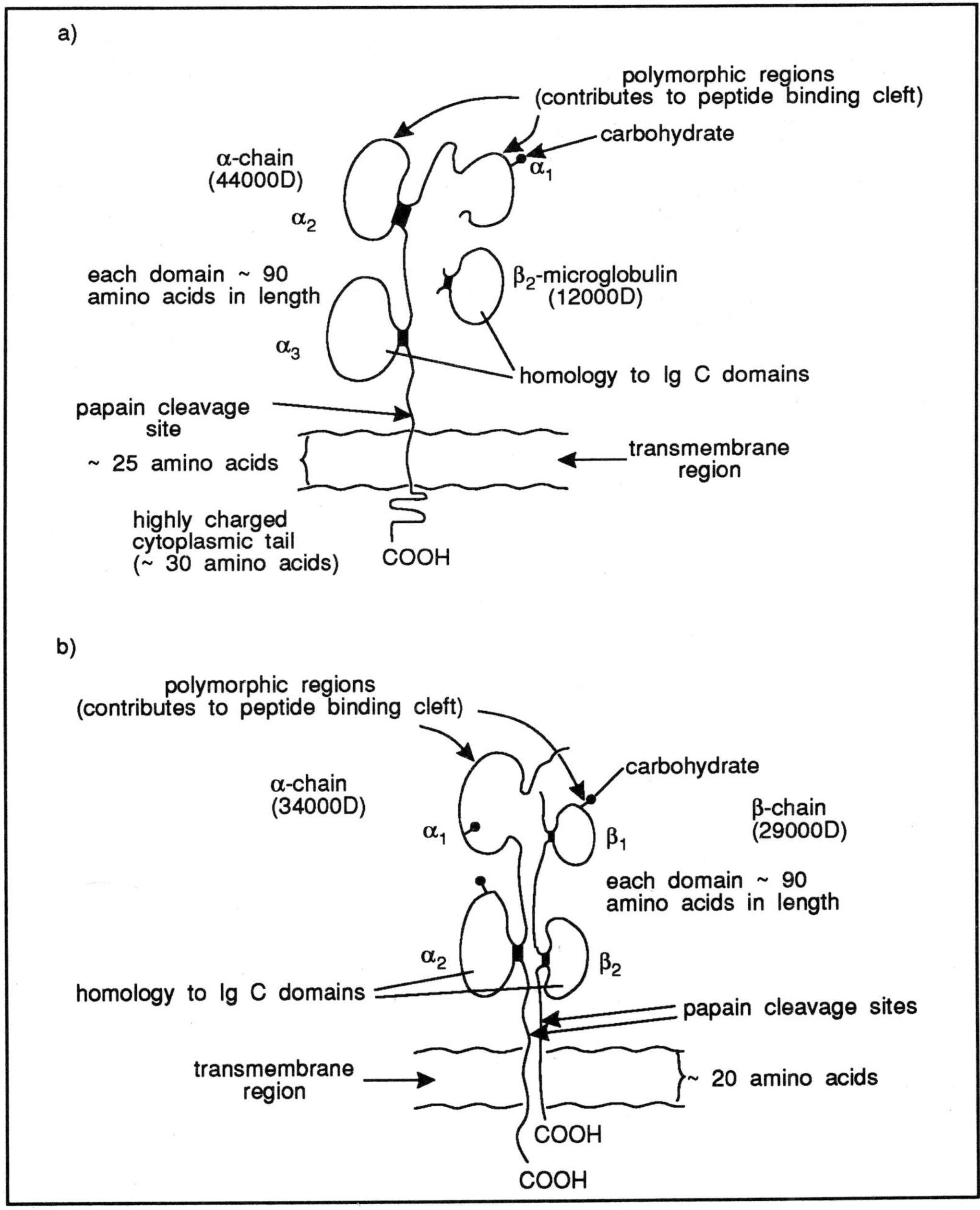

Figure 1.7 Structural features of MHC Class I and II molecules. a) Class I molecules, b) Class II molecules. ■ disulphide bonds.

1.3.2 Characteristics of peptide binding to MHC molecules

The whole sequence of each antigenic molecule is not converted into peptides but probably just a few short sequences. Because of the allelic variation between different MHC Class II molecules which will be reflected in the ability of the binding cleft to bind a particular peptide, some of these peptides will be bound, others will not. It is possible that more than one MHC allelic product eg DQ3 and DR2, can bind the same peptide.

Π Each MHC molecule can bind many different peptides, even unrelated peptides. Can you explain why this is necessary?

Since each individual only possesses a few, possibly no more than 20, different MHC Class II molecules (and only 6 different MHC I molecules) and there are potentially thousands of different foreign peptides which can result from degradation of the myriad of antigens in the environment, many unrelated peptides must bind to a single MHC molecule. This, of course, means that the binding affinities of MHC molecules are much lower than those of antibodies.

The last point we will make is that MHC molecules do not normally appear on the cell surface without a peptide attached to it. Since it is not likely that all APCs in an individual contain sufficient antigen to generate enough peptides to fill all the binding sites, we need another source of peptides. We now know that the binding sites of most MHC molecules, both Class I and II are occupied by self peptides generated by degradation of host proteins.

self peptides

Everything we have said about peptide binding to MHC Class II molecules also applies to MHC Class I molecules.

Π To emphasise the salient features of this section, make a list of the criteria for peptide binding.

You should have included:

- only parts of antigens become peptides;
- different MHC molecules may bind the same peptide;
- many different peptides can bind to each MHC molecule;
- self peptides compete with foreign peptides.

1.3.3 Antigen processing and presentation

Although we are mainly concerned with MHC Class II presentation as is the case for B cells, we will deal with MHC Class I as well as it is useful to compare the two processes. Current ideas are depicted in Figure 1.8. You should use this figure to follow the description in the text. You will find processing leading to MHC Class I presentation on the left of the diagram and that leading to MHC Class II presentation on the right.

target cells

endogenous antigens

proteasomes

peptide transporters

In cells presenting Class I-peptides which are called target cells, various cytoplasmic host (self) proteins and foreign proteins derived from, say, viruses replicating inside the cell (these are called endogenous antigens), are degraded into peptides in the cytoplasm. It is thought this takes place in complexes called proteasomes, the resulting peptides being transported to the endoplasmic reticulum by peptide transporters which then attach them to freshly synthesised MHC Class I molecules. As we said earlier, some peptides may not bind to the available MHC molecules. The MHC Class I-peptide complexes are then moved via the Golgi apparatus to the cell membrane for presentation to cytotoxic T cells.

exogenous antigens

In APCs such as B cells a different mechanism operates. Antigens outside the cell (exogenous antigens) become attached to the APC by a variety of mechanisms. These include antigens which have been opsonised by C3b particles as we have already seen.

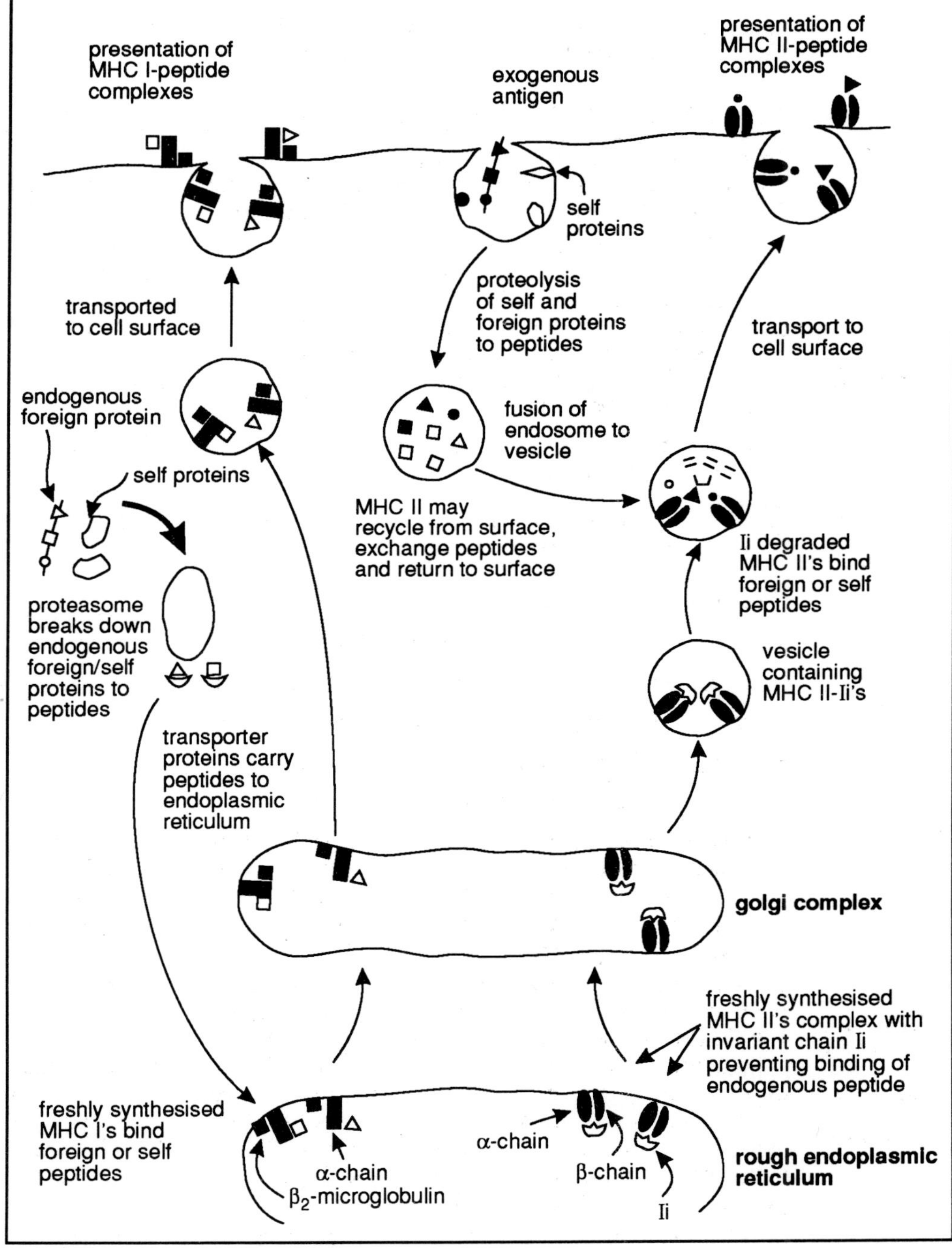

Figure 1.8 Antigen processing mechanisms. On the left hand side of the figure, we have shown the processing of endogenous antigens involving MHC I. On the right hand side, we show the processing of exogenous antigen involving MHC II. Ii = invariant chain.

We shall find out that antibodies can also opsonise antigens and the antibodies then become attached to some APCs. In the case of B cells the antigen specific antibody receptors attach to their respective epitopes on the antigen and the whole antigen-antibody complex is endocytosed into the cell by invagination of the cell membrane thus forming an endosome. Both membrane host proteins including the antibody receptors and foreign proteins are then degraded into peptides.

antigen-antibody complex is endocytosed

MHC Class II molecules are synthesised alongside MHC Class I molecules and are thus exposed to the same endogenous peptides. However, attachment of these peptides is prevented by invariant chains Ii which bind to the binding clefts of MHC Class II molecules. Vesicles now carry these MHC Class II-Ii complexes to the endosomes where the Ii is cleaved off and exogenous self or foreign peptides bind. The MHC Class II-peptide complexes are then expressed on the cell surface for presentation to T helper cells.

invariant chains Ii

Π How do you think the T cells distinguish between MHC I and II molecules binding foreign peptides from those with self peptides?

the thymus and auto-reactive T cells

auto-reactive cells

T cells start their development in the bone marrow like B cells but mature in the thymus. In this environment they are exposed to MHC Class I and II molecules expressing the range of peptides derived from host proteins which APCs and other cells normally produce. These auto-reactive cells are, for the most part, destroyed in the thymus so that only T cells which respond to MHC molecules binding foreign peptides develop into mature T cells. In some cases, we think that some auto-reactive T cells are present in individuals but these are suppressed by T cells so they do not respond and cause tissue damage.

SAQ 1.5

Which of the following statements is/are correct?

1) Transfection of human macrophages with the mouse serum albumin gene will result in presentation of mouse peptides on MHC Class II molecules.
2) Self peptides compete with foreign peptides for binding sites on both MHC Class I and II molecules.
3) T cells will always respond to MHC molecules if they are presented on cells with foreign peptide attached.
4) B cells are called antigen presenting cells because they are antigen specific.
5) Each peptide derived from endosomal processing in a B cell will only activate a single helper T cell.

1.3.4 The T cell receptor

TCR-2

Human T cell receptors (designated TCR-2) from a majority of T cells, including all T_H cells, are heterodimers of 2 polypeptide chains α and β of molecular weight 43-49kD and 38-42kD respectively joined together by a disulphide bridge. Each chain is divided into two sections of equal length called variable and constant domains. There is great amino acid sequence variability in 3 short stretches of each variable domain which accounts for the wide spectrum of specificities possessed by the TCRs of different T cells for the many MHC-peptide complexes.

1.3.5 Other signalling and adhesion molecules involved in the T-B interaction

All TCRs are always found associated with a complex of 5 polypeptides, the so-called CD3 complex which transmits activating signals to the cell when the TCR binds to MHC-peptide.

There are many different CD molecules found on various lymphoid cells and the classification CD (cluster of differentiation) simply means that they have been identified by many antibodies from different laboratories around the world. There are presently about 80 such molecules but for the purpose of this text you need to know very few. Two important ones are CD4 and CD8 which are found on T_H and T_C respectively.

During binding of TCR molecules to MHC II-peptide complexes, CD4 molecules on the T helper cells also attach to the invariant parts of MHC II molecules on the interacting B cell. This CD4 - MHC interaction may stabilise the TCR-MHC II-peptide binding especially if it is of low affinity or the CD4 itself may have a signalling role. CD8 plays a similar role when cytotoxic T cells bind cellular MHC I-peptide complexes.

Although these are the major players in the interaction of the T and B cells, it is now known that there are many other so-called adhesion molecules which serve to hold the cells close together and others which contribute to the activation process. It is beyond the scope of this short review to describe these in any detail although we have included some of them in Figure 1.9 for completeness. We should emphasise that most of these additional molecules are not unique to T and B cells. (Further details of these adhesion molecules are given in the BIOTOL text 'Cellular Interactions and Immunobiology').

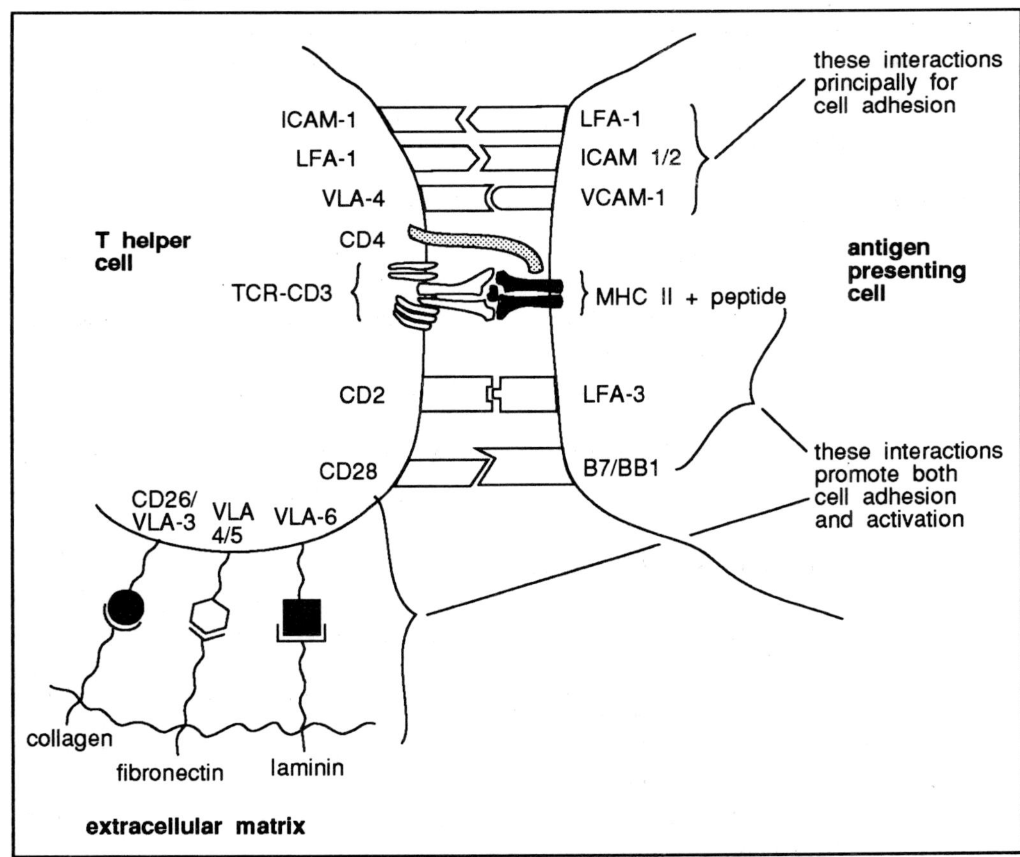

Figure 1.9 Molecules involved in cell interactions between T helper cells and antigen presenting cells. ICAM 1/2 intercellular adhesion molecule 1/2; LFA-1/3 lymphocyte function associated antigen 1/3; VCAM-1 vascular cell adhesion molecule-1; VLA 3-6 very late antigen 3-6.

1.4 Structure of antibodies

We shall now take a rest from T and B cells interactions for a while and examine the structure, function and genetics of antibodies in some detail. Armed with this knowledge we can then return to our study of cell interactions and complete the story on the humoral response.

1.4.1 The four chain monomeric unit of antibody structure

There are 5 classes of antibodies, IgG, IgA, IgM, IgE and IgD which can be distinguished by unique structural features and physiological properties.

binding sites for two identical epitopes

However, all antibodies are structurally similar being composed of one or more monomers of two identical heavy chains (50-70 000 Daltons) and two light chains (25 000 Daltons) joined together by disulphide bridges as shown in Figure 1.10. There are binding sites for two identical epitopes at the N-terminal end and the carboxyl end activates innate immune mechanisms such as complement and phagocytosis.

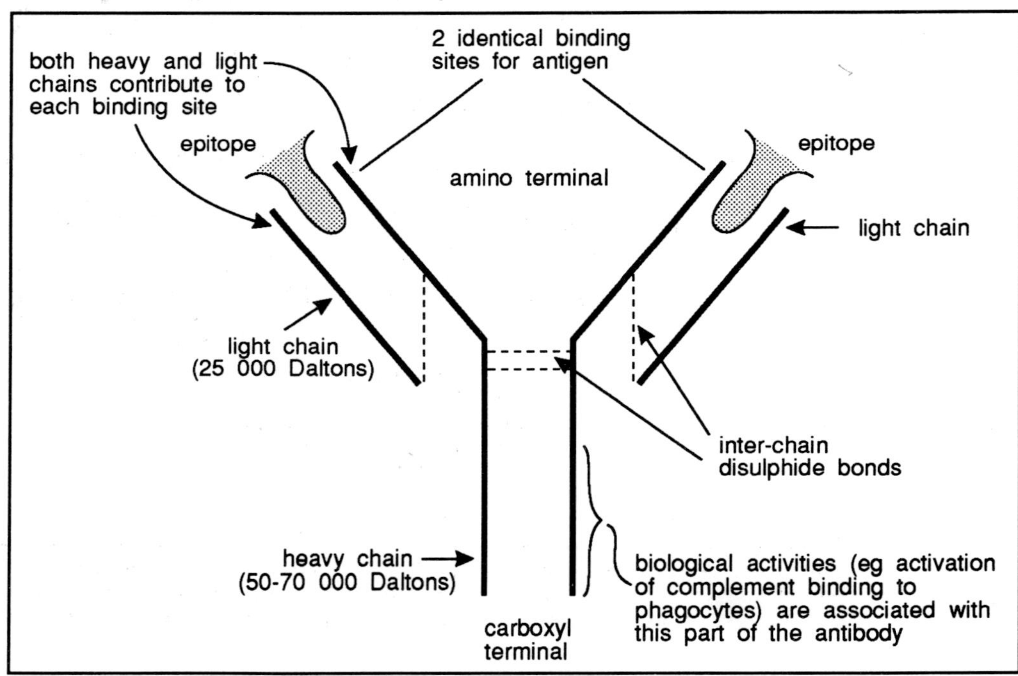

Figure 1.10 Basic structural and functional parts of an antibody molecule.

Π Close the text and construct a diagram of the monomeric antibody molecule including the following labelling: heavy chain, light chain, H-L S-S bridge, H-H S-S bridge, antigen binding site, N-terminal end, carboxyl terminal end. Check your drawing with Figure 1.10.

1.4.2 The domain structure of antibodies

Now use the stylised drawing of the IgG molecule in Figures 1.11 to follow the more detailed description of antibody structure given below. You will notice that both heavy and light chains are divided into repeating units of about 100-110 amino acid residues called domains each of which contains an intrachain disulphide bridge enclosing a loop of about 60 amino acids. The heavy chain has 4 such units, the light chain two. The majority of the domains interact with their opposite numbers on adjacent chains thus contributing to the tertiary structure of the molecule.

domains

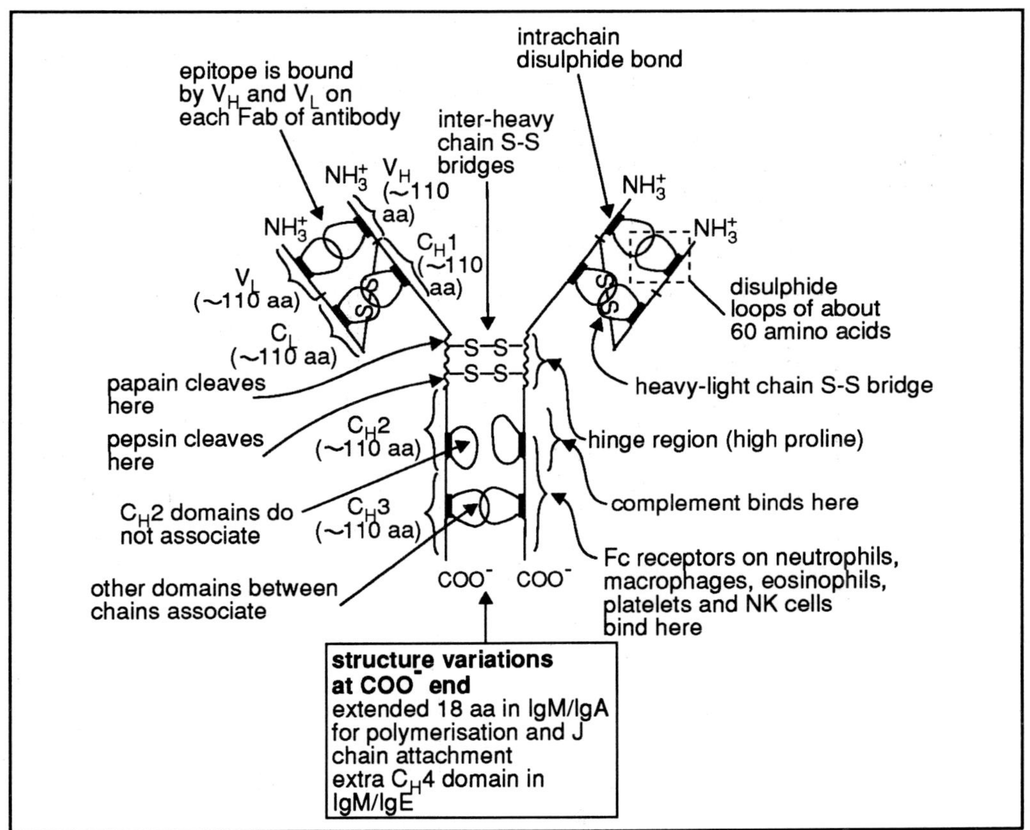

Figure 1.11 Major structural features of IgG. The repeating domain structure is shown and the disulphide links including the most common arrangement for the heavy-light chain link. The binding sites for complement and cellular Fc receptors which result in expression of biological function are indicated.

variable domains

Sequence analysis of light and heavy chains from antibodies of varying specificities has showed there is a high degree of sequence variability in the N-terminal domains and these have been designated V_L and V_H respectively.

Π How would you explain this sequence variability with respect to the function of the antibodies?

You will recall that we said earlier that the N-terminal ends of the antibodies bind the antigen. Hence the sequence variability within the variable domains from one antibody to another accounts for the antibody specificity for a particular epitope.

constant domains

The remaining domains show little sequence variability from one antibody to another and have been called constant domains, that of the light chain being designated C_L, the 3 constant domains of the IgG heavy chains C_H1, C_H2 and C_H3.

hinge region

Situated between C_H1 and C_H2 domains is a nonglobular hinge region high in proline and possessing varying numbers of disulphide bridges between the heavy chains. The hinge region allows a certain degree of independent movement of the N-terminal ends of the molecule containing the antigen binding sites.

Π Close your text and construct a fully labelled diagram of IgG. Check it against Figure 1.11.

1.4.3 Reduction and enzyme digestion of antibodies

The basic structure of antibodies was determined by Porter and Edelman using reducing agents and the enzymes papain and pepsin. A summary of the results of these treatments are shown in Figure 1.12 using IgG antibodies.

reducing agents

Treatment with reducing agents such as 2-mercaptoethanol or dithiothreitol leads to the cleavage of the interchain S-S bonds resulting in isolated light and heavy chains which can be separated chromatographically.

papain hydrolysis of IgG

Fab

Fd

When IgG is treated with the enzyme papain at neutral pH, two identical fragments called Fab (fragment-antigen binding) and one called Fc (fragment-crystallisable) each about 50kD, could be identified. The Fc is further degraded, with prolonged exposure to papain, to a fragment Fc′ which consists of the two C_H3 domains and small peptides derived from the C_H2 domains. Reduction of the Fab fragments results in Fd as indicated in Figure 1.12.

Pepsin cleavage of IgG in acid pH results in two major fragments $F(ab')_2$ and a pFc′ fragment. The $F(ab')_2$ consists of the two Fab fragments joined together by part of the hinge region and the pFc′ is similar to the Fc′ product from papain cleavage.

Π Examine Figure 1.12 carefully and then construct yourself a summary figure omitting the diagrammatic representations of the molecular structures.

Use the following format:

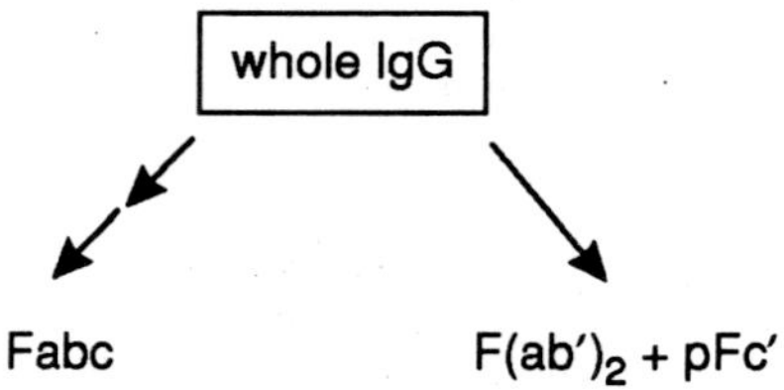

Also write on to your drawing the conditions which are needed for each of the steps.

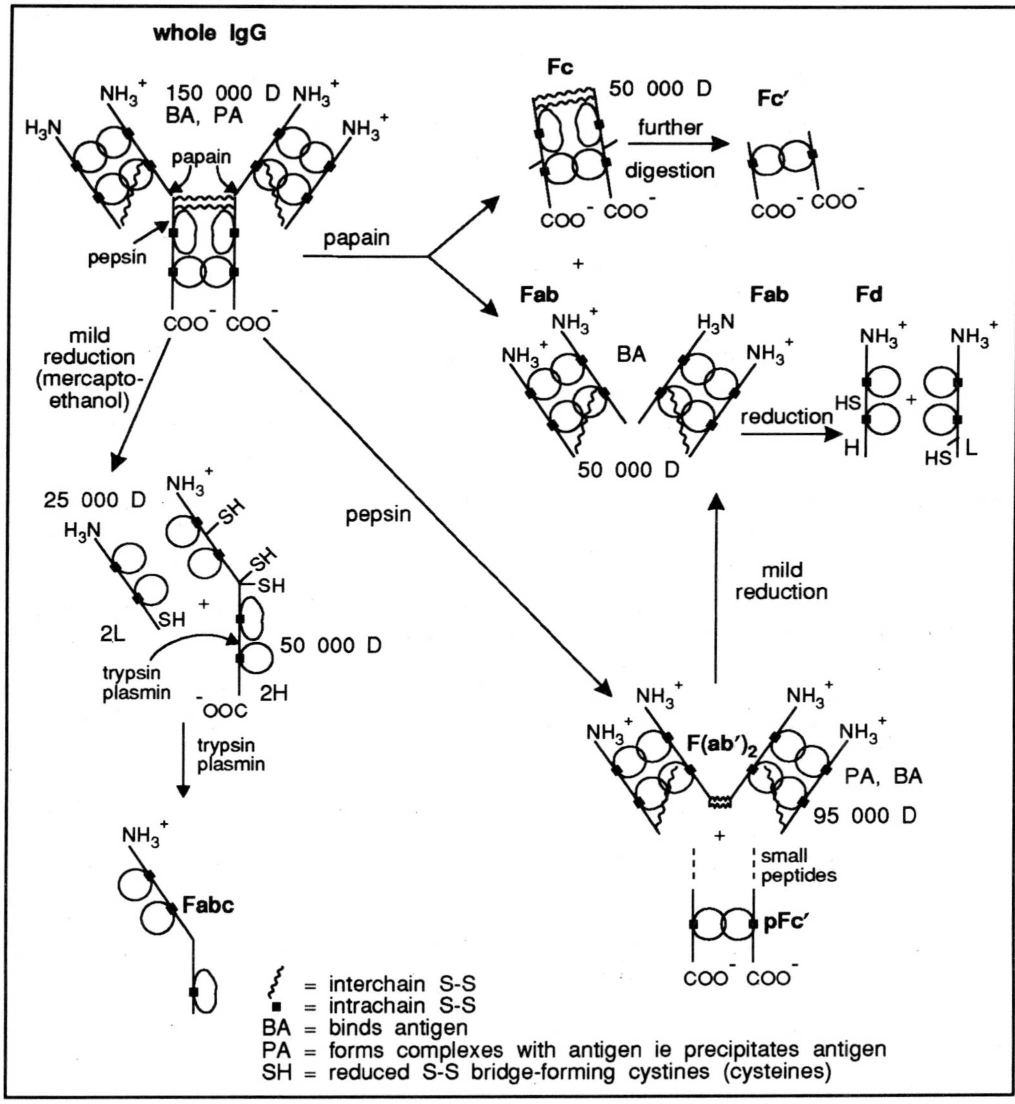

Figure 1.12 Reduction and proteolysis of antibodies.

SAQ 1.6

Immunoprecipitation is dealt with in the next chapter and involves the formation of large antigen-antibody complexes which precipitate out of solution. Can you decide from first principles which one of the following molecules would precipitate antigen?

1) Fab fragment.
2) IgG treated with pepsin.
3) Fc.
4) IgG treated with reducing agent.
5) Reduced and plasmin treated antibodies.

1.4.4 Heavy and light chain isotypes

kappa and lambda light chains

Based on the amino acid sequences of light chains they can be divided into two major groups called kappa (κ) and lambda (λ) isotypes; the lambda chains can be further divided into 4 subtypes λ1, λ2, λ3 and λ6. These light chains are common to all the five antibody classes ie an IgG and an IgM molecule can possess the same light chains. However, each antibody molecule can only possess one type of light chain as any single B cell can only produce one type of light chain.

isotypes

In a similar manner, 5 major classes (isotypes) of antibodies based on major differences in the amino acid sequences of heavy chains, have been identified in most mammals. They are IgG in which the heavy chain is designated with gamma (γ), IgM (μ), IgA (α), IgE (ε) and IgD (δ).

anti-isotype antibodies

These heavy chain isotypes and the kappa and lambda light chains will induce the production of anti-isotype antibodies which are specific for the particular antibody chain if injected into, say, a rabbit. For instance, immunisation of a rabbit with human gamma chain will result in antibodies which bind only to IgG and not to the other antibody classes.

SAQ 1.7

Match each item from the left column with one of the features in the right column using each item only once.

1) Hinge region	A) Isotype
2) Antigen combining site	B) 110 amino acids
3) Domain	C) V_H V_L
4) F(ab')$_2$	D) Proline
5) Heavy chain	E) 2 (V_H V_L)

1.4.5 Major structural features of IgG, IgM, IgA, IgE and IgD

IgG, IgD and IgE are monomeric in form ie they consist of two heavy and two light chains. IgM is a pentameric macroglobulin consisting of 5 monomers linked together. IgA exists in various forms but principally as a dimer.

IgG antibodies

four subclasses of IgG

IgG antibodies are the most common antibodies in serum representing about 75% of the total serum pool. About 50% of the total body IgG is found in extracellular fluids, the remainder in the blood. There are four subclasses of IgG (IgG1-4) bearing about 90% homology to each other, IgG1 being the most common (2/3 total) and IgG2 representing about 1/4 of IgG antibodies. The molecular weight of IgG antibodies is about 146 000 Daltons except for IgG3 which is 170 000 Daltons due to an extended hinge region. There are single oligosaccharide moieties attached to each $C_\gamma 2$ domain.

interheavy S-S bonds

The major structural difference between the IgG subclasses is in the number of disulphide bonds joining the two heavy chains in the hinge region. This varies from 2 for IgG1 to 15 in IgG3. The IgG subclasses also differ in their ability to mediate various physiological responses. We shall deal with some of these properties later.

IgM
extra constant domain
tailpiece
J chain

IgM constitutes about 10% of serum antibodies and the majority is intravascular. The pentamer has a molecular weight of about 900 000 but a monomeric form is found as an antigen receptor on young B cells. The subunits of IgM possess an extra constant domain $C_{\mu}4$ and also an octadecapeptide tailpiece at the carboxyl end of each heavy chain and are joined together by disulphide bridges as shown in Figure 1.13. IgM possesses an additional polypeptide called the J chain of about 15kD which links 2 of the 5 subunits by a disulphide bridge. This chain is thought to regulate the formation of IgM polymers in the B cell.

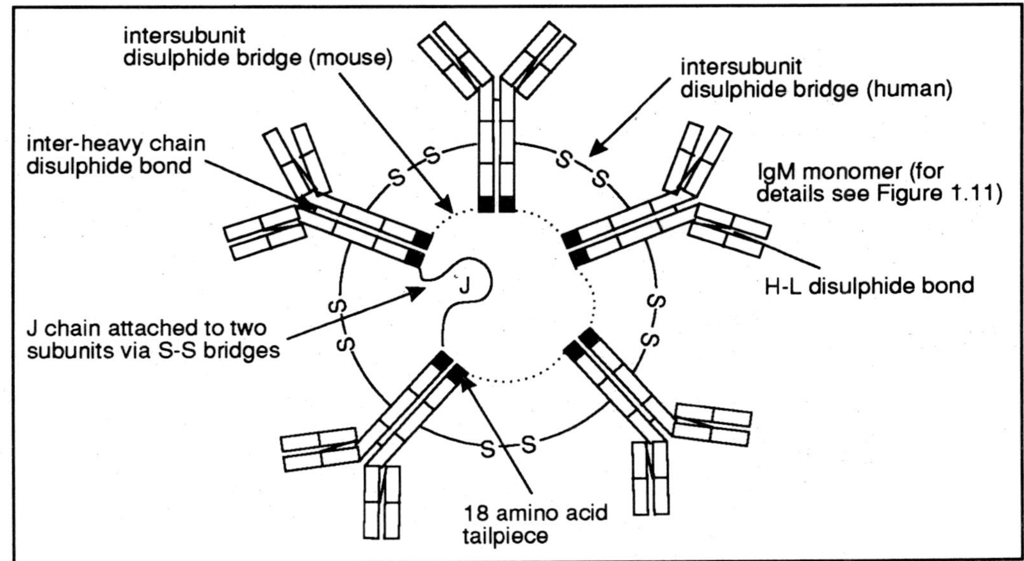

Figure 1.13 IgM molecule (stylised).

About 12% of IgM is carbohydrate, this comprising 5 oligosaccharides attached to each heavy chain by asparagine residues distributed along the four constant domains. These may prevent the domain interactions we see in IgG molecules.

IgM is functionally pentavalent

Although possessing 10 antigen binding sites, IgM is functionally pentavalent using one binding site per monomer. There is no defined hinge region in IgM.

IgA
external body secretions
secretory IgA
secretory component

IgA is the second most common antibody in serum existing as a monomer, dimer or trimer in serum. It possesses 3 constant domains in each alpha heavy chain and the carboxyl terminal tailpiece found in IgM. The main function of IgA is to protect the external mucosal surfaces and is found in the external body secretions as a dimer called secretory IgA (sIgA) shown in Figure 1.14. You can see from the figure that it is made up of an IgA dimer, a J piece and an additional polypeptide of 70 000 Daltons, the secretory component (SC). SC transports IgA from the submucosa where it is made by B cells across the epithelial cells to the mucosal surface.

IgE
reaginic antibody

IgE is the so-called reaginic antibody present in normal serum at extremely low levels and also found on the surfaces of tissue mast cells. IgE is responsible for atopic reactions to pollens, bee stings etc. It is monomeric but possesses an extra constant domain like IgM having a molecular weight of about 190 000 Daltons. There is no hinge region making IgE a very rigid molecule with little flexibility.

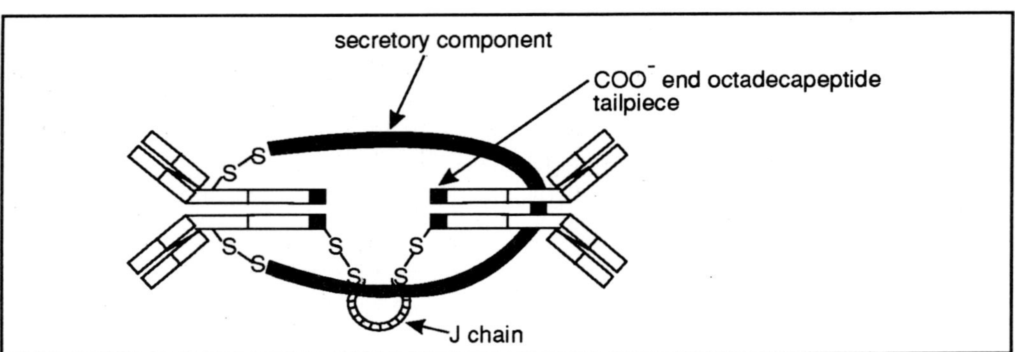

Figure 1.14 Secretory IgA (stylised) showing J chain attachment to the tailpiece of each subunit. The favoured model of secretory component attached to one subunit rather than both subunits is shown.

IgD

The final antibody class IgD is only found in trace amounts in serum. Human IgD has a molecular weight of about 180 000 Daltons, possesses an extended hinge region, 3 constant domains and a unique octapeptide tailpiece at the carboxyl ends of the heavy chains. Murine IgD does not possess a $C_\delta 2$ domain and the hinge region is shorter than that of human IgD. In both species, IgD is rich in carbohydrate.

1.4.6 Allotypes

As the name suggests (allo = allele), individuals of the species exhibit minor differences (generally one or two amino acids) in amino acid sequences in some of the antibody constant domains. Since these are inherited in Mendelian fashion not all members of a species express any one allotype. In Man, these allotypes have been found in IgG (called Gm allotype) and IgA (Am allotype) heavy chains and kappa (Km allotype) chains. As an example of an allotype, Gm4+ individuals possess arginine at position 214 in IgG1 whereas Gm4- individuals do not. Since any one B cell can only produce one heavy chain sequence, a process known as allelic exclusion, an individual heterozygous for any one allotype will have some antibodies bearing the allotype, while others will not.

allelic exclusion

SAQ 1.8

Complete the comparative table below.

Characteristic	IgG	IgM	sIgA	IgD	IgE
No of domains/molecule	12				
Possess allotypes (+/-)	+				
Isotype (Greek letter)					
Monomers/molecule	1				
Molecular weight (kD)	146				
J chain? (+/-)	-				
Extended COO⁻ end? (+/-)	-				
Domains/heavy chain	4				

1.4.7 The antigen binding site

Complementarity Determining Regions (CDR)

Framework Regions (FR)

The heterogeneity we referred to earlier in variable domains is confined to just three short regions in the proximity of residues 30-35, 50-65 and 95-100 in both heavy and light chains which interact with the antigenic epitope, thus defining the specificity of the antibody. They are called Complementarity Determining Regions (CDR) and are numbered 1-3 in each heavy and light chain variable domain as shown in Figure 1.15. The remaining amino acid residues are relatively invariant from antibody to antibody and have been called Framework Regions (FR) as they contribute to the overall shape of the antigen binding site. You will notice from Figure 1.16 that, due to the folding of the polypeptide, these CDRs are arranged close together and both sets from heavy and light chain domains come together to form the binding site.

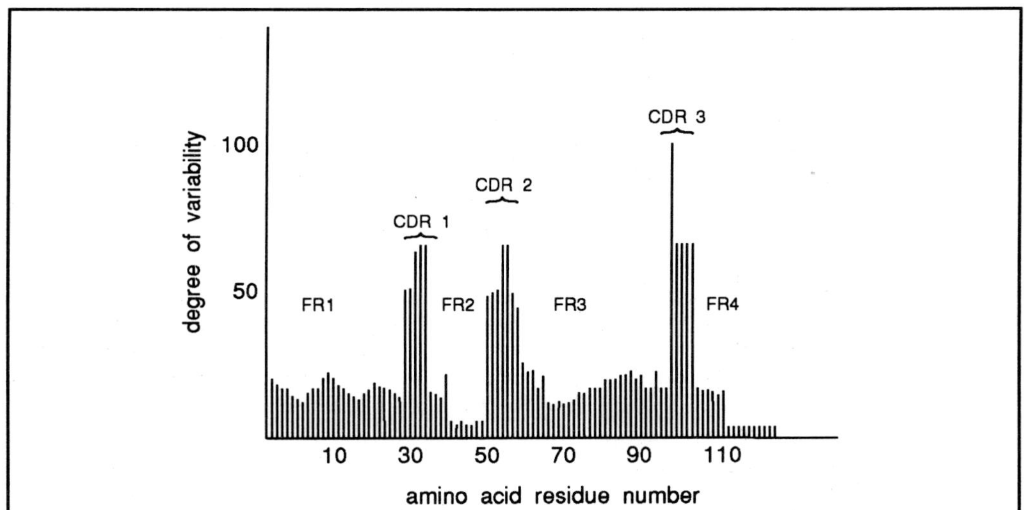

Figure 1.15 Variability plot of amino acids in V_H. FR = Framework Regions; CDR = Complementarity Determining Regions (see text for details).

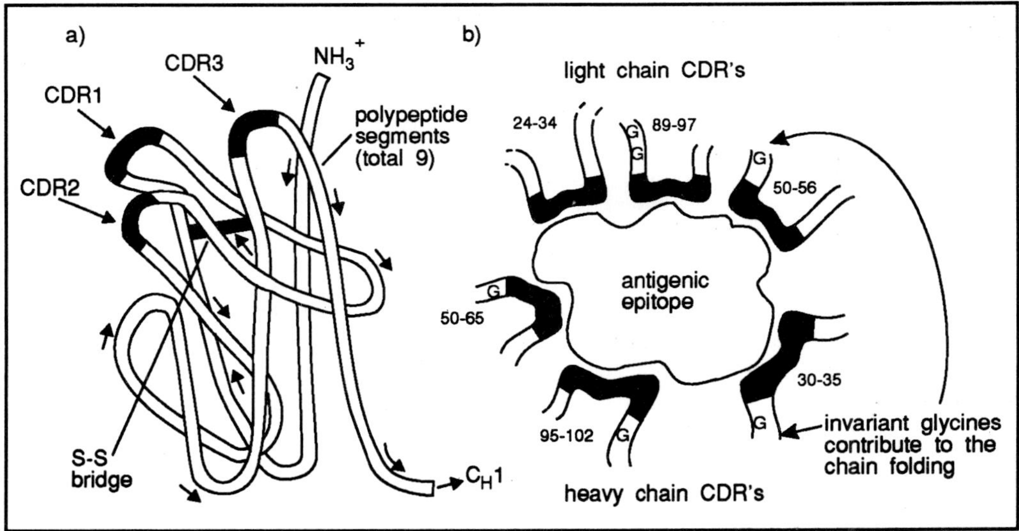

Figure 1.16 a) V_H folding of polypeptide segments showing apposition of CDRs. b) Interactions of CDRs from H and L chains with antigenic epitope.

1.4.8 Idiotypes

idiotypes

idiotope

paratope

anti-idiotype antibodies (anti-Ids)

Along with isotypes and allotypes, idiotypes complete the heterogeneity in the general structure of antibodies. Idiotypes describe the unique amino acid sequences found on the V domains of an antibody molecule. Each of the sequences is called an idiotope and all the idiotopes found on the V_H and V_L of the antibody are called the idiotype. Many of these idiotopes are found within the CDRs of the actual binding site of the antibody (site-associated) and make up the paratope. Other idiotopes are within the V domains but outside the actual binding site (non-site associated). These idiotopes act like antigenic epitopes and induce the production of antibodies called anti-idiotype antibodies (anti-Ids). If the antibodies are directed against the actual paratope they will block the interaction of antigen with the antibody.

Π Can you think of any possible structural relationship between anti-Ids and the antigenic epitope and how this could be useful?

internal image

vaccines

Some anti-Ids which fit within the binding site of Ab1, will have the same structural characteristics as the actual epitope. We say that the binding site of Ab2 is the internal image of the epitope. You could therefore use anti-Ids as vaccines instead of antigen! This would be useful if the actual antigen was toxic. The main ideas on idiotypes are illustrated in Figure 1.17. Note that Ab2 anti-Id competes with the epitope for the paratope. Ab3 anti-Id does not compete with the epitope (antigen).

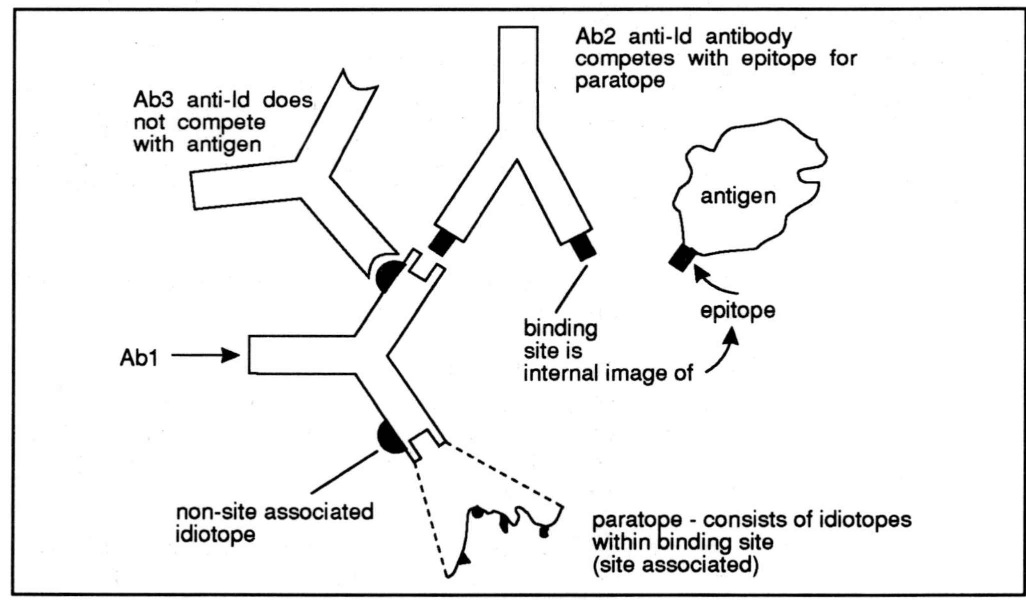

Figure 1.17 Idiotypes and anti-idiotypes. Note that the paratope consists of a set of idiotopes within the binding site made of V_H and V_L and these, together with the non-site associate idiotopes, comprise the idotype of the antibody.

SAQ 1.9

Which the following statements are true or false?

1) Heterogeneity in antibodies is due equally to different isotypes, allotypes and idiotypes.

2) IgM, IgA and IgE have the same allotypes but different isotypes.

3) You can generate anti-Ids to human antibodies allogeneically (in other individuals) or xenogeneically (in other species, say rabbits).

4) Complementarity determining regions (CDRs) are amino acid sequences in the epitope which interact with the paratope.

5) Allotypes have been found in IgG and IgA subclasses and on the constant domains of light chains.

1.5 Biological functions of antibodies

Π Before you read on, it might be useful to go back over our comments at the end of Section 1.2.7 on the inadequacies of the innate immune system and how antibodies enhance this system.

Antibodies can only be effective if the pathogen remains extracellular since antibodies cannot enter cells which are infected with pathogen. Once they gain entry into cells, the host has to activate cell mediated immune mechanisms involving other cells including T cells, not antibodies. Having said that, let us now examine ways in which antibodies fulfil a protective role in the host.

Π Before you read on, can you think of any protective role that is achieved by antibody simply binding to the antigen?

1.5.1 Fab-mediated activities

The principal benefit of Fab binding to antigen is that the antibody may block some essential function of the antigen or pathogen. Many toxins mediate their damaging effects by entering host cells via specific receptors. High affinity IgG is particularly good as an antitoxin in blocking these activities.

antitoxin

sIgA in external secretions and in the mucous lining of the epithelial layer encounters antigens before they breach the physical barriers. It may bind to bacteria preventing colonisation and infection or to viruses thus blocking viral entry into the epithelial cells. sIgA has a similar effect on food antigens preventing their passage across the intestinal wall. The overall result is that the sIgA coated pathogens/antigens are finally expelled from the body by a variety of mechanisms (refer to Figure 1.4) and fail to cause infection or damage to the host.

1.5.2 Fc-mediated functions

If the pathogen manages to gain entry to the host in spite of sIgA, then the specific antibodies have to initiate its destruction to eliminate it from the host. This is achieved by phagocytosis and/or killing of the pathogens by various cell types or through the activation of complement. These mechanisms are dependent on the Fc portion of the antibody molecule once the latter has specifically bound to the antigen.

Fc-mediated phagocytosis

opsonisation

The professional phagocytes, neutrophils and macrophages, express receptors for IgG in Man and particularly for IgG1 and IgG3. Once the antibodies specifically bind to their epitopes on micro-organisms, foreign mammalian cells or soluble antigens such as proteins, a process known as opsonisation, receptors on phagocytes bind the Fc regions of the bound antibodies and the whole antigen-antibody complex is engulfed by the cell and destroyed.

Fc-mediated lysis

antibody-dependent cellular cytotoxicity

Cells or pathogens, when coated with specific antibodies such as IgG or IgE, may be killed by a non-phagocytic mechanism. Various cells, possessing Fc receptors, bind the antibodies and secrete a variety of extracellular enzymes or toxic mediators. Examples of this method are destruction of transplanted tissues by macrophages, neutrophils or natural killer cells, the killing of liver fluke larvae by eosinophils and measles virus-infected cells by macrophages. This process is known as antibody-dependent cellular cytotoxicity (ADCC).

Mast cells and IgE

A third use of Fc receptors is that used by mast cells and the closely related basophils. Mast cells are situated throughout the body including the skin, the submucosal tissues and the lungs and express high affinity receptors for locally synthesised IgE. Binding of antigen by IgE induces these cells to release a number of mediators including histamine, leukotrienes and chemotactic factors which promote local inflammatory responses as a defence against the antigen. The consequences of this mast cell activation are discussed in Figure 1.18.

Π Examine Figure 1.18 to ensure that you can distinguish between the different uses of Fc receptors on various cell types.

SAQ 1.10

Match each item in the left column with those in the right column using each item only once.

1) Fab	A) Cell lysis
2) IgE	B) Antigen-antibody complex
3) Opsonisation	C) Neutrophil Fc receptor
4) IgG1	D) Virus-receptor inhibition
5) ADCC	E) Mast cell

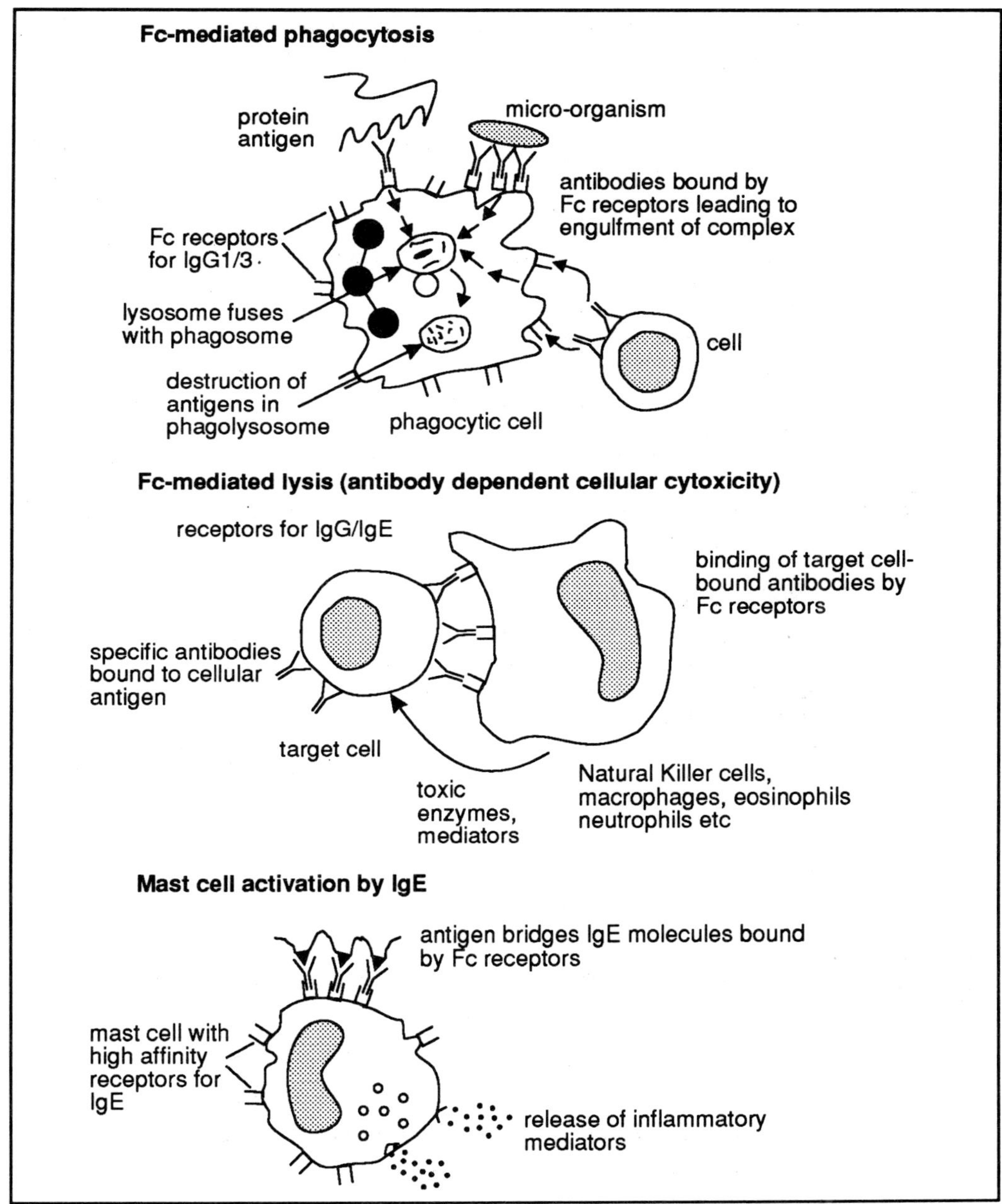

Figure 1.18 Fc-mediated defence mechanisms.

Fc-mediated complement activation

Complement proteins are found in inactive forms in body fluids and are produced at inflammatory sites by cells such as macrophages. We have summarised the pathways of complement activation in Figure 1.19 and you should follow this closely while you read this section. May we remind you once more that this is only a review of a very complex topic so that you will particularly understand the biological implications of complement activation by antibody Fc's. (For further details, you may find the BIOTOL text 'Cellular Interactions and Immunobiology').

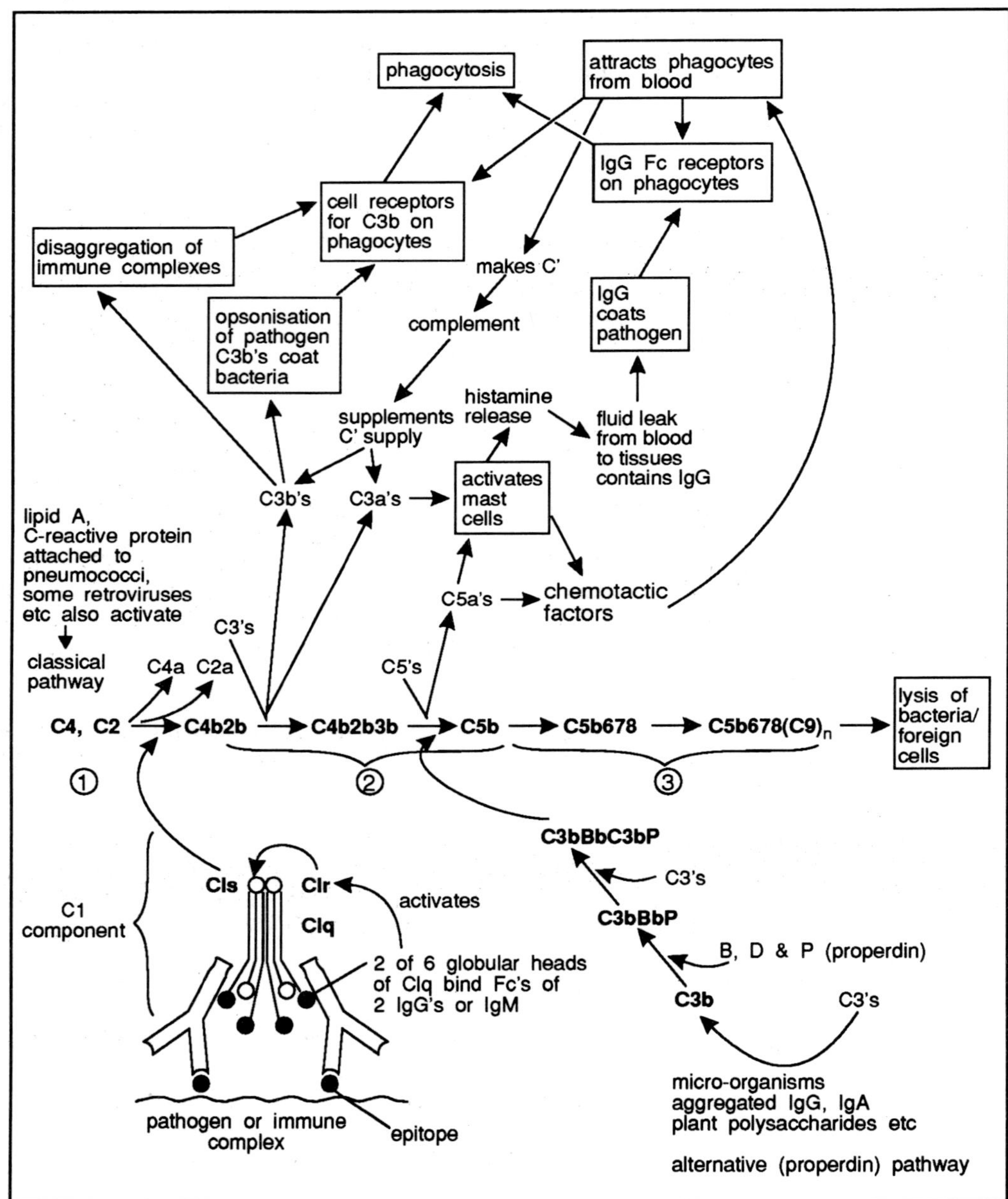

Figure 1.19 Complement activation and the biological consequences: ① = activation step; ② = amplification step; ③ = membrane attack complex.

Complement (C′) is activated in the classical pathway when the first component of complement C1q is bound by two Fc regions derived from two IgG molecules or one IgM molecule. Only IgM and IgG (except IgG4) activate complement. Most of the reactions described below take place on the pathogen surface.

direct killing of pathogens by complement

membrane attack complex

C1q binding then activates C1r which has attached to C1q. This then cleaves C1s to produce an active enzyme which cleaves the next two components C4 and C2 resulting in C4b and C2b particles which unite to form an enzyme C4b2b. This can cleave many C3 particles leading to the formation of the last enzyme C4b2b3b which then cleaves many C5's resulting in C5b molecules being produced. Each of these then combine with C6, C7 and C8 to form C5b678 in a nonenzymic manner. This complex then binds a number of C9 particles which forms a pore through the cell membrane of the pathogen resulting in lysis and death. The C5 to C9 sequence is called the membrane attack complex (MAC).

You will have noticed that, during the early enzymatic reactions, other reactive particles were produced such as C3b, C3a and C5a. These possess important biological reactivities.

C3b and phagocytosis

You will remember from our discussions on innate immunity that C3b particles opsonise the antigen which is then engulfed by phagocytes expressing receptors for C3b.

anaphylatoxins

histamine

neutrophil chemotactic factor

Let us now deal with C3a and C5a which are called anaphylatoxins. They are so-called because they attach to mast cell receptors mediating the same effects as IgE (Section 1.4.5). Histamine and neutrophil chemotactic factor (NCF-A) are released from the mast cells and diffuse to the nearest blood capillary where histamine promotes loss of fluid from the blood and NCF-A promotes neutrophil migration from the blood stream to the inflammatory site. The histamine effect may result in more antigen specific IgG being brought to the site from the blood thus enhancing the response. The neutrophils will be used for phagocytosis of the antigen which is coated with antibodies or C3b particles. C5a is also a chemotactic factor for phagocytes and is instrumental in activating these cells at the site.

As you will see from Figure 1.19, complement can be activated by other means and there is an alternative pathway which can be activated by some micro-organisms (see our earlier comments on innate immunity).

Π This has been a very brief review of a complex series of reactions. We suggest it would be very useful for you to re-read this section and then make a list of the major consequences of complement activation.

You should include the following:

- destruction of pathogens using the membrane attack complex (C1-9);
- attraction of phagocytic cells to the site by mast cell and C5a chemotactic factors;
- promotion of phagocytosis using C3b opsonisation of antigens;
- possible enhancement of Fc-mediated phagocytosis using serum derived IgG due to histamine effects.

Other Fc related functions

The Fc regions confer on 3 out of 4 of the IgG subclasses resistance to proteases. These antibodies, therefore, give long-term protection to the individual due to their long half lives. Additionally, the Fc region promotes the passage of IgG across the placenta thus providing protection to the infant during foetal and neonatal (newborn) life when the child's own immune system is deficient.

As we have seen, the Fc region is instrumental in the transfer of sIgA to the mucosal surface. Additionally, the attachment of secretory component to a site on the Fc region results in its transport into mother's milk which provides protection to the infant against gut pathogens.

SAQ 1.11

Which of the following are true or false?

1) Complement activation results only from interaction of Fabs with epitopes.

2) C3a is an anaphylatoxin that activates mast cells and is chemotactic for neutrophils.

3) The membrane attack complex is dependent on the generation of C5b.

4) Opsonisation only involves the coating of antigens by C3b particles.

5) The membrane attack complex can result from both classical and alternative pathways of complement activation.

1.6 Antibody genes

We are now going to briefly review the mechanisms by which a very large repertoire of antibody specificities is derived from a relatively limited number of antibody genes. We have to emphasise that this section is a brief review of a very complex topic and you are advised to refer to other immunology texts including the BIOTOL text 'Cellular Interactions and Immunobiology') in which the topic is fully described.

1.6.1 Antibody gene chromosomal organisation

You should refer constantly to Figure 1.20 as you read this section. There are unlinked gene families encoding the kappa, lambda and heavy chains on separate chromosomes. On each of these chromosomes you will find distinct gene segments which encode the variable and constant domains.

V, D and J segments

If we begin at the 5′ end of the Igh locus (encodes the heavy chain) in Man we find a family of about 100 individual variable (V) segments then a cluster of up to 30 diversity (D) segments and further downstream 6 joining (J) segments. A considerable distance downstream at the 3′ end of the locus we find the constant (C) segments which encode individual constant domains of the antibody molecule. Each segment is separated from its neighbours by noncoding sections of DNA called introns.

introns

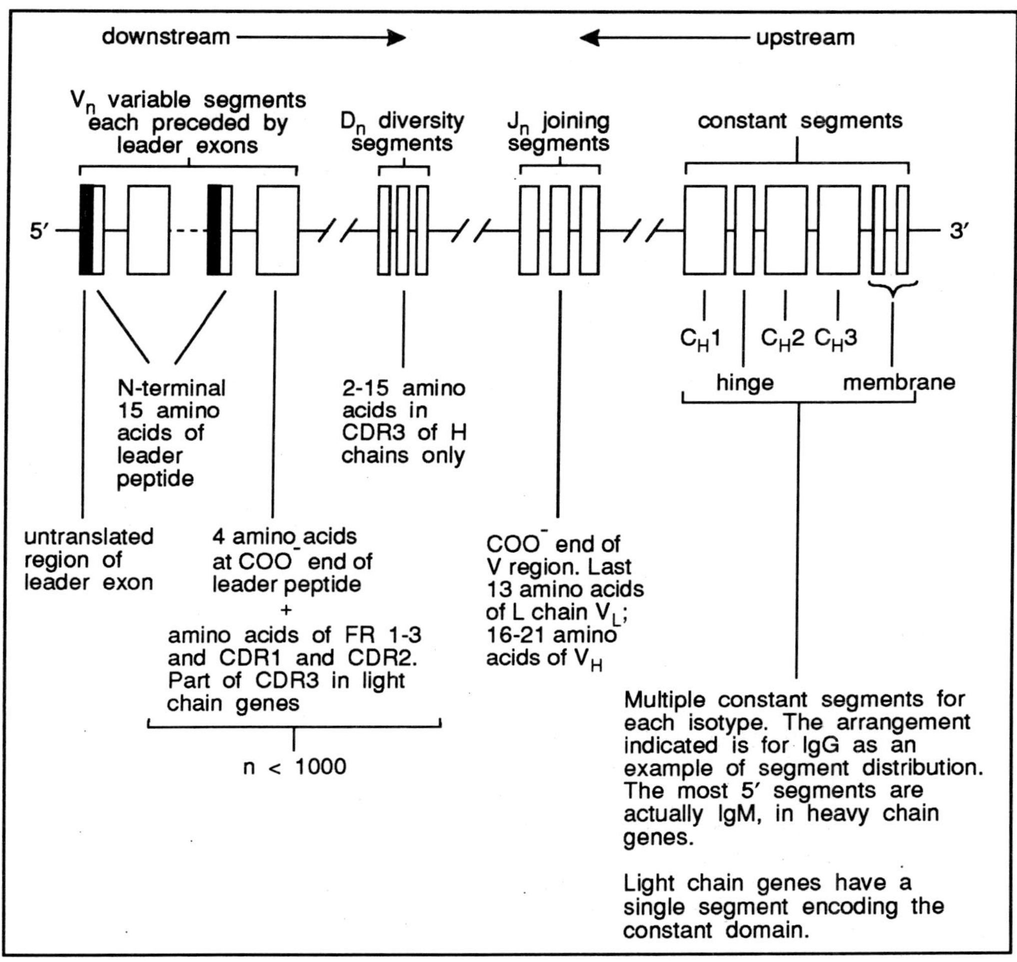

Figure 1.20 Immunoglobulin exons (segments) encoding the variable and constant domains of antibodies showing the general arrangement on the chromosomes. Notice that D segments are only in heavy chain genes and the arrangement of segments in lambda (λ) light chain genes differs from this arrangement. The values of n are only estimates based on Ig genes of Man and mouse and include nonfunctional genes (pseudogenes). Note that exons are represented by boxes, introns by a single line.

basic DNA rearrangement mechanism

During development of each B cell, a functional variable gene is created by a process known as DNA rearrangement. The heavy chain V gene is formed by recombination of one of the D segments with any one of the J segments with the loss of the intervening DNA. One of the V segments then joins to the DJ to form a VDJ or variable gene which encodes the whole 110 amino acid sequence of the heavy chain variable domain. Light chain variable genes are formed by a single step mechanism when a V segment joins to a J segment since there are no D segments in light chain genes.

Π What is the maximum number of variable genes that can be generated by DNA rearrangement in Man?

If we assume that any of the V, D and J segments can be selected, the total number will be 100 (V) x 30 (D) x 6 (J) = 18 000 variable genes.

Π Close your text and construct a diagram of a heavy chain gene rearrangement to $V_4D_3J_4$ from families of 6 variable segments (number them 1-6), 4 diversity segments and 6 joining segments. Indicate in your diagram the DNA sequences which are deleted in the process. Check your diagram against Figure 1.21.

At the 5′ end of each V segment and a short distance from it is located a leader exon that encodes a translation initiation signal (not translated) and the first 15 amino acids of a 19 amino acid residue leader peptide. This peptide is found at the amino end of each antibody chain and facilitates the movement of the Ig chain into the lumen of the endoplasmic reticulum where it is cleaved off before Ig secretion.

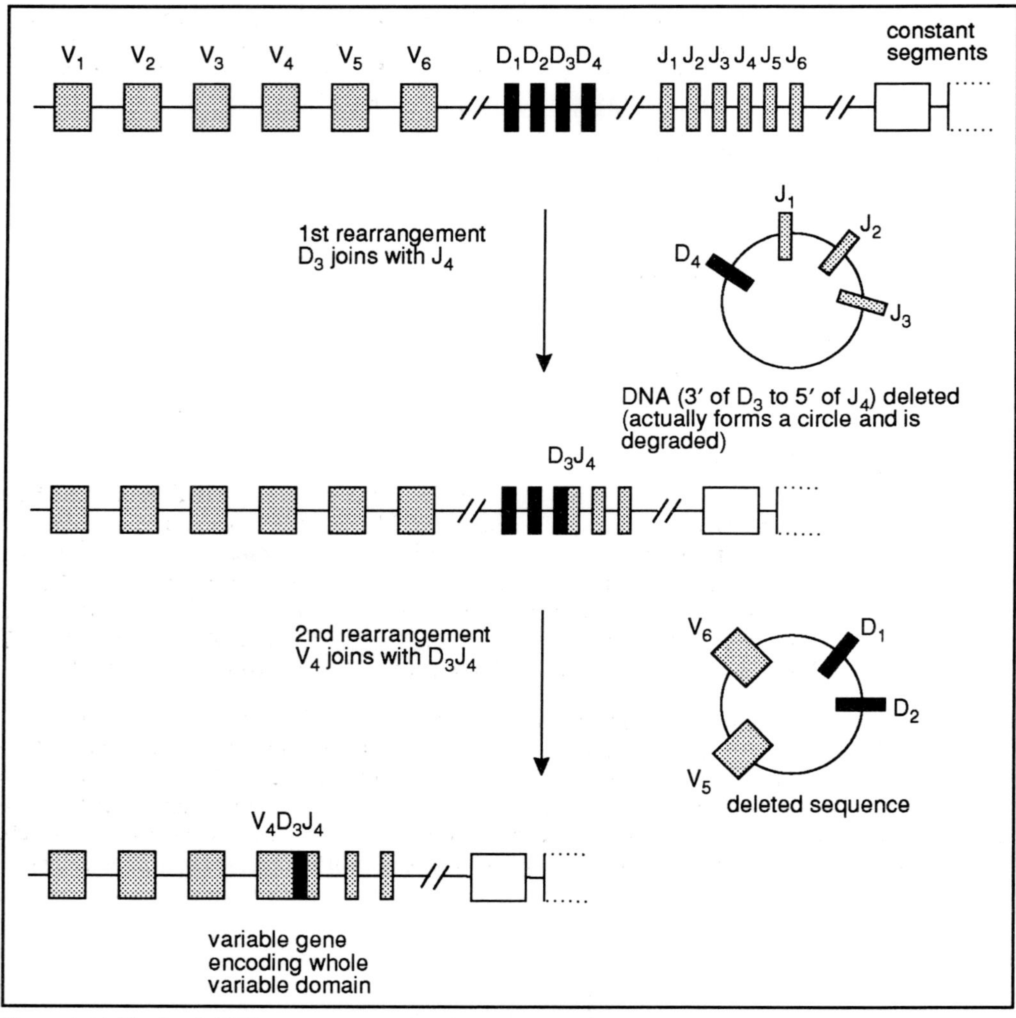

Figure 1.21 The basic DNA rearrangement mechanism in heavy chain genes.

what the V, D and J segments encode

Each V segment encodes the last 4 amino acids of the leader peptide and those which comprise CDR1, CDR2 and framework regions (FR) 1-3. CDR3 in heavy chains is encoded to varying degrees by the D segments with contributions from the J segments. The latter also encodes FR4. In the light chain genes, in addition to CDR1/CDR2 and FR1-3, the V segments also encode part of the CDR3 with the J segment encoding the remainder and FR4.

SAQ 1.12

Let us assume that in the total mouse genome mouse kappa chain genes comprise 250 V segments and 5 J segments. The V segments encode 98 amino acid residues and the J segments 13 residues. If mice also possess in total 1000 V_H segments encoding the first 98 amino acid residues of the V_H domain, 12 D segments encoding, on average, 8 amino acid residues and 4 J segments encoding 13 amino acids, calculate the percentage of the total genome used to encode all the variable segments of mouse antibodies (we are ignoring the few lambda variable segments). Assume that the total mammalian genome is 3.5×10^9 basepairs. Select your answer from those provided below.

1) 17%.

2) 0.004%.

3) 2.1%.

4) 0.01%.

5) 5.9%.

1.6.2 Light chain genes

The detailed arrangements of the lambda and kappa chain genes in Man and mouse are shown in Figure 1.22. In murine DNA there are two clusters of joining and constant segments which are associated with a single V segment. Each V segment is thought to be restricted to rearrangements with the J segments in its own cluster only. This means that there is only one functional rearrangement for $V_{\lambda}2$, that is with $J_{\lambda}2$ since $J_{\lambda}4$ is a pseudogene (a non-function but related sequence of nucleotides). You can see, then, that there is little germline diversity in murine lambda chain genes.

lambda chain genes in Man

There is more potential diversity in lambda chain genes in Man where any one of about 100 V segments can rearrange with one of 6 J_{λ} segments. The resulting VJ must then use the constant segment which is associated with the J segment. This rule applies to the murine lambda genes as well. The multiple constant segments leading to a number of lambda subtypes. Kappa chain genes in Man and mouse possess multiple V segments which can rearrange with up to 5 J segments and there is a single constant segment.

Kappa chain genes

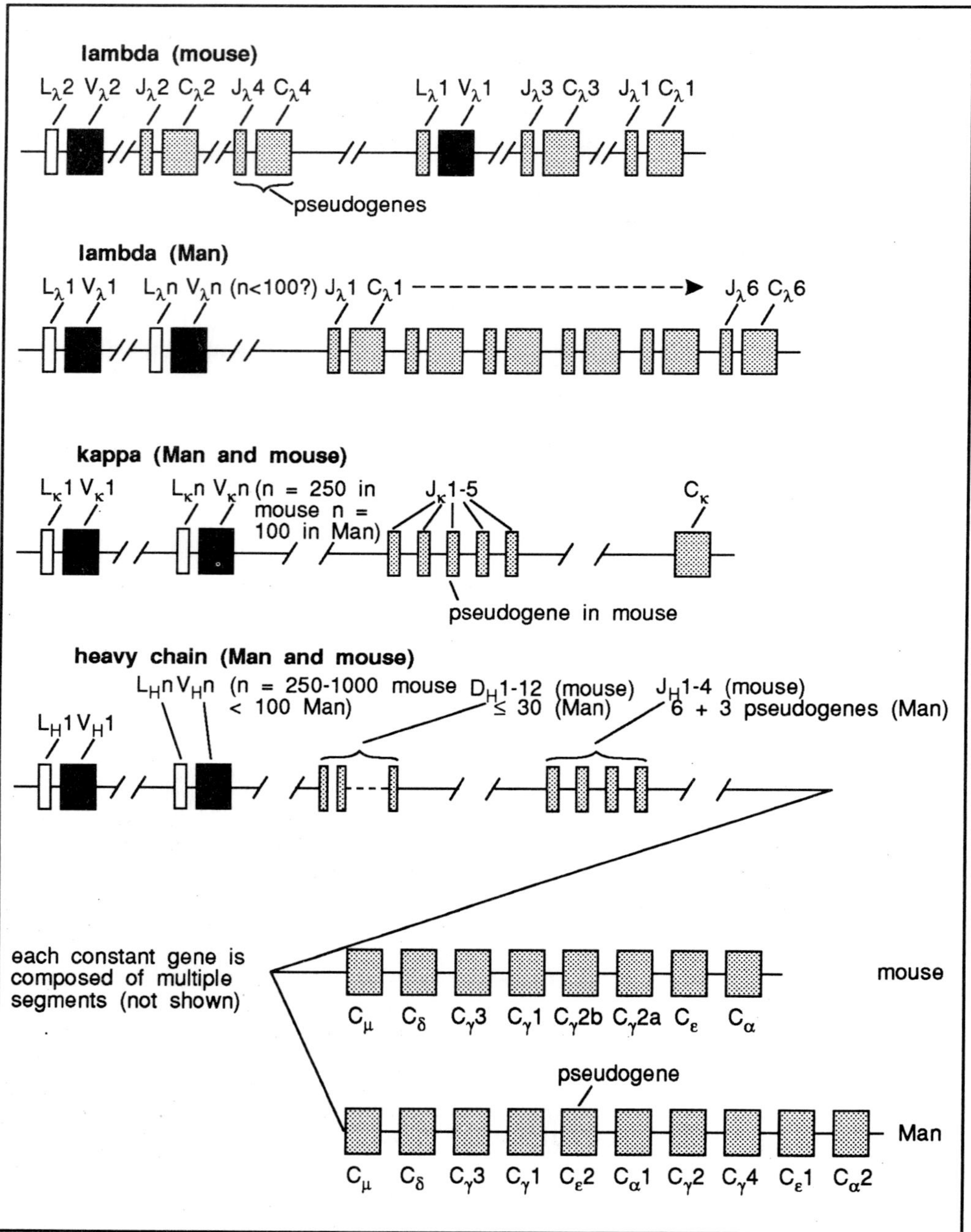

Figure 1.22 Organisation of germline Ig genes at the lambda, kappa and heavy chain loci in Man and mouse. Figure not to scale.

1.6.3 Heavy chain genes

There are up to 1000 V segments in mouse of which about 2/3 are functional and probably not more than 100 V segments in Man. The D segments may encode as little as two amino acids or 11 or more. In some antibodies they encode the whole of the

CDR3, in others they share this task with the J segment. As you can see in Figure 1.22 there are 4 J segments in mouse and 9 in Man of which 3 are pseudogenes.

Quite a distance downstream (6-8Kb) from the J cluster are the constant segments encoding the constant domains of all the isotypes.

In Figure 1.22 we have shown the sequential arrangement of the genes encoding the isotypes but remember that each of the boxes represents multiple segments as we indicated in Figure 1.20 for the IgG constant genes.

SAQ 1.13

Which one of the following is correct?

1) Introns comprise a major portion of the eukaryotic genome and code for the protein products of the cell.

2) DNA rearrangements lead to VDJ joining in all antibody genes.

3) The leader peptide is encoded by the leader exon and is involved in movement of the Ig chain into the lumen of the endoplasmic reticulum.

4) J segments invariably encode longer amino acid sequences in light chains than heavy chains.

5) V segments encode the complete amino acid sequence of the heavy chain except CDR3 and FR4.

1.6.4 Ordered rearrangement of Ig genes

ordered rearrangement of Ig genes

There is an ordered rearrangement of Ig genes in pre-B cells during their development. Heavy chain rearrangement occurs first with a D-J joining followed by a V to DJ rearrangement. If the resulting sequence is readable (in frame) then we say that it is a productive rearrangement. It is thought that only one in three rearrangements are productive. If the rearrangement is nonproductive then the Igh locus on the other chromosome goes through the same process. If this is also unsuccessful, then the B cell is nonfunctional and does not develop.

IgM and IgD act as antigen receptors

If there is a successful rearrangement, then the cell produces the IgM heavy chain which remains in the cytoplasm. This somehow activates rearrangement at the kappa locus; if this results in a productive rearrangement on either chromosome, the cell then makes membrane IgM containing kappa chains ($\mu_2\kappa_2$). If both kappa chain gene rearrangements are nonproductive, then the cell attempts rearrangements at the lambda locus. If successful, the cell expresses membrane IgM containing lambda chains. The cell then synthesises IgD with the same specificity and this is expressed on the cell surface. Both IgM and IgD act as antigen receptors on the young B cells.

The mature B cell then leaves the bone marrow and settles in lymphoid tissues. If it meets the antigen for which it is specific, it will secrete polymeric IgM. During the primary response it may switch to making IgG of the same antigen specificity (see Section 1.6.7).

1.6.5 Germline antibody diversity

The diversity of the antibody pool of any individual is primarily based on the number of V segments in the germline. However, Man possesses no more than 100 V segments which encode the first two CDR's and FR1-3. It will be obvious to you, then, that much of the diversity in the germline is generated by joining of D and J segments to the limited numbers of V segments but that this results in sequence variation in CDR3, not in CDR1/2.

CDR3 is the major source of antibody diversity

What we are saying is that whereas there is only about 100 sets of CDR1/2 sequences, there are many more potential CDR3 sequences. At the Igh locus, there are about 30 D segments and 6 functional J segments providing 180 potential DJ joinings. If we assume that any of these can rearrange with any of the V segments we end up with a total of 18 000 different heavy chains which, however, possess only 100 different CDR1/2 sets.

In kappa genes, each of the 100 or so V segments provides part of the sequence for the CDR3 along with any of the 5 J segments resulting in 500 potential kappa chain sequences.

Π Before you read on, calculate how many different antibodies containing kappa chains can be derived from the germline genes.

random association of heavy and light chains

Additional diversity is generated by random association of heavy and light chains. The total number of antibodies containing kappa chains, therefore, is $18\,000 \times 500 = 9 \times 10^6$ antibodies.

1.6.6 Non-germline generation of diversity

junctional diversity

N-regions

Three additional mechanisms greatly increase the total number of antigen binding sites in antibodies. The first, which we call junctional diversity is due to imprecise joining of the V, D and J segments at the heavy and light chain loci resulting in changes in the nucleotide sequence of CDR3 which can result in amino acid changes at this site. This may result in antibodies of higher or lower affinity. A second mechanism involves the insertion of up to 30 nucleotides called N-regions during the segment joining in heavy chain genes resulting in a much extended CDR3 and obviously changes in antibody affinity or even specificity!

somatic hypermutation

maturation of the response

The third mechanism is that of somatic hypermutation and it represents point mutations occurring in all CDR regions by some unknown mechanism. These mutations occur during immune responses and especially after the B cell has switched from making IgM and IgD to IgG and other isotypes. It is thought that this process is to produce antibodies with improved affinity for the antigen and would explain the so-called maturation of the response. It is commonly found that with each response to a single antigen, the overall quality or affinity of IgG increases dramatically and is thought to be due to the preferential selection of B memory cell clones expressing high affinity antibody receptors for the antigen.

SAQ 1.14

Match the items in the left column with those in the right column.

1) Somatic mutation	A) Nonfunctional rearrangement
2) Junctional diversity	B) Immune response
3) N regions	C) CDR3
4) Vλ2Jλ4Cλ4	D) Oligonucleotides

1.6.7 Class switching

Early B cells, as we have said, express both monomeric IgM and IgD as antigen receptors on their surfaces. After stimulation by antigen some of their progeny switch their isotype expressing on their surface IgG, IgA or IgE instead of IgM or IgD and secrete IgG, IgA or IgE antibodies. Using probably a similar mechanism to that which created the VDJ gene, the cell moves the VDJ downstream to link to other heavy chain constant segments thus creating a different antibody class with the same binding specificity. As we observed in Sections 1.4 to 1.5 various antibody classes possess unique biological functions so this mechanism is a means of supplying the various antibodies needed.

Π Can you think of the possible reasons for switching the heavy chain gene from IgM to 1) the gamma 1 chain and 2) the alpha chain gene?

Switching to the gamma 1 chain gene would result in IgG1 and possibly in somatic mutation resulting in high affinity IgG antibodies. Since these can be distributed to the tissues as well as the blood and have a long half life, they serve a protective role in the animal which is not served by IgM. Switching to the alpha chain gene results in production of IgA which would be distributed to the mucosal surfaces, again as a protection against invasion by pathogens.

1.6.8 Membrane and secreted forms of antibodies

The heavy chains of secreted and membrane IgM differ in the amino acid sequence of the carboxyl terminal ends, secreted IgM possessing a 18 amino acid tailpiece used in polymerisation while the membrane form possesses a 26 amino acid transmembrane portion and 3 amino acid cytoplasmic section. The tailpiece gene segment is attached to the 3′ end of the $C_\mu4$ segment. A short distance downstream are two small segments encoding transmembrane and cytoplasmic sequences. The B cell can simultaneously produce both forms of IgM by alternative RNA processing of a primary transcript which encompasses all segments. Membrane forms of other isotypes can be similarly synthesised.

1.7 The B cell response revisited

You should now have a reasonable picture of the main events which occur from the time when B cells encounter antigen to the development of memory cells.

Π Close the text and try to construct a diagram illustrating all the events involved in the humoral response. Check your drawing with Figure 1.23.

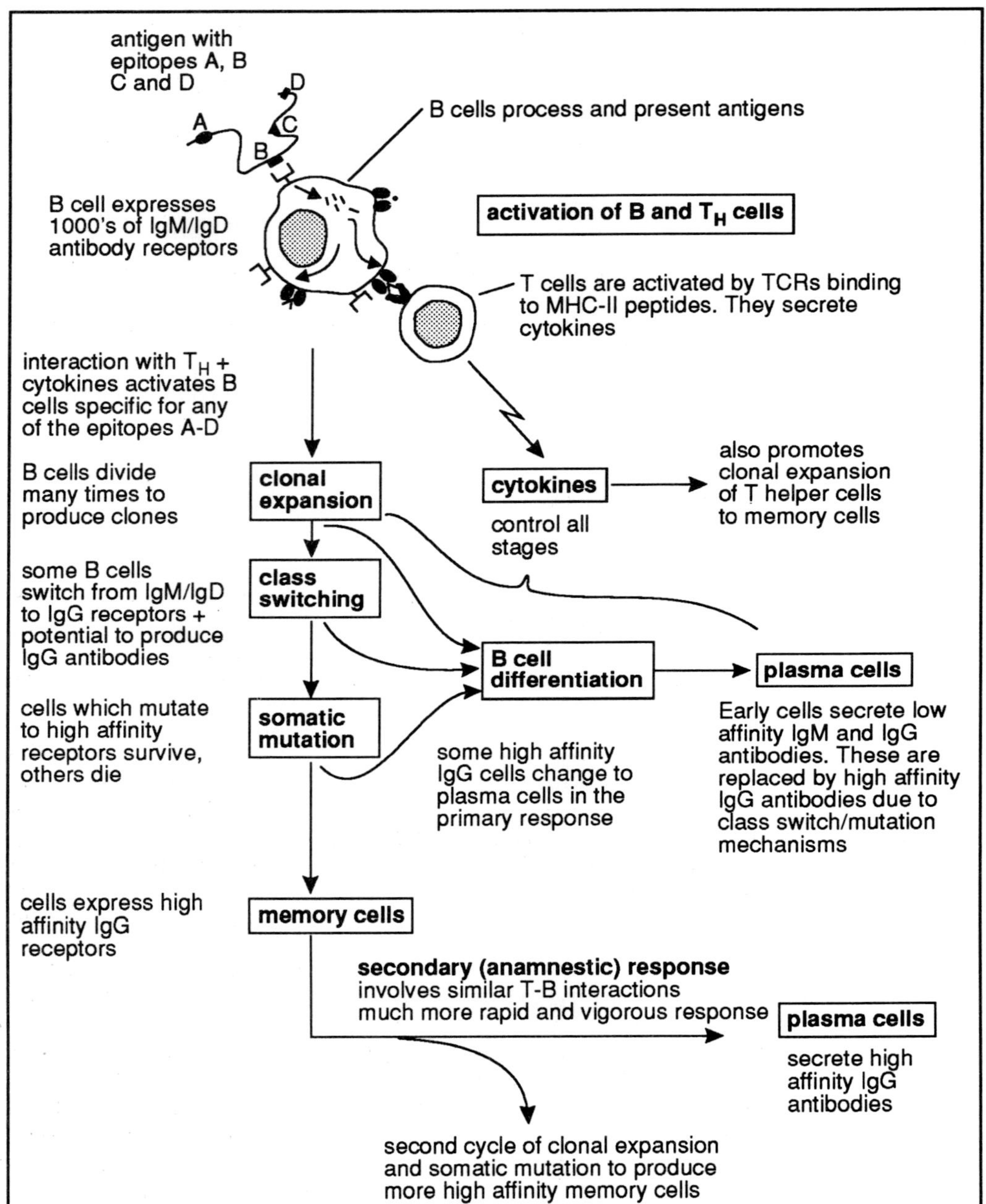

Figure 1.23 T-B collaboration and the humoral response. Description of the events leading from the first encounter of B cells with antigen (primary response) via the maturation of the IgG response (improvement of IgG affinity with time), memory cells to the characteristics of the secondary response. Note that at some sites in the body such as the submucosa, class switching would result in IgA and IgE B cell receptors and antibodies.

SAQ 1.15

Which of the following statements are correct?

1) Cytokine activity is confined to the primary response.

2) Somatic mutation has no effect on the quality of antibodies in the primary response.

3) Class switched progeny maintain the specificity of the parent B cell.

4) Somatic mutational events require the presence of antigen.

5) The secondary response is more rapid and vigorous due to increased numbers of antigen specific T and B cells and the production of better quality antibodies.

1.8 The physiological response

1.8.1 Sites of antibody production

You have learnt quite a lot about the way in which B cells are activated resulting in the production of antibodies and memory cells and you may well question as to how all these events relate to what happens in the body (*in vivo*) during an immune response and what the memory cells do from day to day when they are not responding to antigen.

lymphatics

lymph node

Antigen usually enters the body through breaks in the skin or across the epithelial layer of the respiratory and urinogenitary tracts and the gut and finds itself in the tissues. If deposited in the skin, specialised cells called Langerhans cells and macrophages may engulf some of the antigen and then transport it along vessels called lymphatics to the local lymph node. These lymph nodes are small encapsulated organs packed with cells, a large majority of which are T cells, B cells, other antigen presenting cells and phagocytic cells. Most of the antigen is destroyed in the node by phagocytic cells.

interdigitating dendritic cells

In a primary response, specialised antigen presenting cells called interdigitating dendritic cells, which are derived from Langerhans cells, process and present antigen to the naive T_H cells (those which have not been activated previously) resulting in clonal expansion and production of memory cells. These T_H cells can now interact with naive B cells (young B cells undergoing a primary response) which have processed and are presenting antigen and the events described in Figure 1.23 occur in the order shown. Some of the B cell progeny will develop in the lymph node into plasma cells, produce antibodies for a few days and then die; other progeny will move out of the node into interconnecting lymphatic vessels and be transported to other lymph nodes and even to the spleen and bone marrow where they will produce antibodies. The antibodies produced by the lymphoid tissues will be carried by the lymphatics to the blood. Up to half of the IgG antibodies will finally diffuse back into the tissues where it can bind to antigen if it gains access a second time.

Antigens crossing the epithelial layer of the alimentary, respiratory and urinogenitary tracts will be deposited in the submucosa where they will activate local B cells, with the help of T cells, to produce IgA and IgE.

recirculatory pool

The primary response results, as we have seen, in production of memory T and B cells and these migrate out of lymphoid tissue during the primary response and are carried by the lymphatic vessels to the blood stream. These cells are now part of the recirculatory pool of lymphocytes which regularly exit from the blood into the tissues, and subsequently they find their way up the lymphatic system back to the blood. Since these cells are constantly moving around the body, they can be recruited at any time to respond to a second invasion of the antigen at any site in the body. As we said earlier, memory cells respond more rapidly than the cells in a primary response and the higher affinity antibodies produced are much more effective at intercepting antigen thus promoting its destruction.

1.8.2 The kinetics of IgM and IgG production

Let us imagine that we immunise a rabbit with antigen and withdraw serum samples daily over the next four weeks. We then give a second injection to induce the secondary response. Antibody levels in the serum will follow the patterns shown in Figure 1.24. There is a lag phase where no antibody is detected followed by a logarithmic increase in antibody levels. After a time, which varies with the type of antigen, the concentrations of antibody in the serum level off and then slowly decline.

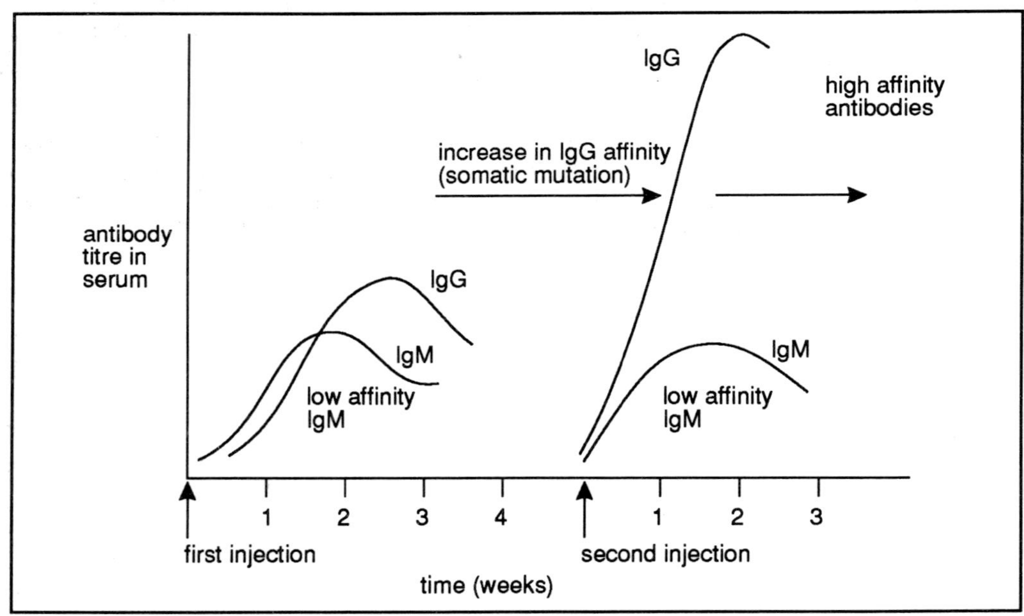

Figure 1.24 Serum IgG and IgM antibodies during primary and secondary responses. Note the change in affinity of IgG as well as the change in the titre levels.

We often find that the first antibodies are IgM which are of poor quality and this remains so throughout the response. IgG antibodies usually appear in serum about a day after IgM and initially these are also of low affinity. As the response progresses, however, the affinity of IgG antibodies produced increases for the reasons illustrated in Figure 1.23.

After a second injection, we observe changes in IgG production. There is a much shorter lag time between injection and detection of IgG, its first appearance often coinciding with that of IgM, the amount produced is much greater and the average affinity of the

IgG antibodies is much higher. In contrast, the IgM response mirrors that of the primary response.

Π How do you explain the differences in IgM and IgG production?

The production of IgM observed in the secondary response is probably due to fresh B cells from the bone marrow undergoing essentially a primary stimulation by the antigen and hence this IgM production has the same characteristics as IgM produced in a primary response. In contrast, many cells from the primary response have clonally expanded and class switched to IgG. Some of these have undergone somatic hypermutation with the capability to produce high affinity IgG. The overall result is that many more cells capable of producing IgG and higher affinity IgG.

1.8.3 Major features of the physiological response

You should by now be able to put together a reasonable picture of the physiological response, identify various mechanisms at the different stages and make some conclusions as to the final results in terms of antibody production and its contribution to protection of the host including enhancement of innate immunity. We have summarised these events in Figure 1.25.

Π We suggest that you go through this figure carefully and write down what you think you know about each stage of the response depicted in the figure. Then check your information from the relevant sections in this chapter. It might be helpful to redraw Figure 1.25 onto a large piece of paper.

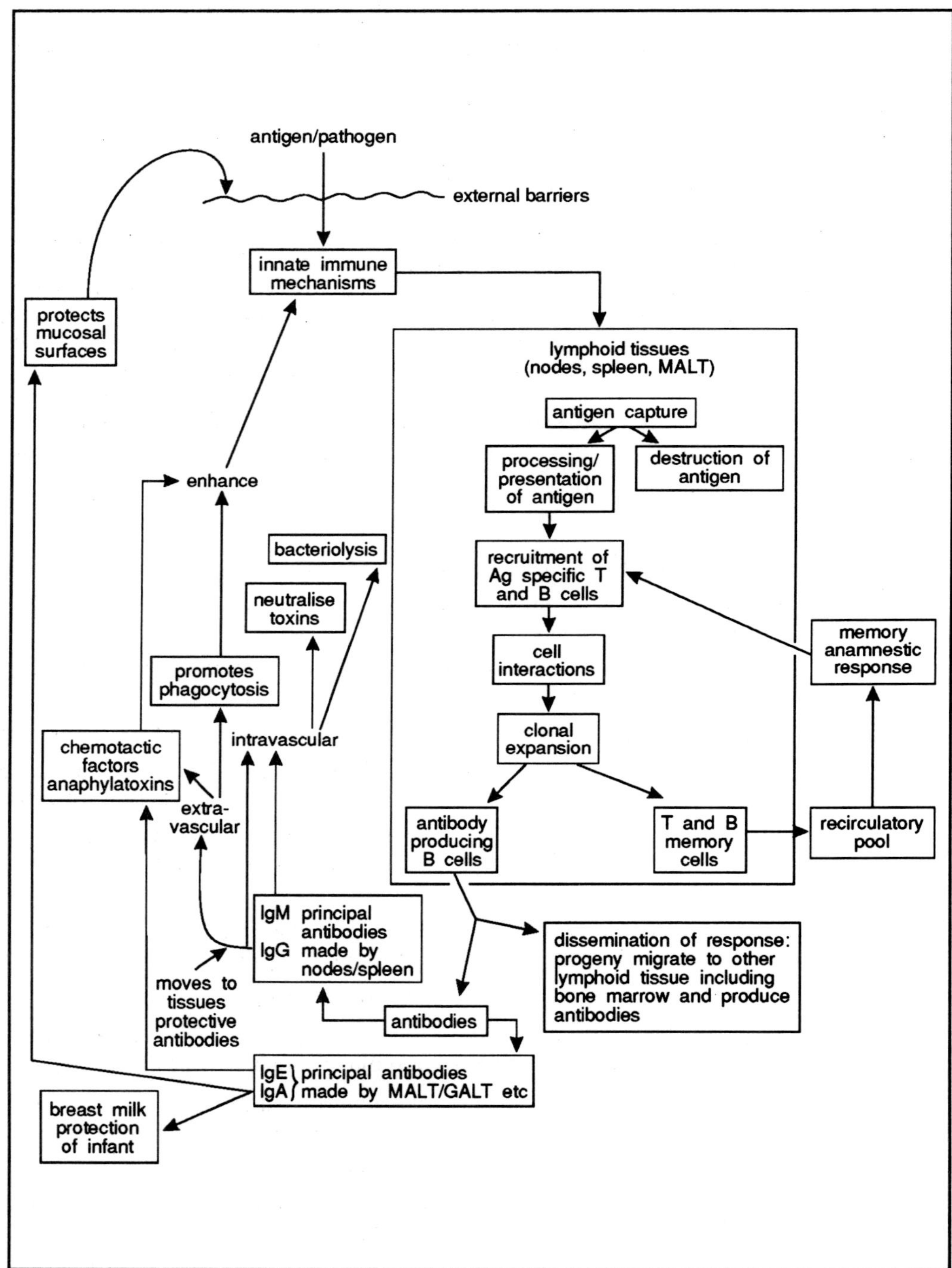

Figure 1.25 The physiological response. MALT = mucosa associated lymphoid tissue. GALT = gut associated lymphoid tissue.

Summary and objectives

In this chapter we have provided you with an overview of innate and acquired immunity. We then directed your attention to a detailed description of the structure and function of antibodies and have also analysed the genetic basis of antibody diversity. This knowledge will provide a solid background for much of the remainder of the text. We also examined the role of cell interaction molecules in the immune response. Antigen is processed into peptides which become bound to MHC II molecules which are expressed on the surface of antigen presenting cells including B cells. Antigen specific T cell receptors on T helper cells interact with these complexes and become activated, clonally expand and produce cytokines. The cytokines promote the activation and clonal expansion of the B cells and some of the progeny differentiate to plasma cells and produce antibodies whereas others become memory cells.

Now you have completed this chapter you should be able to:

- explain what is meant by self and nonself, describe the major features of antigens and define epitope, antigen, immunogen;
- describe the major features of humoral and cell mediated immunity and the roles of B cells and the different T cell subpopulations;
- list the major components of innate immunity and explain its deficiencies;
- construct a detailed diagram of the human major histocompatibility complex and list the major structural features of the MHC I and II molecules;
- describe exogenous and endogenous antigen processing and presentation mechanisms and list the major criteria governing binding of peptides to MHC molecules;
- construct a detailed diagram of an IgG molecule and identify structural and biological differences between antibody isotypes;
- define the terms isotypes, allotypes and idiotypes;
- list the biological functions of antibodies and describe the steps that result in complement mediated lysis, chemotaxis, opsonisation and phagocytosis;
- describe the chromosomal arrangement of antibody light and heavy chain gene segments and the basic DNA rearrangement mechanism;
- list the major features of light and heavy chain genes;
- explain the terms functional rearrangement, junctional diversity, N regions, somatic mutation and class switching;
- describe the major features and mechanisms of the physiological humoral response and explain how it results in protection of the host.

Antigen-antibody reactions

Antigen-antibody reactions

2.1 Introduction

We are now going to examine the many different ways to detect the presence of antibodies and means of quantifying them. At the same time, of course, since antibodies bind specifically to complex molecules such as proteins and simpler compounds such as haptens, availability of antibodies with specificity for any such molecule provides us with tools to detect and measure antigens. The choice of the method will obviously depend on the nature of the antigen ie whether it is a soluble protein, a bacterium or a mammalian cell and we shall give you some guidance on what method is suitable. Another important consideration is the sensitivity of the particular assay. If the levels of antigen or antibodies you wish to detect or measure are extremely low, then some of the methods we shall describe are of no use since they are not sensitive enough.

Within the confines of a single chapter it is obviously not possible to deal with all the methods used to measure antibodies and antigens. However, we shall introduce you to examples of all the major approaches which can be divided into:

- precipitation methods;
- agglutination methods;
- enzyme-labelled reagent methods;
- radiolabelled immunoassays;
- immunohistochemical assays;
- Western blotting.

Before we begin our studies of these various methods, let us examine the nature of the binding of antigen by antibodies.

2.2 The molecular basis of antigen-antibody reactions

2.2.1 The non-covalent nature of antigen-antibody reactions

There are no covalent bonds formed in antigen-antibody complexes, these being held together entirely by non-covalent interactions similar to those which stabilise enzyme-substrate interactions. However, complex formation differs in one important aspect from enzyme-substrate interactions in that neither participant ie antibody or antigen undergoes any change. Having said that, there are antibodies called abzymes which do catalyse reactions in the same manner as enzymes and we will deal with these later.

Π From your knowledge of protein-protein interactions can you list the types of non-covalent interactions which could be involved in formation of antigen-antibody complexes?

The forces holding the epitope within the antibody binding site are the same ones which are involved in protein-protein interactions and include hydrogen bonds, electrostatic or ionic interactions, van der Waals forces and hydrophobic associations. These interactions only operate effectively when the fit between the antigen and antibody is precise. For instance, van der Waals forces are only effective at physiological temperatures when there are several atoms from each molecule involved in the binding and hydrogen bonding only occurs between atoms in very close apposition to each other. You can imagine, then, that the interacting antigens and antibodies must have closely fitting complementary structures like a lock and key model. Within the binding site, there may be a predominance of ionic interactions in some antigen-antibody complexes whereas in others hydrophobic interactions may play a significant role. Having said that, it is thought that hydrophobic interactions may contribute 50% of the binding force in most interactions. In some interactions, the energy generated by these non-covalent interactions may be low and, in others, high; the antibodies involved are said to be low affinity and high affinity antibodies respectively. You will also conclude that minor changes in antigen structure will profoundly affect the binding energy of the interaction and this explains the extreme specificity of most antibodies.

complementary structures

2.2.2 Antibody affinity and avidity

affinity

avidity

The strength of binding between a single antibody combining site and the antigenic epitope is referred to as the affinity of the antibody. You must be able to distinguish this term from avidity which is used to describe the strength of binding of a whole antibody rather than a single binding site or even that of an antiserum which may consist of many different antibodies of varying specificities.

For example, if an IgG antibody is bound to a single epitope on an antigen with the other combining site unoccupied, then the avidity is the same as the affinity. If however, both IgG combining sites have interacted with two identical epitopes on the antigen, the binding energy (the avidity) is considerably greater (up to 1000-fold) than the sum of the two binding energies (affinities) of the individual sites because for dissociation to occur, both binding sites have to simultaneously break from the two epitopes. With IgM, the difference between affinity and avidity is even more striking (more than 10 million fold) since there may be up to 5 IgM binding sites attached to epitopes.

2.2.3 The law of mass action and affinity determination

association rate

dissociation rate

The formation of antigen-antibody complexes is reversible and obeys the law of mass action which states the rate at which the complexes form (the association rate) is proportional to the concentrations of antigen and antibody ie [Ag] x [Ab] and can be expressed as k_a [Ag] [Ab] where k_a is the association rate constant. The rate at which the complexes dissociate (the dissociation rate) can be similarly expressed as k_b [AgAb] where k_b is the dissociation rate constant. At equilibrium, we would expect that both rates are equal.

$$k_a [Ag] [Ab] = k_b [AgAb]$$

$$\text{And } K = \frac{k_a}{k_b} = \frac{[AgAb]}{[Ag] [Ab]} \qquad \text{Equation 2.1}$$

Let us consider monovalent antibodies ie an antibody using only one binding site, reacting with an epitope, say a hapten. We can call the total antibody concentration $[Ab_t]$, the free antibody concentration at equilibrium $[Ab_f]$ and the concentration of antibodies bound to antigen $[Ab_b]$. Then the concentration of free antibodies is:

$$[Ab_f] = [Ab_t] - [Ab_b] \qquad \text{Equation 2.2}$$

Substituting the value for $[Ab_f]$ into Equation 2.1 we get:

$$K = \frac{[AgAb_b]}{[Ag]\,[Ab_f]} = \frac{[AgAb_b]}{[Ag]\,([Ab_t] - [Ab_b])}$$

$$K\,[Ag]\,([Ab_t]-[Ab_b]) = [AgAb_b]$$

$$K\,[Ag]\,[Ab_t] = [AgAb_b] + K\,[Ag]\,[Ab_b] = [AgAb_b]\,(1 + K\,[Ag])$$

Rearranging:

$$\frac{K\,[Ag]}{1 + K\,[Ag]} = \frac{[AgAb_b]}{[Ab_t]} \qquad \text{Equation 2.3}$$

The right hand side of the equation represents the fraction of antibody molecules bound by antigen and we will now call this r. We thus obtain the following equation:

$$r = \frac{K\,[Ag]}{1 + K\,[Ag]} \qquad \text{Equation 2.4}$$

If we now also consider antibodies with n identical binding sites then it can be shown that Equation 2.4 becomes:

$$r = \frac{n\,K\,[Ag]}{1 + K\,[Ag]} \qquad \text{Equation 2.5}$$

If we now call the antigen concentration c and rearrange Equation 2.5, we can get:

$$\frac{r}{c} = nK - rK \qquad \text{Equation 2.6}$$

(Try to do this rearrangement for yourself)

If we know values of r and c over a range of antigen concentrations we can plot a graph of r/c against r. This is called a Scatchard plot (Figure 2.1) and it will be a straight line of slope -K. From the intercept with the x axis, the value of n, (the number of binding sites per antibody molecule) can be deduced and from the intercept with the y axis, a value for nK is found. As noted below, only monoclonal antibodies with specificity for a hapten will give a straight line plot and a calculation of the affinity constant using the slope is only applicable to these antibodies.

For a divalent antibody, when half the antibody binding sites are occupied (r = 1) and n = 2 then it follows from Equation 2.6:

$$K_0 = \frac{1}{[Ag]} \text{ or } \frac{1}{c}$$

average association constant

K_0 here is the average association constant which is the reciprocal of the free antigen concentration. For a divalent antibody such as IgG you could obtain this value by reading the r/c value when r = 1 from a Scatchard plot.

The unit of K are litres per mole (1 mol^{-1}) and that of c is moles per litre. Low affinity antibodies invariably have K values of about 10^5 1 mol^{-1} whereas high affinity antibodies may have K values as high as 10^{11} 1 mol^{-1}.

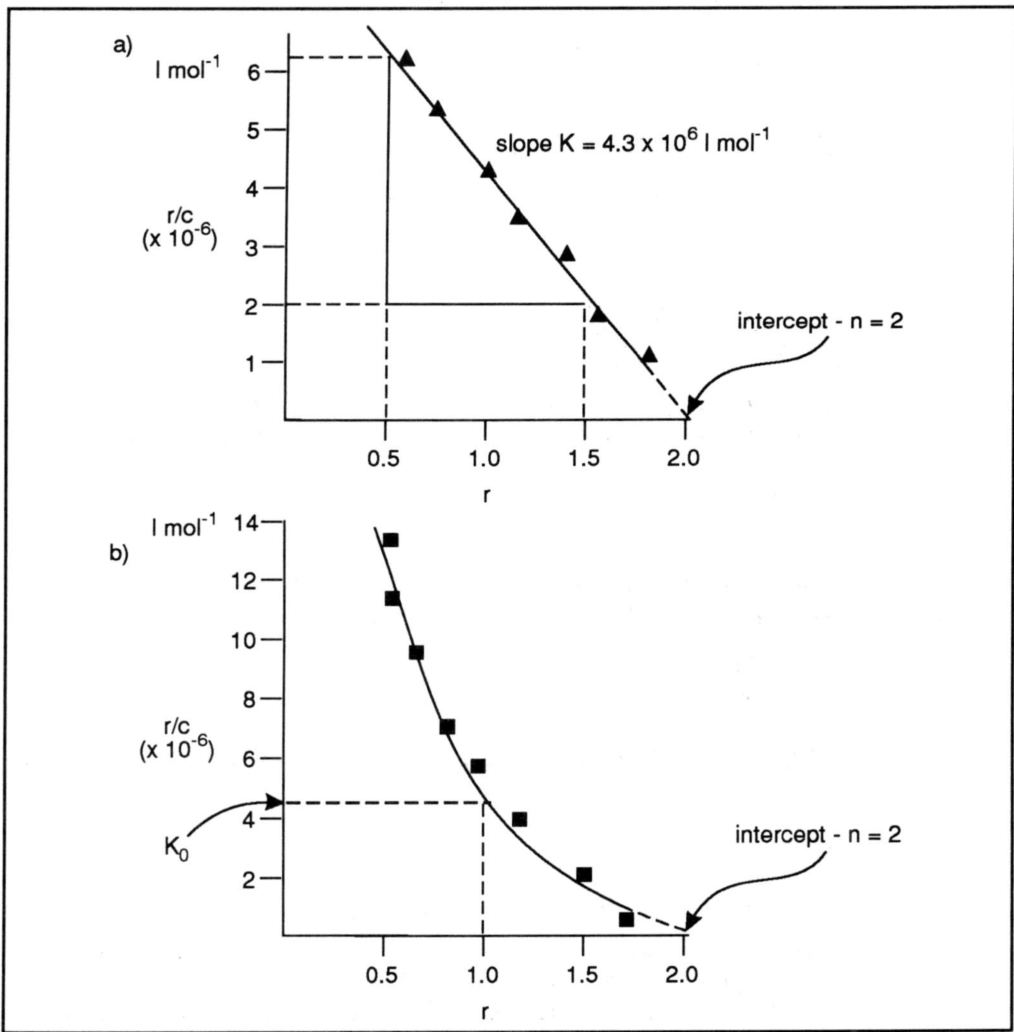

Figure 2.1 Scatchard plots of hapten (ligand) binding by a monoclonal antibody a) and heterogeneous antibodies b). Plot a) is of monovalent haptens binding to monoclonal antibodies where all antibody combining sites possess the same association constant. The plot of r/c = nK - rK is therefore a straight line. Plot b) shows results obtained using polyclonal antibodies, the non-linearity being due to the variety of binding sites possessing different K values amongst the antibody population. The average value for K, designate K_0, is the reciprocal of the hapten concentration when r = 1 ie when, on average, each antibody molecule is binding one hapten molecule (one out of two binding sites occupied). For the hypothetical results shown, the intercept on the x axis is 2 indicating the antibodies are divalent, eg IgG.

Antibody affinity can also be calculated using the Langmuir plot (Figure 2.2) using an equation derived from Equation 2.5 where:

$$\frac{1}{r} = \frac{1}{nK} \cdot \frac{1}{c} + \frac{1}{n}$$

If we plot 1/r (y axis) against 1/c (x axis) the intercept on the y axis is 1/n and the slope of the line is 1/nK.

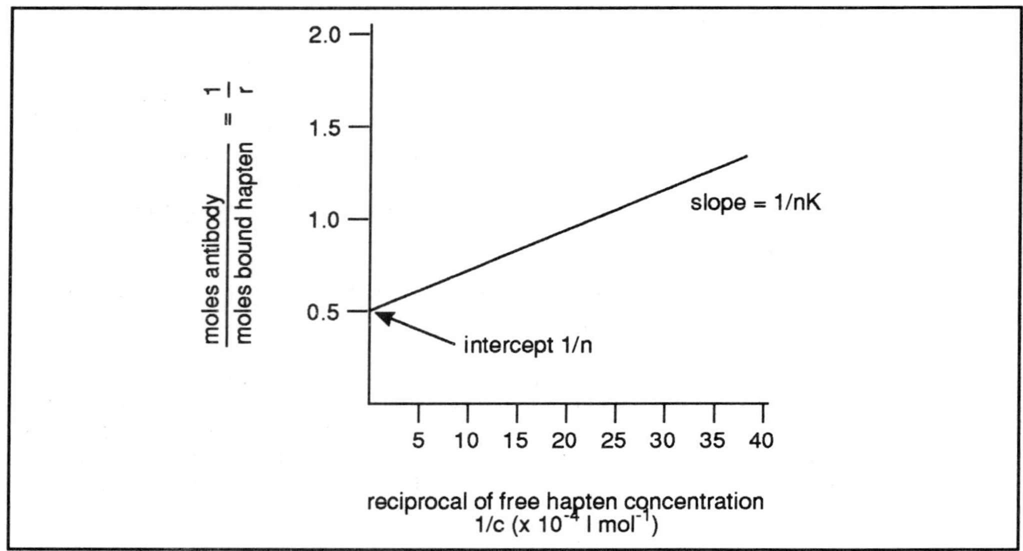

Figure 2.2 The Langmuir plot. The idealised plot using homogenous antibodies and monovalent antigen.

Π From the Langmuir plot in Figure 2.2 determine the valency of the antibody and the approximate value of K.

Since the intercept on the y axis is 1/n, then n = 2 and the valency of the antibody is 2. If you estimated the slope by reading off the x and y axis values as was done in Figure 2.1 you should have found K to be about 2.5×10^5 1 mol^{-1}. (Slope = $0.8/40 \times 10^{-4} = 1/nK$).

Π Can you write down a simple rule of thumb describing the correlation between antigen concentration and antibody affinity?

The rule is simply that the higher the affinity of an antibody, the lower the concentration of free antigen (ligand) needed for the binding sites to be occupied.

Homogenous and heterogenous antibodies

In reality, only homogenous antibodies give a straight line plot in affinity determinations. This is because all the antibodies possess the same affinity for the epitope (hapten) and the Scatchard plot will be a straight line with slope of -K. However, with heterogenous antibodies such as those present in an antiserum, there are many antibodies with different affinities for the one epitope and the plot will deviate from a straight line (Figure 2.1). In such cases, an average association constant (K_0) is determined assuming that each antibody in the population is using one binding site only as shown in the figure.

The heterogeneity index

This heterogeneity is often evaluated by the Sips distribution function which leads to the equation:

$$\log \frac{r}{(n-r)} = a \log K + a \log c$$

where a is the heterogeneity index which gives an estimate of the deviation of the antibodies from the average affinity. If you assume divalence of the antibody (n = 2) and plot log (r/(2-r) against log c you will obtain a straight line and K can be directly read from the plot. When half the binding sites are filled (r = 1) then log (r/(2-r) = 0 and K is the reciprocal of c. The slope of the graph is the heterogeneity index a.

Π A Sips plot gave a slope of 1. Is the antibody a monoclonal or a polyclonal antibody preparation?

The slope of 1 means that the heterogeneity index = 1. In other words, all of the antibodies bind with the same affinity. This indicates that the antibody preparation is a homogeneous one and contains a single type of antibody molecules. In other words, it is a monoclonal antibody preparation. With polyclonal antibodies, the heterogeneity index is less than 1 reflecting multiple affinities for the ligand with an average affinity of K.

2.2.4 The experimental determination of affinity

There are a host of methods by which affinity or association constants can be measured. Within the confines of this chapter we cannot review them all and we shall concentrate on the used methods most often used.

Equilibrium dialysis

Dialysis chambers consist of two compartments separated by a semipermeable membrane which allows free movement of small molecular weight substances such as haptens but not of large molecules such as antibodies. An alternative and less expensive method is to use dialysis sacs containing the antibody solution suspended in a fixed volume of hapten solution at various concentrations. Figure 2.3 illustrates the principals of this technique. At the start of the experiment, a number of chambers contain a fixed amount of antibody in one compartment and varying amounts of the radiolabelled hapten in the other compartment. During the experiment, the hapten will freely diffuse into the antibody compartment until equilibrium is reached. If the antibody is not specific for the hapten, then the final concentration of hapten will be equal in both compartments. However, with specific antibodies in the compartment, more hapten will be found in the antibody compartment due to binding of hapten by the antibodies. The amount of free hapten and bound hapten can be determined by measuring the radioisotope in each chamber and these figures can then be used to determine K and n using Scatchard analysis.

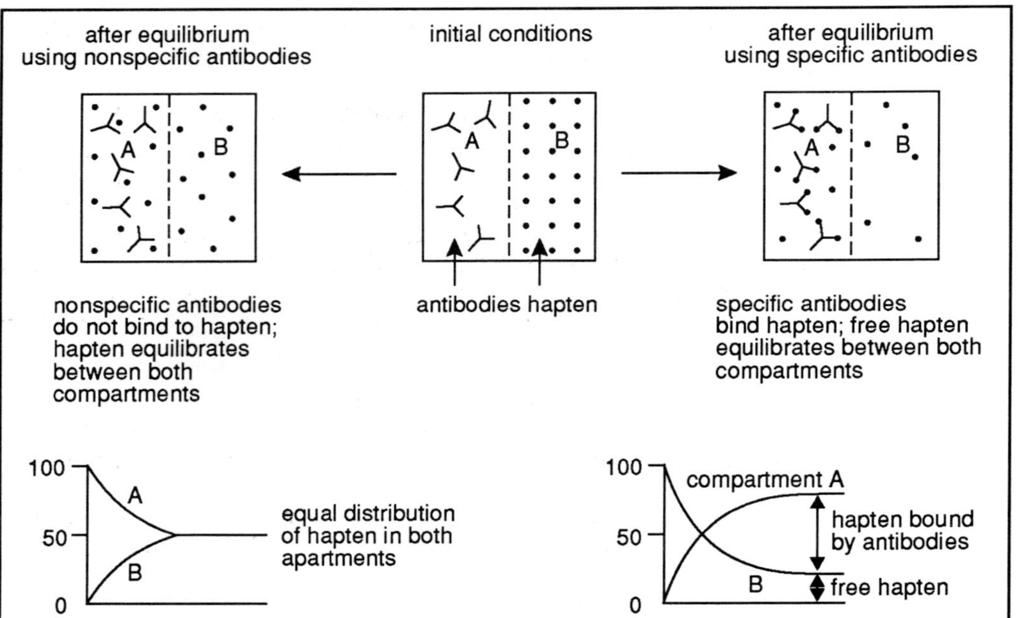

Figure 2.3 Measurement of affinity using equilibrium dialysis. Dialysis chambers are set up with a fixed amount of antibodies in one apartment and varying amounts of ligand (hapten) in the other. At equilibrium in the left chamber the concentration of radiolabelled hapten is equal in both compartments A and B as the antibodies are not specific for the hapten. At equilibrium in the right chamber the concentration of hapten in compartment A is greater than in B as some hapten has bound to the specific antibodies and cannot freely move between compartments. However the free hapten concentration will be equal in both compartments. The amount of radiolabelled hapten bound to the antibodies will therefore be the total cpm in compartment A - the total cpm in compartment B. In practice, many chambers are set up containing different amounts of radiolabelled hapten so that many ratios of free/bound ligand are obtained.

SAQ 2.1

Construct a graph from the following data and determine the valency of the antibody and the value of K.

r/c (l^{-1} μmol)	r (= fraction of Ab molecules binding hapten)
4	0.20
3	0.60
2	1.15
1	1.60

Ammonium sulphate globulin precipitation

This technique is used to determine the relative levels of specific antibodies in sera but can be adapted to estimate association constants. The basis of the technique introduced by Farr is that 40% saturated ammonium sulphate solution precipitates antibodies and antigen-antibody complexes but not free antigen in most cases. The procedure, then, involves radiolabelling the hapten and then incubating various amounts of this with a fixed amount of antibody followed by ammonium sulphate precipitation of the antibody-hapten complexes. The amount of bound and free hapten can then be estimated and the association constant determined in a way similar to that described for equilibrium dialysis.

Fluorescence quenching and fluorescence enhancement

aromatic amino acid residues

Since antibodies are proteins they fluoresce in the ultraviolet region because of the presence of aromatic amino acid residues (mainly tryptophan). The absorption is maximal at 290nm and the emission at about 345nm. When antibodies bind some haptens or antigens, some of the excitation energy is transferred to the hapten resulting in a reduction in the fluorescence of the antibody - hence the name fluorescence quenching. Thus, by varying the amount of hapten one can produce a similar graph to that for equilibrium dialysis since the amount of quenching relates to the amount of bound hapten and the association constant can be determined.

This method is not as useful as equilibrium dialysis since controls have to be introduced to account for unequal quenching by the first and second ligand with a bivalent antibody but it is a more rapid technique and can be used for non-dialysable ligands.

Also some of the emitted light may be adsorbed by some free reagents in the reaction mixture. Thus controls need to be run to check and allow for non-specific quenching.

fluorescence enhancement

Some small organic molecules are themselves fluorescent but have a different fluorescence spectrum when bound by antibody. For instance, free folic acid exhibits maximum fluorescence at 350nm which shifts to 362nm when bound to antibody. This change can be measured and is dependent on the number of antibody-combining sites occupied by the fluorescent molecules. Thus the affinity can be determined. This process is called fluorescence enhancement.

SAQ 2.2

Which one of the following statements is correct?

1) An antibody with a K_0 value of 10^6 1 mol^{-1} is of a higher affinity than one with a K_0 value of 10^{10} 1 mol^{-1}.

2) Antibody avidity describes the binding strength of each individual binding site of an antibody molecule.

3) Equilibrium dialysis is a generally applicable method to determine association constants for antibodies to most antigens.

4) Fluorescence enhancement can be used with antigens which fluoresce themselves.

5) The higher the antibody affinity, the higher the concentration of free ligand needed for occupation of all the antibody binding sites.

We have now completed our review of the primary interactions between antibodies and antigens. These describe the actual interaction between the antibody binding site and the epitope. We are now going to look at the secondary reactions or consequences which result from these primary reactions and we shall begin by looking at the precipitation of antigen-antibody complexes.

2.3 Antibody-antigen reactions - precipitation reactions

2.3.1 The precipitin reaction

Interactions between antibodies and antigens can result in the formation of an insoluble precipitate. However, there are various criteria to be met to achieve precipitation of immune complexes. For instance, hapten-antibody complexes remain soluble in solution and no precipitate is observed. This is due to there being only a single determinant (epitope) on most haptens and only one antibody can bind to it; hence no large complexes are formed.

Π Can you think of the conditions which are necessary for immunoprecipitation in solution to occur?

Precipitation only occurs when antibodies can link together antigen molecules to form a large complex. This obviously implies that the antigen molecule must have at least two epitopes and the antibody must be at least divalent.

Π However, some monoclonal antibodies do not precipitate antigens. Can you think why?

For monoclonal antibodies to precipitate antigens there must be at least two identical epitopes on each antigenic molecule so that the antigens can be linked together by the antibodies. You can imagine that it is easier to form large complexes with polyclonal antibodies since there will be many opportunities for cross linking of multiple epitopes.

2.3.2 The precipitin curve

Not all antigen-antibody complexes are insoluble because the solubility of the complexes is dependent on the ratio of antibody to antigen. This can be examined very simply as follows. A series of test tubes is set up each tube containing a constant amount of antibody solution. Increasing amounts of antigen are now added to each consecutive tube and the tubes are then incubated for 24hr followed by centrifugation. The pellets (precipitates) consisting of antigen-antibody complexes are quantitated and the amount of free antibody or antigen in the supernatant is also determined. There are various methods for quantitating the samples; these include UV absorption, colour reactions of proteins such as using ninhydrin or the Folin-Ciocalteau reagent, ELISA (Section 2.6.2) or radiolabelling of either the antibody or the antigen.

prozone

Figure 2.4 shows a typical precipitin curve after such an experiment. You will notice that the curve can be divided into three zones. The first zone, called the prozone represents an excess of antibody over antigen. In the test tubes you can detect free antibody but no free antigen and there are few if any large complexes. This is because there is an excess of antibodies so that they tend to bind to single antigenic molecules rather than linking them together. You can see the types of complexes we would expect in Figure 2.4.

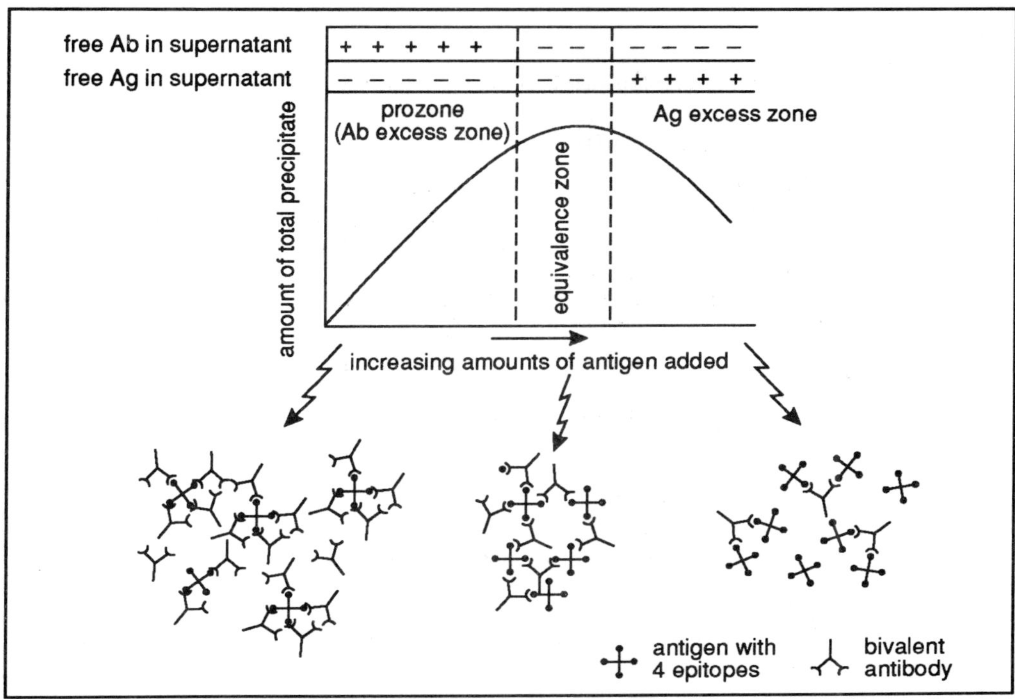

Figure 2.4 The precipitin curve: the amount of precipitate plotted against the amount of antigen added. The figure also shows the types of complex predominant in each zone. In the prozone, the predominant complex will consist of multiple antibody molecules bound to single antigen molecules. Most of the complexes are soluble. In the equivalence zone antibody molecules are crosslinked with antigen molecules forming large complexes; there is no free antibody or antigen molecules left in the supernatant. In the antigen excess zone antibody molecules are relatively scarce and may be used to crosslink two antigen molecules giving a ratio of 2:1 (Ag:Ab). We show the antigen molecules as having 4 identical epitopes and the antibodies having identical specificities. However, a similar plot would result if polyclonal antibodies (say in serum) were used against antigen molecules bearing different epitopes.

equivalence zone

The next zone, called the equivalence zone, is characterised by the absence of both antibodies and antigens in the supernatant since they are all contained in the large complexes which have precipitated out of solution. In the test tubes represented by this zone, the antibodies and antigens have crosslinked forming a stable lattice as shown in Figure 2.4. The third zone, called the antigen excess zone is characterised by a lack of free antibodies and the presence of free antigen molecules in the supernatant. In these test tubes, the excess antigen molecules have bound to individual antibody molecules preventing crosslinking as shown in Figure 2.4.

antigen excess zone

There is an alternative to the so-called lattice theory. It has been suggested that antigen and antibody may combine to form both soluble and insoluble complexes. The latter form the core of the precipitate but substantial quantities of the soluble complexes coprecipitate due possibly to interactions between the Fc regions of the participating antibodies.

2.3.3 Determination of the amount of specific antibodies in antiserum

You should be able to see that the precipitin reaction provides a method for determining the amount of specific antibodies in a sample of, for example, antiserum. Look at Figure 2.5 and read the legend for the experimental details. Since we added a known amount of antigen to the test tube which contains the highest amount of precipitate (representing equivalence) we can calculate the amount of antibodies present in the precipitate using optical density data as follows. The extinction coefficient of 1mg ml^{-1} of IgG in sodium hydroxide is 1.439 and that of albumin is 0.530. Since we know that 85µg of albumin are present in the precipitate, this would result in an extinction of 0.045.

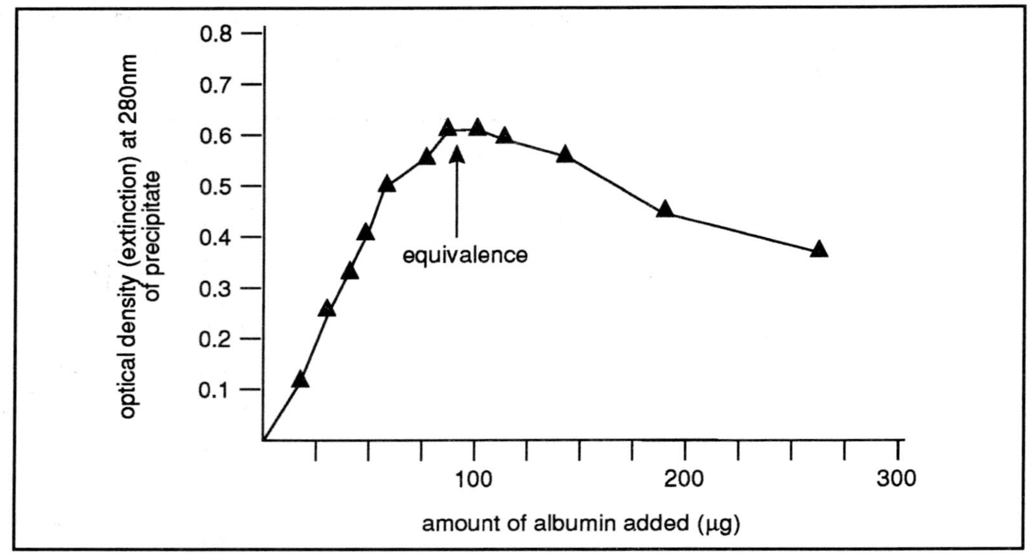

Figure 2.5 Precipitin curve of albumin with anti-albumin antiserum. Increasing amounts of albumin were added to 0.1ml aliquots of antiserum and the extinction of the resulting precipitates redissolved in sodium hydroxide were determined. The equivalence point (when 85µg albumin had been added) was determined by ELISA analysis of the supernatant; no free antibodies or antigens were found in the supernatant at this point.

Π Before you read on, calculate from this information and data from Figure 2.5 the amount of albumin specific IgG in the antiserum.

Since the extinction of the precipitate dissolved in sodium hydroxide from the graph (Figure 2.5) is 0.61, the amount of IgG in the precipitate has an extinction of 0.61 - 0.045 = 0.565. This represents 0.565/1.439 mg IgG = 0.393mg. This was contained in 0.1 ml so the concentration of IgG anti-albumin antibodies is 3.93mg ml of antiserum.

SAQ 2.3

Assuming that the molecular weights of albumin and IgG are 68 000Da and 150 000Da respectively, calculate the molar ratio of IgG to albumin in the precipitate at equivalence from the above data.

2.3.4 Determination of antigenic valency

Heidelberger plots

The same approach can be used to determine how many epitopes per antigenic molecule can be bound by antibodies. At extreme antibody excess you would expect that all epitopes will be bound by antibodies and an analysis of the antibody in the precipitate at this point will give us the molar ratio of antibody to antigen. By determination of the ratios of antibody to antigen in precipitates after addition of various amounts of antigen, we can construct plots similar to those in Figure 2.6. Such plots are called Heidelberger plots. The intercept on the y axis (infinite antibody excess) will indicate the number of antibody molecules which bind to one molecule of antigen and hence the number of available epitopes per molecule of antigen. Figure 2.6a shows the type of plot one would expect to obtain when a protein molecule with 5 different epitopes is reacted with a mixture of 5 monoclonal antibodies each with specificity for one of the epitopes. In antibody excess you would expect that all epitopes would be bound by antibodies resulting predominantly in Ab_5Ag complexes. However, a linear plot is not always obtained as is shown in Figure 2.6b. A curved plot may be due to heterogeneity in affinity of the reacting antibodies and also may arise through differences in the relative availability of epitopes on the antigen.

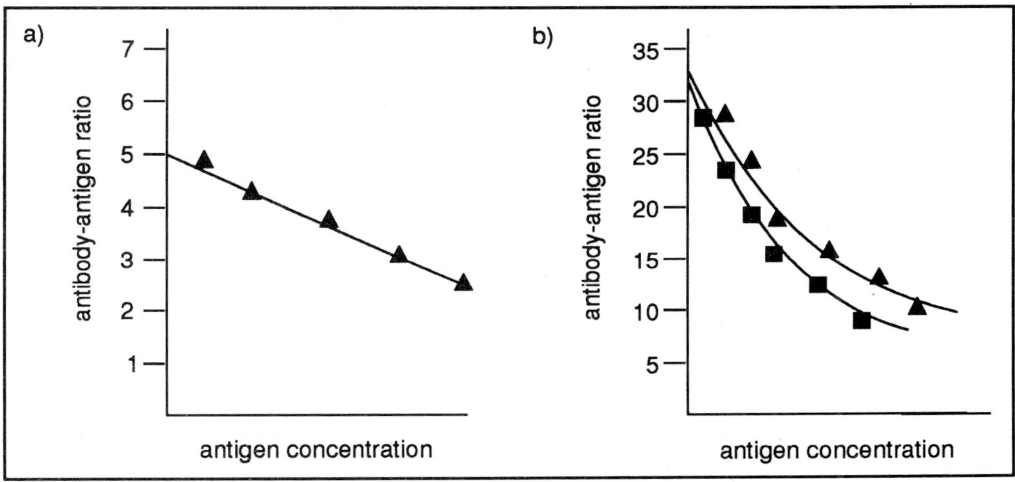

Figure 2.6 Determination of antigen valence. Figure 2.6a shows an idealised Heidelberger plot extrapolated to antibody excess giving an antigen valency of 5. Thus the majority of complexes would consist of Ab_5Ag. In many experiments using heterogeneous antibodies a curved plot is obtained as shown in Figure 2.6b. These two curves extrapolate to a valency of about 33 indicating the presence of 33 epitopes on each antigen molecule. Some of these could represent repeat epitopes at different sites.

2.3.5 Precipitation methods for detection of antigen-antibody reactions

Precipitation methods are performed in solution or in semisolid media such as agar gel and are designated single diffusion methods if only one of the two participants (antigen or antibody) actively diffuses or double diffusion if both participants diffuse towards each other. Examples of the single diffusion technique are the Oudin and Mancini methods and examples of the double diffusion are the Oakley-Fulthorpe tube technique and the Ouchterlony plate technique.

Single diffusion

Oudin technique

Antiserum is added to molten agar and then allowed to set in either tubes (Oudin technique) or glass plates (Mancini technique). In the Oudin technique a solution of antigen is then layered on to the top of the agar and the antigen is allowed to diffuse into the agar. A concentration gradient forms in the agar which decreases from the antigen solution to the bottom of the tube. A precipitate forms when the concentrations of antigen and antibody are equal (the equivalence zone). As more antigen diffuses into the gel, however, this precipitate dissolves due to antigen excess and reappears further down the tube. The movement of the precipitin band down the tube will depend on the concentration of antigen on top of the agar relative to the concentration of antiserum in the agar and on the size and shape of the antigen since smaller molecules will diffuse faster in the agar. Obviously, if there are several antigens in the solution of different molecular weights, then several bands will appear in the agar.

Π In an experiment, a monoclonal antibody was mixed with agar and poured into a tube. A mixture of antigens with common epitopes was overlayered on the agar and diffusion allowed to take place. After 24hr, 4 precipitin bands were observed. Does this experiment tell us how many antigens were present in the mixture?

The number of precipitin bands tells us the minimum number of antigens with which the antibody was able to interact to form precipitates but it is not an estimate of the total number of antigens in the mixture. This is because some precipitates may be due to the antibody precipitating with different antigens which diffuse at about the same rate resulting in overlapping precipitin bands.

single radial immunodiffusion (SRID)

The Oudin tube technique is an example of diffusion in one dimension. The Mancini technique, also called single radial immunodiffusion (SRID) is an example of single diffusion in two dimensions and is the basis of an important quantitative assay. In this technique, holes are punched into the agar and serial dilutions of known quantities of the antigen and samples of test antigen solution (concentration unknown) are inserted into the wells. The slides are then kept in a damp atmosphere (to prevent drying out of the gels) for a few hours. The antigens radially diffuse into the agar surrounding each well and will form a precipitin ring with the antibody in the agar at equivalence. The distance of the ring from the well is directly proportional to the concentration of the antigen originally present in the well.

The diameter of each ring is measured. With uniform antibody concentration and gel thickness throughout the plate, the area enclosed by the precipitin ring is directly proportional to the antigen concentration, A. The practice is to plot the square of the diameter of the ring against antigen concentration and, if the diffusion has gone to completion, which may take a few days, then the plot is a straight line according to the equation:

$$d^2 = (K \times A) + w$$

where w is the intercept, this being dependent on the volume of antigen in the well and its diameter. The slope K is inversely proportional to the antibody concentration in the gel. Look at the example given in Figure 2.7.

This method is frequently used in hospital laboratories for determining the concentrations of plasma proteins in serum samples, particularly IgG, IgM and IgA. It cannot be used for assessing IgE as the method is not sensitive enough (about 1.5mg l^{-1}).

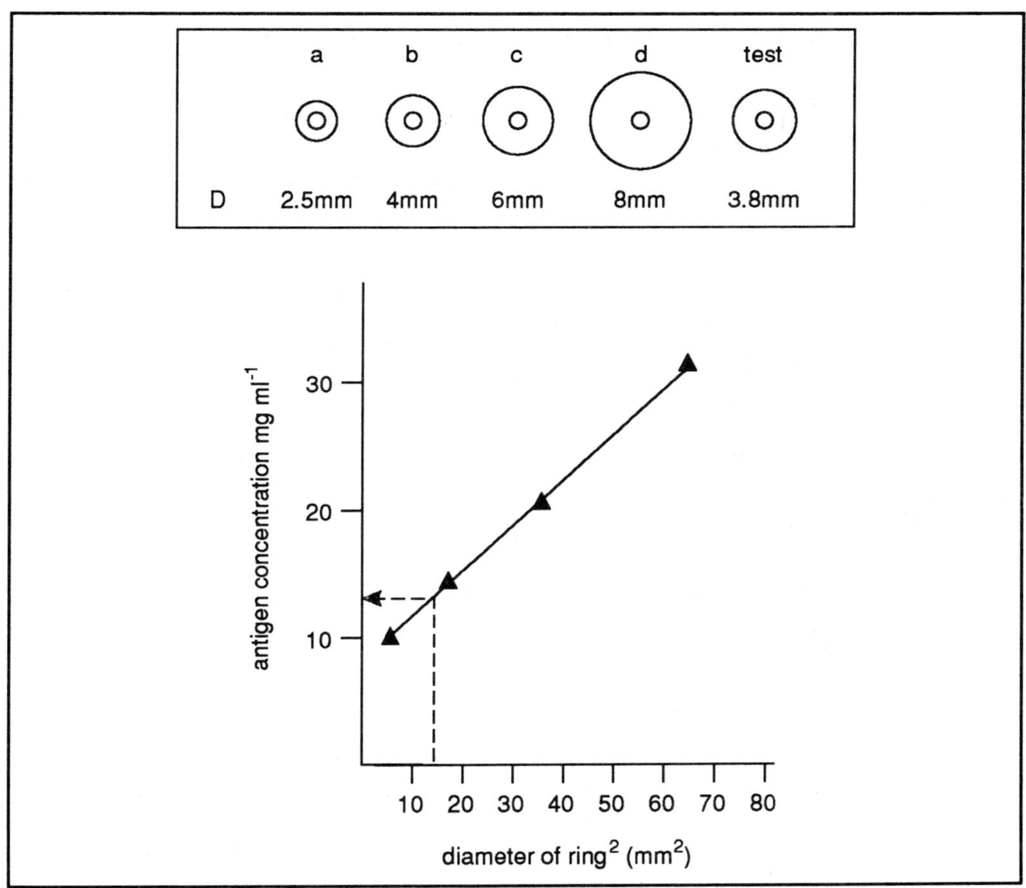

Figure 2.7 Single radial immunodiffusion measurement of antigen. Standard concentrations of antigen (a-d) ranging from 10 - 32mg ml^{-1} and a test sample of unknown concentration were subjected to SRID and the diameters of the precipitin rings measured on completion of diffusion. These values were squared and plotted against the standard antigen concentrations. From the graph the concentration of the antigen in the unknown sample is about 13mg ml^{-1}.

SAQ 2.4

Plot a graph of IgG concentration against diameter of the results of a Mancini assay shown below and determine the concentration of the IgG test sample which was diluted once in saline prior to the test and which gave a ring diameter of 5.24mm. Select your answer from those provided below.

IgG concentration (mg ml^{-1})	Ring diameter (mm)
0.45	5.91
0.345	5.0
0.29	4.47
0.241	3.87
0.130	2.24

1) 0.24mg ml^{-1} 2) 1.0mg ml^{-1} 3) 0.74mg ml^{-1} 4) 0.145mg ml^{-1} 5) 0.56mg ml^{-1}.

Double diffusion

Oakley-Fulthorpe technique

In this technique, both antibody and antigen diffuse towards each other forming a precipitin band at equivalence. The Oakley-Fulthorpe technique is a modified version of the Oudin technique. A layer of agar containing antibody is at the bottom of the tube being overlaid with a layer of agar. The antigen solution is then placed on top of the agar layer. The antigen diffuses down into the agar and the antibody diffuses from the lower layer into the middle agar layer to form precipitin band(s).

Ouchterlony technique

reaction of identity

The more commonly used Ouchterlony technique is performed on plates of agar in which are punched wells in which is placed either antigen or antibody. As you can see from Figure 2.8 antibody and antigen diffuse radially from the wells and where they meet at equivalence a precipitin band appears. This technique enables us to discern antigenic relationships between molecules using polyclonal antibodies present in antisera. Examine Figure 2.8. You will notice that there are 3 major patterns which can be obtained. In Figure 2.8a, the precipitin line is continuous between the antibody well and the two antigen wells. This is called a reaction of identity indicating that the antibodies present in the antiserum are binding to the same epitopes on both antigens. This is what you would expect if the antigens were identical as they are in the figure.

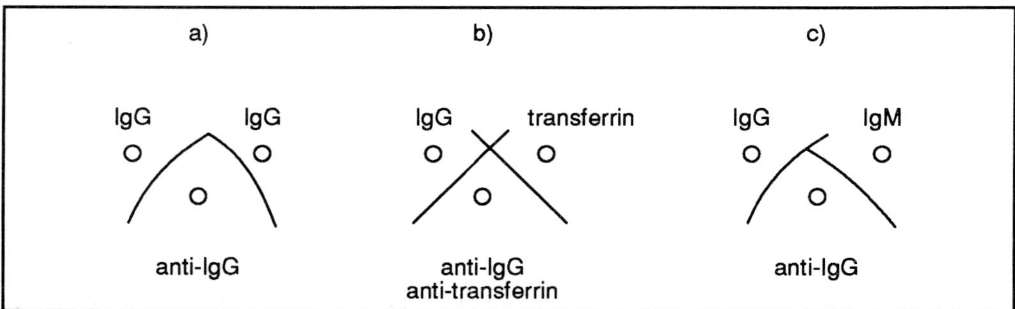

Figure 2.8 Ouchterlony technique showing the reactions of identity a) non-identity b) and partial identity c). Note the spur in c) which points to the IgM well (see text).

reaction of non-identity

If the antigens in the adjacent wells are not structurally related ie they have no common epitopes, then the precipitin lines would cross rather than form a continuous line (Figure 2.8b). This is the sort of result you would expect if you had anti-IgG and anti-transferrin in the centre well and IgG and transferrin in adjacent separate wells. This is called a reaction of non-identity.

reaction of partial identity

If the antigens are not identical but structurally related ie they possess some common epitopes then a different pattern emerges. This time you will observe a continuous precipitin line as with the reaction of identity but there will be an extension of the precipitin line, called a spur, at the intersection of the two lines. This spur will point in the direction of the well which contains the antigen which lacks some of the epitopes. This is the result you would expect if the antibody well contained anti-whole IgG antiserum and the antigen wells contained IgG and IgM. There would be a continuous precipitin line formed by anti-light chain antibodies which would form a precipitate with both IgG and IgM and a spur formed by precipitation of IgG by antibodies to the heavy chain of IgG (epitopes not present in IgM). This is called the reaction of partial identity.

If there are multiple antigens in the sample such as is the case for precipitation of whole serum by anti-serum antibodies, then you will observe multiple precipitin lines due to precipitation of antigens of different molecular weight precipitated by specific antibodies in the antiserum. For instance, you would see a line for IgG-anti-IgG, one for albumin-anti-albumin and one for IgM-anti-IgM. If the antibodies are all IgG they will diffuse at the same rate through the gel but albumin, IgG and IgM will diffuse at different rates. The albumin precipitin band would be neared to the antibody well (molecular weight of albumin 65 000 Da), the IgG precipitin band would be next (IgG molecular weight 150 000 Da) and the IgM precipitin band would be nearest to the antigen well due to the slow diffusion of IgM (molecular weight 900 000 Da). You should remember that the number of precipitin bands observed represents the minimum number of antigens present in the mixture, not necessarily the actual number since some precipitin bands may contain heterologous complexes due to a similar diffusion rate by different antigens.

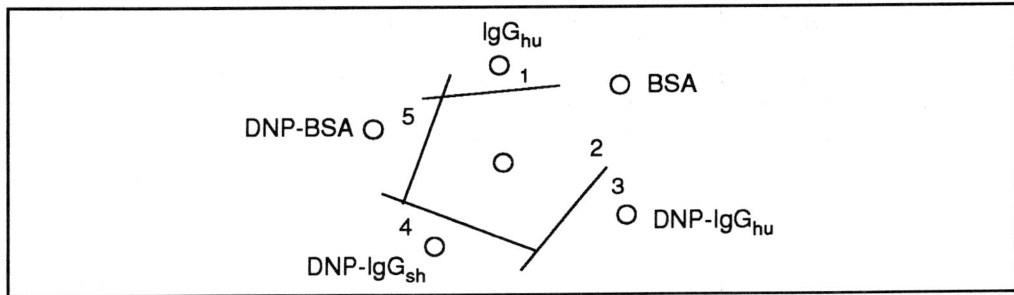

Figure 2.9 Results of an Ouchterlony test (SAQ 2.5). Antiserum in centre well is anti-DNP-IgG_{hu}. IgG_{hu} = human IgG; IgG_{sh} = sheep IgG.

SAQ 2.5

Which of the following statements correctly explain the precipitin patterns found in Figure 2.9? Refer in your answers to the numbers assigned to each precipitin reaction.

1) Line 5 shows that the antiserum is specific for BSA.

2) There are epitopes on human IgG which are not present on sheep IgG (line 3).

3) Lines 4 and 5 show a reaction of partial identity.

4) Lines 1 and 5 show a reaction of non-identity between human IgG and BSA.

5) The lack of a precipitin line in 2 is most probably due to a low concentration of BSA.

immuno-electrophoresis

An important qualitative test based on double diffusion is immunoelectrophoresis in which antigens are first separated by electrophoresis and then identified by precipitation using specific antisera. Microscope slides or glass plates are coated in warm agar and this is allowed to gel. Wells and troughs are then punched in the gel to produce a pattern such as that shown in Figure 2.10. A small volume of the antigen mixture (minimal concentration 50-100 $\mu g\ ml^{-1}$) is then electrophoresed across the slide using electrophoretic buffer of about pH 8 -8.6 at which pH most proteins possess an overall negative charge and will move at various rates towards the anode. The troughs are then filled with suitable antisera and the slide placed in a water - saturated atmosphere overnight. During this period, both antigens and antibodies will diffuse towards each other and form precipitin lines at equivalence. Using this technique, it is possible to identify about 30 different proteins in serum using an antiserum to whole human serum prepared in rabbits.

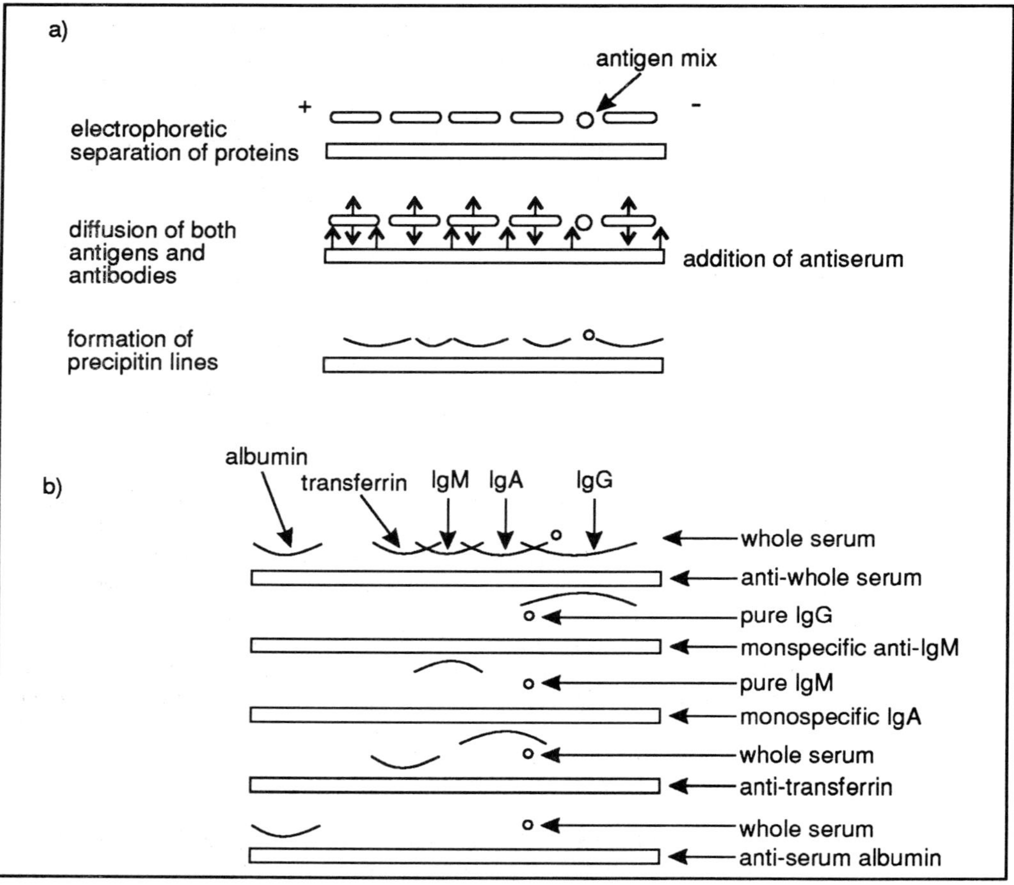

Figure 2.10 Immunoelectrophoresis. Figure 2.10a shows the principle of the method. A well in the agar is filled with antigen and the gel subjected to electrophoresis. Antiserum is added to the trough and the antigens and antibodies allowed to diffuse towards each other forming precipitin arcs. Figure 2.10b shows stylised results of immunoelectrophoresis of whole serum. The majority of the precipitin lines have been omitted for clarity. Each of the precipitin lines can then be identified using either purified antigens and anti-whole serum or whole serum and monospecific antisera.

Individual antigens can be identified using monospecific antisera as shown in Figure 2.10. In such an analysis, it is usually observed that the IgG precipitin line is on the cathodic side of the application well as shown in Figure 2.10. This is due to electro-endosmosis which results in movement of buffer fluid towards the cathode ie in a direction opposite to that of the anodically migrating proteins. This can partly be corrected by subjecting the gel to electrophoresis for 30 minutes before adding antigen to the well.

Π Examine the precipitin pattern in Figure 2.10. What does it tell you about the overall charge on albumin?

Since albumin moves the farthest distance towards the anode it must be the most electronegative of the two proteins.

two dimensional immuno-electrophoresis

This method can be modified to produce a semiquantitative method called two dimensional immunoelectrophoresis or crossed electrophoresis. In this technique, the strip of gel containing the electrophoresed proteins is laid on another plate and covered with warm antibody-containing gel. This is allowed to set and then an electric current is applied perpendicular to the direction of the initial protein migration. A series of precipitin peaks results each of which is proportional to the initial concentration of antigens in the mixture. This method is restricted to proteins of anodic migration.

counter immuno-electrophoresis

Other electrodiffusion methods include counter immunoelectrophoresis and rocket immunoelectrophoresis. The first method makes use of the observation that IgG antibodies tend to move to the cathode during electrophoresis. Thus a modification of the Ouchterlony method is to apply a current to gels in which wells contain IgG antisera (which will move towards the cathode because of electro-endosmosis) and anodically migrating antigens. This technique is called counter immunoelectrophoresis. The resulting precipitin lines appear in a much shorter time than in the Ouchterlony method.

Rocket immunoelectrophoresis is a modification of the single radial immunodiffusion method based on the same observation on IgG as the above method. In this method wells to contain the antigens are cut into antibody-containing agar on a plate. Once again the antigens have to be anodically migrating proteins (not IgG). On applying an electric current, the antigens migrate towards the anode precipitating with the antibodies as they migrate through the gel. When all the antigen is used up from each well, the precipitation stabilises generally forming a rocket shaped precipitin arc. There is a linear relationship between peak height and antigen concentration and a standard curve can be constructed using standard antigen solutions. This method has moderate sensitivity measuring down to about $1 \mu g \, ml^{-1}$.

Π Examine Figure 2.11 and make sure you know the principle of each method. If the concentration of antibody in the gel used for rocket electrophoresis was doubled would the precipitin arcs have been larger or smaller?

The answer is that they would have been smaller. As the antigen migrated into the gel, it would form complexes with the antibody and would precipitate. The more antibody present in the gel, the shorter the distance the antigen would have to migrate before it would be removed from migration by forming complexes.

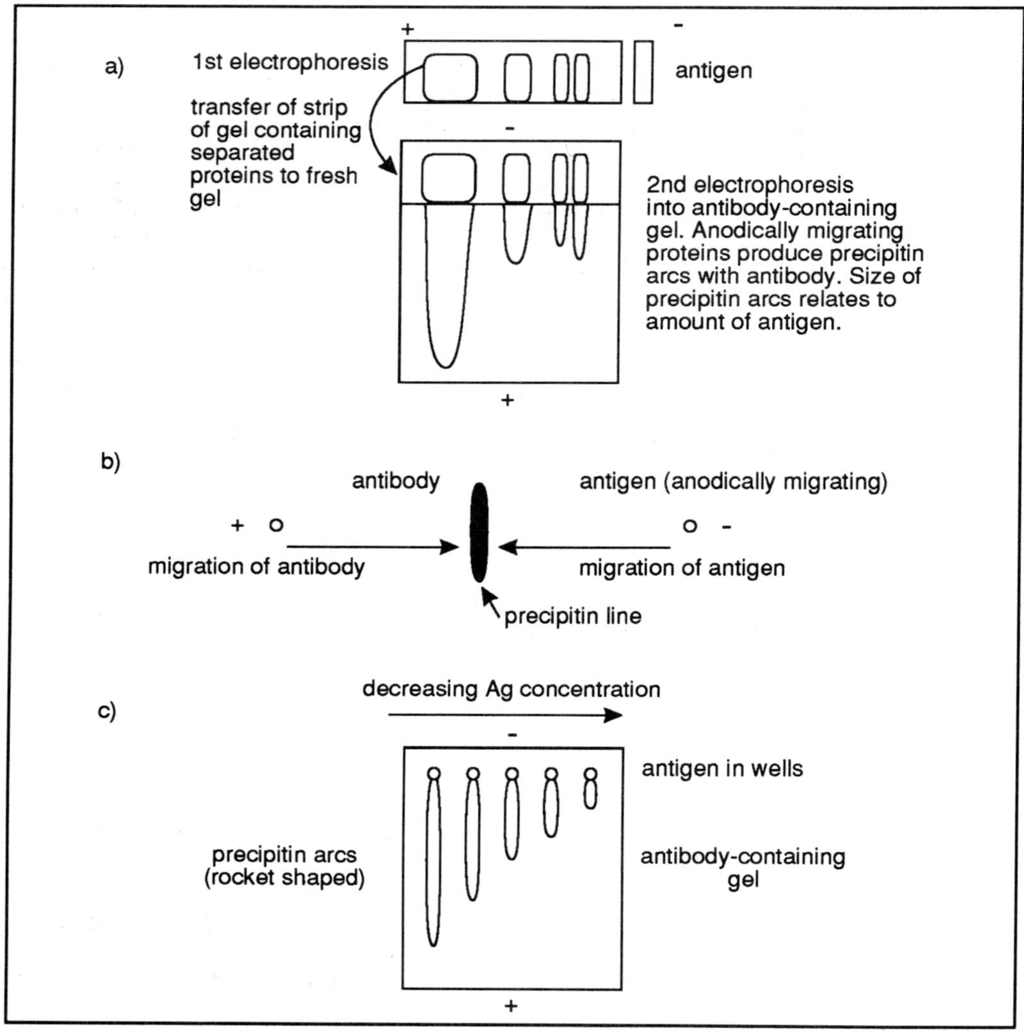

Figure 2.11 Electrodiffusion method for antigen-antibody reactions. a) Two dimensional immunoelectrophoresis. The mix of antigens are separated in the first electrophoresis and the gel strip containing the separated proteins is placed on a glass plate and covered with antibody containing gel. Electrophoresis of anodically migrating proteins results in precipitin arcs in the gel, each of which covers an area proportional to the concentration of antigen. b) Counterimmunoelectrophoresis. Antibodies and antigens are placed in adjacent wells. On electrophoresis, the anodically migrating antigens and cathodically migrating antibodies form a precipitin line at equivalence. c) Rocket immuno-electrophoresis. Various concentrations of antigen are placed in wells of antibody-containing gels. On electrophoresis rocket shaped precipitin arcs form, the length of the precipitin arcs are proportional to the concentration of antigen.

Observation of precipitin lines in gels

In order to clearly observe precipitin lines in gels, the gels should be held against a dark background using diffuse white light. Alternatively, or if you wish to preserve the sample, the gel can be stained. To do this, the gel is first soaked in saline for about a day to remove all protein which is not in the precipitated complexes and is then stained with protein stains such as Coumassie Brilliant Blue.

Turbidimetry and nephelometry

When many samples have to be analysed, many laboratories employ more sophisticated instrumental techniques such as turbidity and nephelometry and these quantitative immunoprecipitin techniques in a fluid medium are routinely applied for the measurement of albumin, IgG, IgM, IgA and IgD, fibrinogen and complement components among others. In these techniques, monochromatic light is passed through a solution of antigen with antibody and the amounts of complexes formed are measured by either the amount of light scatter (nephelometry) or light absorption (turbidimetry). The latest instruments use laser technology and can detect nanogram quantities of antigen-antibody complexes.

SAQ 2.6

Indicate by a + which of the following techniques are qualitative and which are quantitative methods.

Technique	Quantitive	Qualitative
1) Immunoelectrophoresis (IEP)		
2) Rocket IEP		
3) Mancini		
4) Ouchterlony		
5) Counterimmunoelectrophoresis		
6) 2 dimensional IEP		
7) Nephelometry		

2.4 Antibody-antigen reactions - agglutination methods

Just about any cell, microbial or mammalian, and even inert particles such as polystyrene beads, can be clumped or agglutinated by antibodies specific for surface antigens. This mechanism forms the basis for simple and rapid methods for identifying various microbes and red cells and can also be used for semiquantitative assays of antibodies.

direct agglutination

passive haemag-glutination

If the interaction is between the cells themselves and the antibodies then we call it direct agglutination. It can also be adapted to assay antibodies to other antigens such as proteins which can be covalently or non-covalently attached to cells such as red cells and the anti-protein antibodies will then agglutinate the protein coated red cells. This process, where the red cells are the carriers for an antigen, is called passive haemagglutination.

There have been attempts to replace red cells with inert particles such as those made with polystyrene, polyacrylamide or aminopolystyrenes but agglutination tests using these particles are not as sensitive as those using red cells and we shall concentrate on those methods which use red cells.

The agglutination assays are rapid to perform, relatively cheap and much more sensitive than the precipitation methods: agglutination reactions detect 1-100 ng ml^{-1} of antibodies in a sample, precipitin reactions detect 60 µg ml^{-1} in gels and 20 µg ml^{-1} in liquid media.

2.4.1 The agglutination reaction

The reaction is normally carried out in physiological saline solutions (about 0.15 mol l^{-1} sodium chloride). Because of the overall negative charge on red cells or bacteria at neutral pH it is difficult for them to come close enough for molecules such as IgG antibodies to form a bridge between epitopes on two different cells. IgM, due its larger size, is able to agglutinate the cells much more efficiently than IgG. The presence of salt tends to neutralise the charge effects to allow agglutination.

The agglutination reaction is very similar to the precipitation reaction. The main difference is the size of the antigen. If we have a mixture of two types of red cells such as chicken and mammalian red cells and antibodies specific for each, then clumps formed in the red cell suspension will consist of one type of red cell only agglutinated by the specific antibodies. This is similar to the formation of precipitin lines in gels where each antigen-antibody system precipitates independently of the others.

2.4.2 The titration of agglutinating sera

The principle of this test is very straightforward. As an example, let us examine the titration of antiserum specific for blood group A red cells. One volume of saline (0.1 ml) is dispensed into a series of wells of a microtitre plate. An equal volume of the antiserum to be tested is added to the first well and mixed with the saline and 0.1ml of the diluted antiserum is then transferred to the saline in the second well. After mixing, 0.1ml is removed and transferred to the third well and so on. By this method, called serial dilutions, the antiserum is doubly diluted in each consecutive tube. The practice of serial dilutions is shown in Figure 2.12.

serial dilutions

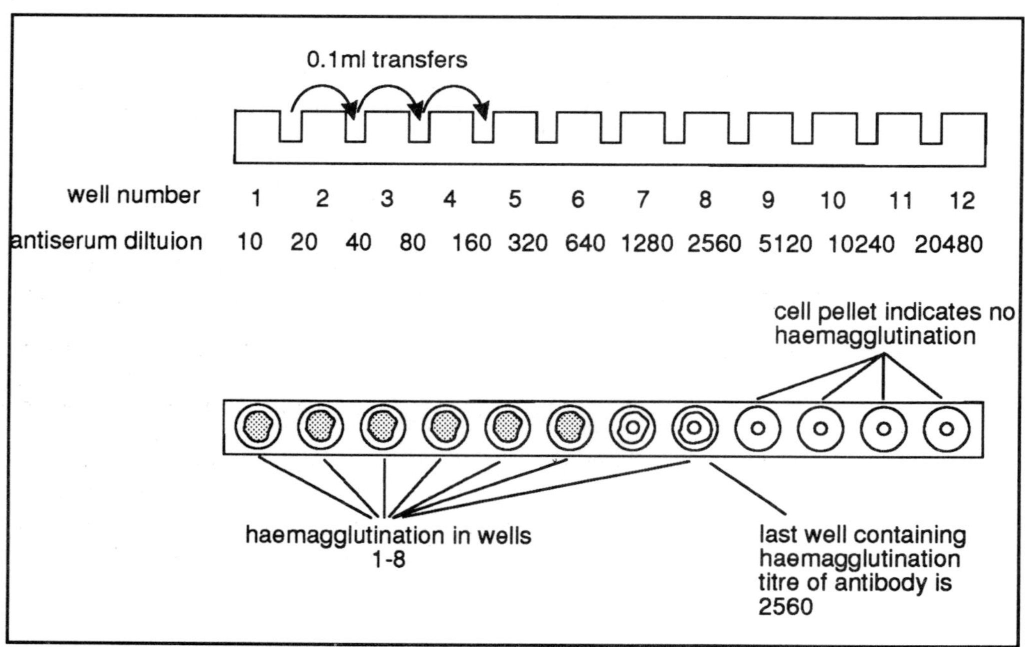

Figure 2.12 The basis of the haemagglutination assay. 0.1ml saline is dispensed into 12 wells of a microtitre plate. 0.1ml of antiserum (diluted 1:5) is added to the first well, mixed and 0.1ml transferred to the second well. After mixing, 0.1ml is transferred to well 3 and so on to give the dilutions of antiserum shown. 0.1ml red cell suspension is added to each well. Patterns of haemagglutination appear in some wells as a carpet of red cells whereas no haemagglutination is indicated by a cell pellet. In the assay shown, the titre of the antiserum is 2560.

A given number of group A red cells in 0.1ml of saline are now added to each well and the microtitre plate is gently shaken to promote mixing and is then left at room temperature for 1 hour. Positive agglutination is seen when the red cells are spread out across the well like a carpet whereas non-agglutinated cells form a tight pellet in the centre of the well.

From such an experiment we can determine the titre of the antiserum. You can see from Figure 2.12 that the last well that contains haemagglutinated red cells is the one representing a 1/2560 dilution. Thus the antibody titre is 2560.

Π What controls would you include in this assay?

heterophile antibodies

You would need to demonstrate non-haemagglutination using serum-free wells and you could also use control sera some of which is nonagglutinating eg normal rabbit serum and a standard haemagglutinating serum. Some normal sera do have background anti-SRBC antibodies in low titre - these are called heterophile antibodies.

prozone effect

It is often observed in haemagglutination reactions that the first few tubes containing the most concentrated antiserum fails to agglutinate the red cells. This is the prozone effect and is most probably due to the excess antibodies binding to single red cells with no tendency to link red cells together. It is similar to what you observed in the precipitin reaction in antibody excess.

Π Can you think of a way in which you could demonstrate that antibodies were attached to the red cells in the prozone?

If you added an anti-isotype antibody (antiglobulin) this would link the antibody coated red cells together resulting in haemagglutination. This is the principal of the Coomb's test discussed below.

2.4.3 Some applications of haemagglutination

There are quite a number of clinical uses based on the direct agglutination of red cells. These include red cell typing in blood banks, the detection of anti-Rh antibodies in haemolytic disease of the newborn and tests for autoimmune haemolytic anaemia.

Coomb's tests

An interesting application is the Coomb's tests for detection of auto-antibodies to red cells. In this test, red cells coated with auto-antibodies in the blood of the patient can be detected (direct Coomb's test) and anti-red cell antibodies found free in the serum can also be analysed (indirect Coomb's test). Examine Figure 2.13. You will notice that some of the patient's red cells are coated with antibody and the red cells can be agglutinated using an anti-globulin reagent ie an anti-isotype antibody. In the indirect Coomb's test, the patient's serum is incubated with reference red cells. The antibodies then attach to the epitopes on the red cell surfaces for which they are specific and the antiglobulin then causes haemagglutination.

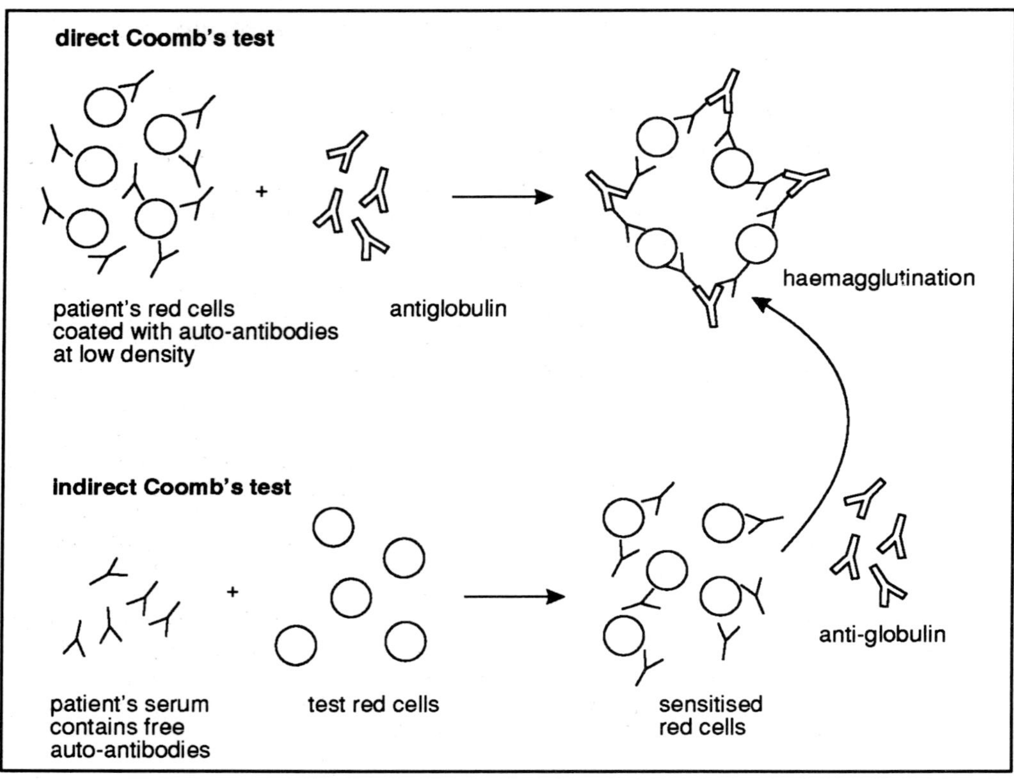

Figure 2.13 The Coomb's antiglobulin test.

Π What would be a simple method to determine the isotype of the auto-antibodies in the patient's serum?

Based on the Coomb's test you could set up a number of tubes containing the sensitised red cells and then add anti-IgG to one tube, anti-IgM to another tube and so on using antibodies to every isotype. The tube showing agglutination would indicate the nature of the isotype of the auto-antibodies.

SAQ 2.7

Some mice in your laboratory become infected with an unknown bacterium and 6 of 10 mice die from the infection. You are able to obtain a culture of the bacteria from one of the dead animals and you decide to investigate why some of the mice survived. What experiments could you do with the serum from one of the survivors which might provide you with information as to why the mice survived?

2.4.4 Passive haemagglutination

An important extention to this assay is in the use of the red cells as indicator cells to assay for a variety of antigens (eg proteins, polysaccharides, antibodies). By simple adsorption of the antigens onto the red cell surface or using covalent linkages, the antigens can be attached to red cells and subsequently the haemagglutination of the sensitised red cells can be used in semiquantitiative assays for the detection of antibodies.

Preparation of sensitised red cells

Many polysaccharides are readily adsorbed by red blood cells and are suitable to test for anti-polysaccharide antibodies in the haemagglutination assays. The coupling procedure simply involves the incubation of the polysaccharides with a suspension of red cells followed by washing.

tanned red cells

Some proteins can also adsorb to red cells but in this case adsorption is greatly improved by first treating the red cells with tannic acid to produce tanned red cells. This is a very simple procedure. Freshly collected or formalinized red cells are suspended in physiological saline and the tannic acid solution is added. After a short incubation period, the cells are centrifuged out of solution and resuspended in fresh physiological saline to which is then added saline containing the protein antigen. After incubation with gentle mixing the cells are centrifuged out of solution, washed and resuspended in saline containing 1% rabbit serum. The red cell preparations are stable in the refrigerator for months. They can be used in haemagglutination assays in the same way as uncoated red cells or can be used in haemagglutination inhibition assays to test for antigens (see below).

chromic chloride

An alternative method for coupling proteins to red cells is to use chromic chloride. This involves coupling the proteins using chromium cations. This is a straightforward techniques simply involving mixing a 50% suspension of red cells in saline with the chromic chloride solution and a solution of antigen. After thorough washing, the sensitised red cells are kept in the cold as a 1% suspension in saline.

homo-bifunctional reagents

There are many reagents which covalently couple proteins to red cells. These are called bifunctional reagents since they provide a covalent link to both the protein and the red cells. They include bis-diazotized benzidine and glutaraldehyde which are called homobifunctional reagents since both reactive groups are the same. However, the disadvantage of these reagents is that the process is rather inefficient due to the formation of protein-protein conjugates.

heterobifunctional reagents

A better approach is to use heterobifunctional reagents where the two reactive groups on the linking reagent are different and therefore will link with distinct functional groups. A good example of such a reagent is N-succinimidyl-3-(2-pyridyldithio) propionate (SPDP) which, in the first step of the coupling process, links 3-(2-pyridyldithio)propionyl (PDTP) groups to amino groups on the protein antigen. The disulphide bonds on the red cell surfaces are then reduced using dithiothreitol resulting in -SH groups. The PDTP-protein conjugate is then added to the reduced red cell suspension resulting in formation of disulphide links between the protein and the red cells. The mechanism is shown in Figure 2.14. We shall meet other heterobifunctional reagents, some of which could also be used to prepare sensitised red cells, later in the text.

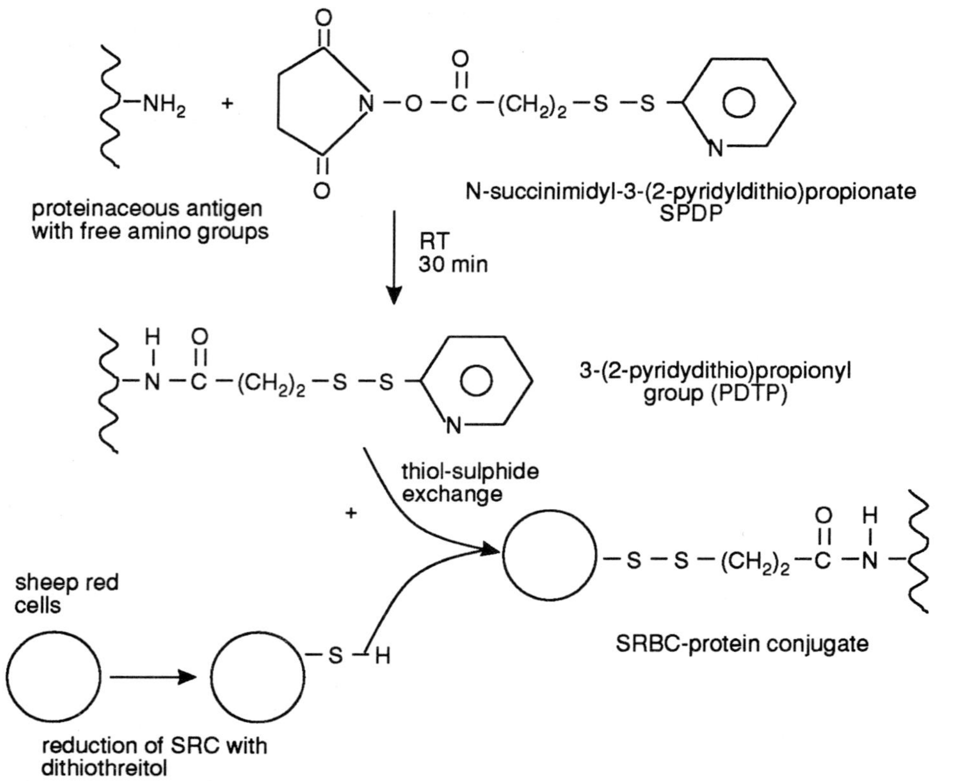

Figure 2.14 An example of the use of heterobifunctional reagents to couple protein antigens to sheep red cells. See text for a description of the mechanism. RT = room temperature.

Some uses of the passive haemagglutination technique

Clinical tests include assays for auto-antibodies to thyroglobulin in the serum of patients with thyroid disease. There are commercial kits available containing thyroglobulin sensitised sheep red cells which will be agglutinated in the presence of anti-thyroglobulin antibodies in the serum. In some kits, the thyroglobulin is coupled to latex particles and agglutination of latex particles is observed in this test. Rheumatoid factors are IgM auto-antibodies which bind to IgG molecules and react across the species barrier. They are usually present in high quantities in rheumatoid arthritis patients. These rheumatoid factors are tested for by agglutination of either human IgG coated latex particles or rabbit IgG coated red cells (Rose-Waaler test). Tests for rheumatoid factors are usually done in the clinical laboratory on a glass slide by adding the patient's serum to a drop of the sensitised latex particles or red cells and observing the agglutination.

You can also link haptens to red cells to assay for anti-hapten antibodies. A clinical application of this is the detection of antibodies to penicillin. This drug readily attaches to red cells and can be agglutinated by the antibodies.

Π Would you expect all antibodies to agglutinate the sensitised red cells?

Not all antibodies do agglutinate the red cells as we suggested earlier. In such a case, an antiglobulin is used to complete the reaction. However, in some instances, antiglobulins do not work with sensitised red cells.

reversed passive haemag-glutination

An important variation of this test is the reversed passive haemagglutination test such as is used to detect a surface component of the hepatitis B virus.

Π Before you read on, can you think why the test is called 'reversed'?

The term 'reversed' points to the use of antibody coupled to red cells to detect the viral antigen which would cause agglutination due to binding to the antibodies coupled to the red cells.

2.4.5 Haemagglutination inhibition assays

Some viruses possess surface ligands called haemagglutinins which bind to red cells causing haemagglutination. This fact can be used in determining the titre of antisera containing anti-haemagglutinin antibodies since, in the presence of such antibodies, no haemagglutination occurs.

secretor status

pregnancy tests

There are many applications for haemagglutination inhibition assays. It can be used to test for secretor status, which is the ability of individuals to secrete soluble ABH blood group specific antigens into body fluids. Obviously, soluble A antigens will inhibit the agglutination of A blood group red cells by anti-A serum. Clinical tests for fibrinogen include inhibition of agglutination of red cells coated with anti-fibrinogen antisera. The principle is also used in pregnancy tests. These are based on increased human chorionic gonadotropin (HCG) in the urine in early stages of pregnancy. This will inhibit agglutination of HCG coated red cells with an anti-HCG antiserum.

2.4.6 Bacterial agglutination

As we mentioned earlier, not only red cells and inert particles such as latex but also, bacterial cells are used in agglutination tests. The latter tests are used to detect anti-bacterial antibodies. The Widal reaction is an agglutination test for antibodies which appear in the course of infections with various *Salmonella* species and standard suspensions of organisms are used to detect antibodies in leptospirosis, brucellosis and rickettsias. Agglutination tests are also used to detect certain bacterial species such as *Haemophilus influenzae*, *Neisseria meningtidis*, *Streptococcus pyogenes* and *Salmonella* and *Shigella* species using diagnostic antibodies. Bacterial agglutination is about 10x less sensitive than haemagglutination; it detects antibodies down to about 0.1 $\mu g\ ml^{-1}$.

Π We have described rather a lot of tests based upon agglutination (haemagglutination). It might be helpful to produce a summary table of these as an aid for remembering them. We suggest you use the format:

Test	**Basis**
auto-antibodies to thyroglobulin	thyroglobulin coated SRCs or latex particles

SAQ 2.8

Examine the following sheep red cell haemagglutination assay. Then match each of the observed results in the wells indicated in the left column with the most plausible explanation in the right column using each explanation only once. Key: H = haemagglutination observed; NH = no haemagglutination. All sera were diluted 1:10 with saline before use and 0.1ml of sera was serially diluted into 0.1ml of saline in each well.

Row number	Well number								
	1	2	3	4	5	6	7	8	9
A	NH	NH	NH	H	H	H	H	H	NH
B	NH	NH	NH	NH	NH	NH	NH	NH	NH
C	H	H	NH	NH	NH	NH	NH	NH	NH
D	H	H	H	H	NH	NH	NH	NH	NH

1) Wells A 1-3 — a) heterophile antiserum (background antibodies)

2) Wells D 1-4 — b) normal rabbit serum

3) Wells A 1-8 — c) low titre anti-SRC antiserum

4) Wells C 1-2 — d) high titre anti-SRC antiserum

2.5 Assays involving complement

If you were to add fresh rabbit or guinea pig serum to most of the above assays the antibodies would bind the first component of complement in the serum and activate the membrane attack complex resulting in lysis of the red cells. This is the basis of the haemolysis assays, yet another method for assaying antibodies and antigens. There are two methods we shall briefly discuss, the haemolysis assay and the complement fixation test.

2.5.1 Haemolysis assay

Titration of haemolytic antibodies

IgM and IgG antibodies

Since only IgM and IgG antibodies are capable of activating complement in the classical pathway as we saw in Chapter 1, this assay measured only these antibodies. It is very similar to the haemagglutination assay we have just described, the only difference being that you add a source of complement to each well of the microtitre plate.

Π To refresh your memory go back to Section 2.4.2 and read about determining the haemagglutination titre.

Since complement is labile in serum on storage, you have to have a source of serum which has been immediately frozen on collection from the animal. This is usually collected from guinea pigs.

Π What precautions do you think you have to take to ensure that the only source of complement is in the serum you add to the assay?

The source of antibodies, eg as an antiserum, may also contain some complement which would affect the titration. To destroy this complement activity you heat the antisera, usually to 56°C for 30 minutes, before adding it to the test wells. The complement source itself may have some heterophile antibodies and these must be removed prior to using the complement. This is done by mixing sheep red cells with the complement at 4°C for 1 hour followed by centrifugation of the red cells and recovery of the absorbed complement solution.

Π What sort of results you would expect to observe in the wells which show haemolysis and those that do not?

haemolytic titre

If the red cells have been lysed there will be no red cell pellet whereas with no haemolysis there will be a red cell pellet similar to that found in the negative haemagglutination wells. The haemolytic titre is defined as the dilution in the last well giving total lysis with no evidence of any whole red cells settling in the well.

Note that IgM is very sensitive to reduction and therefore addition of 2-mercaptoethanol solution to the assay can be used to determine the titre of IgG antibodies alone.

Determination of total haemolytic complement

von Krogh equation

You can determine the total amount of complement in serum by adding the serum to antibody-coated sheep red cells and measuring the amount of lysis. The results can be plotted according to the von Krogh equation resulting in a sigmoid curve as shown in Figure 2.15. As 100% lysis is difficult to measure, it is the usual practice to determine the 50% lysis point (CH_{50}).

The von Krogh equation is:

$$x = K\left(\frac{y}{100-y}\right)^{\frac{1}{n}}$$

where x = amount of complement expressed in ml of undiluted serum, y = % haemolysis and K = CH_{50} value or the number of lytic units at 50% lysis; n is a constant.

In the laboratory, sensitised red cells are first prepared by treating sheep red cells with a standard amount of anti-sheep red cell antiserum (IgM), the number of red cells used being determined by their absorbance at 541nm when fully lysed. The sensitised red cells are then mixed with buffer and varying amounts of complement (diluted guinea pig serum). Ammonia solution is added to one tube of red cells to induce total (100%) lysis. The tubes are then incubated at 37°C for a predetermined time, placed on ice to terminate the reaction and the remaining intact red cells centrifuged out of suspension. The absorbance of each supernatant is subsequently determined at 541nm and the percentage lysis in each tube determined by comparison to the absorbance of the red cells treated with ammonia. The percentage lysis is then plotted against the amount of complement added calculated as ml of undiluted serum added. A plot similar to Figure 2.15 is obtained and the CH_{50} can be read from the graph.

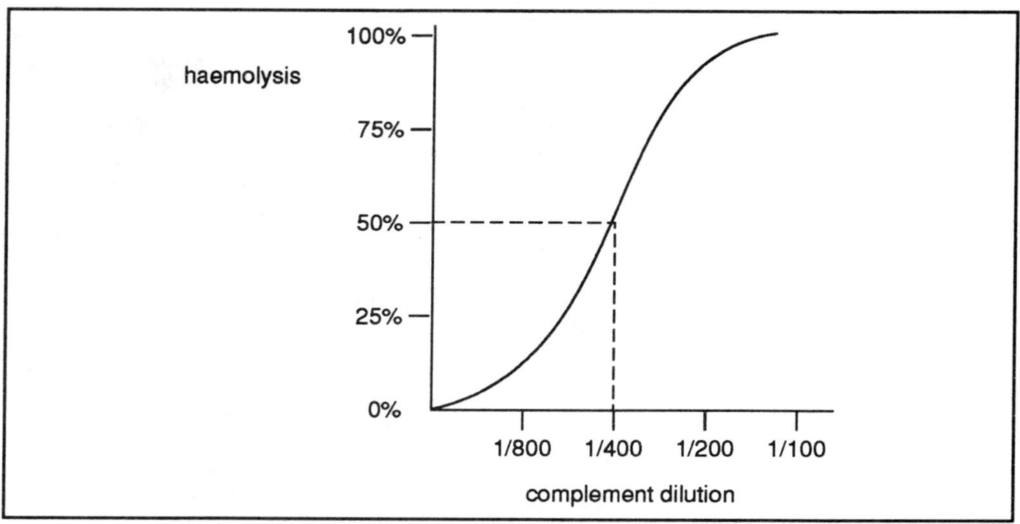

Figure 2.15 Estimation of total haemolytic complement. Various dilutions of serum (as a source of complement) are mixed with a fixed number of sensitised red cells and incubated for a fixed time period at 37°C. Cell control tubes are also set up containing no complement to measure spontaneous lysis and control tubes to measure complete (100%) lysis are set up containing red cells and ammonia instead of serum. After incubation, the optical density of each of the tubes is measured at 541nm. The % haemolysis is then calculated from the optical densities and plotted against either complement dilution or volume of complement added. The CH_{50} is determined from the plot as shown. The CH_{50} value obtained from a graph is expressed in arbitrary units since it is dependent on such factors as the red cell concentration and the amount of antibody used for sensitisation as well as on other factors such as the time of incubation and pH and ionic strength of the reaction mixture.

To obtain a straight line, you can plot log x against log y/100-y and this gives a straight line with a slope of 1/n. The intercept of the line when $\log\left(\frac{y}{100-y}\right) = 0$ is the log dilution giving 50% lysis of the red cells. The complement concentration is normally expressed per ml of serum (CH_{50} units ml^{-1} serum).

SAQ 2.9

In an experiment to determine the haemolytic content of a serum sample various amounts of the serum were incubated with a fixed number of sensitised red cells and the reaction allowed to proceed. At the end of the experiment, the optical densities of the reaction mixtures and that of a 100% lysis control tube were determined. Determine the value of CH_{50} for this sample by determining the percentage haemolysis from the data and plotting this against amount (ml) of serum added. Choose the best answer from the list below.

Tube Number	Serum added (ml)	Optical density
1	control 100% lysis	0.69
2	0.1	0.62
3	0.05	0.46
4	0.03	0.21
5	0.02	0.04
6	0.01	0.01

1) 250 2) 25 3) 15 4) 134 5) 57 units ml^{-1} serum.

The complement fixation test

The haemolytic assay using sensitised red cells can also be used to determine the concentrations of antibodies or antigens in samples or complexes in the serum of patients. Look at Figure 2.16. Let us assume that you have a serum sample from a patient which is suspected of containing immune complexes indicative of some disease. If a standardised amount of complement (so many CH_{50} units) is added to the serum some of the complement will become attached or fixed (hence the name) to the antibodies in the immune complexes and the amount of complement available for lysis will be reduced. After a suitable incubation time, sensitised red cells are added to the reaction and the amount of lysis is recorded. The reduction in lytic activity corresponds to the levels of complex in the serum.

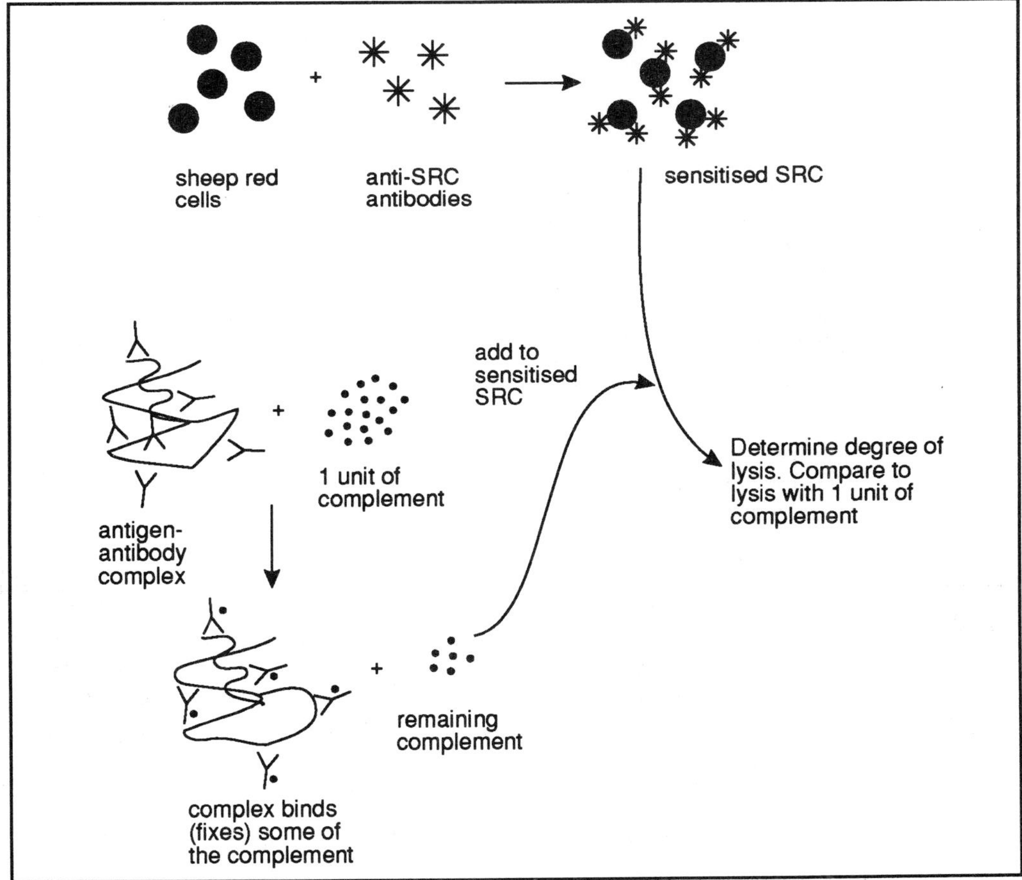

Figure 2.16 The complement fixation test. Sheep red cells (SRC) are mixed with anti-SRC antibodies to produce sensitised SRC. A fixed number of these cells are lysed by a standard amount of complement, say one unit. To detect immune complexes in serum, 1 unit of complement is added to the serum. After incubation, the remaining complement is titrated against the same number of sensitised SRC and the amount of lysis determined spectrophotometrically. The amount of complement fixed by the complexes can then be derived and this correlates to the amount of complexes (antibodies) in the serum.

A more sensitive test is to label the sensitised red cells with ^{51}Cr (as sodium chromate) by incubation of the cells with a solution of the radio-isotope. Lysis of the cells results in release of the radio-isotope from the cytoplasm of the cells and the supernatants of the lysed cells can be analysed for the amount of radio-isotope on a gamma scintillation counter. The counts in the supernatants are compared with those in controls of 100% lysis as before.

This test has a wide applicability, especially in clinical laboratories. Typical examples of its use are in detection of hepatitis antigen in blood, anti-platelet antibodies and the Wassermann test for syphilis.

The plaque-forming cell assay

enumeration of antibody producing plasma cells

Another application of haemolysis is in the detection and enumeration of antibody producing plasma cells. For instance, if you are following the immune response to sheep red cells in mice you may wish to determine the kinetics of plasma cell formation and production of different antibody classes during the physiological response. You can produce cell suspensions from spleens removed from mice and mix these with SRBC in an agarose solution. The agarose forms a loose gel in which are entrapped plasma (spleen) cells and SRBCs. The plasma cells from the spleens will continue to secrete antibodies. The anti-SRBC antibodies secreted by some of the plasma cell will bind to the red cells surrounding it and, in the presence of complement, will cause lysis. Under the microscope, the lysed cells appear as a halo (plaque) in a lawn of red cells and the number of plaques can be counted under low power microscopy to provide an estimate of the numbers of plasma cells per spleen. Alternatively, this assay can be carried out in specially prepared slide chambers instead of in agar.

Π Which antibody class do you think is mainly detected by this method?

The plaque forming assay carried out as described mainly detects IgM producing cells. This is because the amount of antibody secreted is low and only single IgM molecules activate complement. IgG, which also activates complement, is in too low a concentration on the cell surface to bind complement. From Chapter 1 you will remember that you need two IgG molecules situated close together to activate complement and you can do this by linking the IgG molecules with anti-globulin. Following the incubation period needed to allow secretion of antibodies by the plasma cells and subsequent attachment of the antibodies to the SRBC, you can overlay the agar with a solution of anti-globulin. After washing away the excess, complement is added and the plaques which develop are enumerated. This procedure is called the indirect plaque assay.

indirect plaque assay

reverse plaque assay

protein A

The final haemolytic assay is the reverse plaque assay which allows you to determine the total antibodies of any specificity released by the plasma cells. In this method, the SRBC are coated with protein A (which binds Fc regions of some antibodies) or anti-globulin antibodies using the chromic chloride method. The sensitised red cells are then mixed with plasma cells and anti-globulin in agarose. After allowing the agarose to set in a dish and a period of incubation for antibody secretion, the agar is overlaid with a solution of complement. The antibodies, secreted by the plasma cells, will have become bound by the protein A or anti-globulin antibodies and will activate complement resulting in plaques.

secretion of various non-Ig proteins

This method can also be used to detect the secretion of various non-Ig proteins (hormones, enzymes etc) from cells if the appropriate antibodies, which can be bound to the red cells, are available. You can also determine the class of antibodies secreted by plasma cells by attaching class specific antibodies to the red cells.

SAQ 2.10

Which method in the right column would be the most appropriate for each of the tasks in the left column. Use each item only once.

1) Detect immune complexes in serum	a) Radial immunodiffusion
2) Measure IgG in serum	b) Counter-immunoelectrophoresis
3) Detect antibodies to influenza virus in serum	c) Haemagglutination
4) Rapid assay to detect anti-albumin antibodies	d) Complement fixation
5) Detect antibodies to red cells	e) Haemagglutination inhibition

2.6 Enzyme immunoassays

2.6.1 ELISA and radioimmunoassays compared

Enzyme Linked ImmunoSorbent Assay (ELISA)

These immunoassays are extremely sensitive methods for detecting antigen-antibody reactions and are quantitative, capable of measuring levels of antibodies or antigens down to picogram levels. The most popular technique is the Enzyme Linked ImmunoSorbent Assay (ELISA) which is almost as sensitive as radioimmunoassays and has replaced the latter in many instances because of its relatively low cost and because it involves much simpler handling techniques and simpler instrumentation. Most importantly, it eliminates the need for using radio-isotopes and large volumes of scintillation fluids. This is an excellent and, indeed preferred, method for hybridoma screening as well as for determing many antigens. If a quantitative assay is required, there are many sophisticated instruments which can read all 96 wells of a microtitre plate in a few seconds. This again has to be compared with a few hours for scintillation counting using a radio-isotopic method.

There are many variations of the basic method depending on whether you wish to measure antigens or antibodies. We shall briefly examine approaches to measure antigens in unknown samples and then turn our attention to methods that measure antibodies. We shall then look at the practical details of the ELISA and the enzyme systems used for detection. For the moment we shall concentrate on the method using microtitre plates since this is the most common method used in laboratories. Keep referring to Figure 2.17 while reading the text so that you fully understand the principals of the ELISA technique.

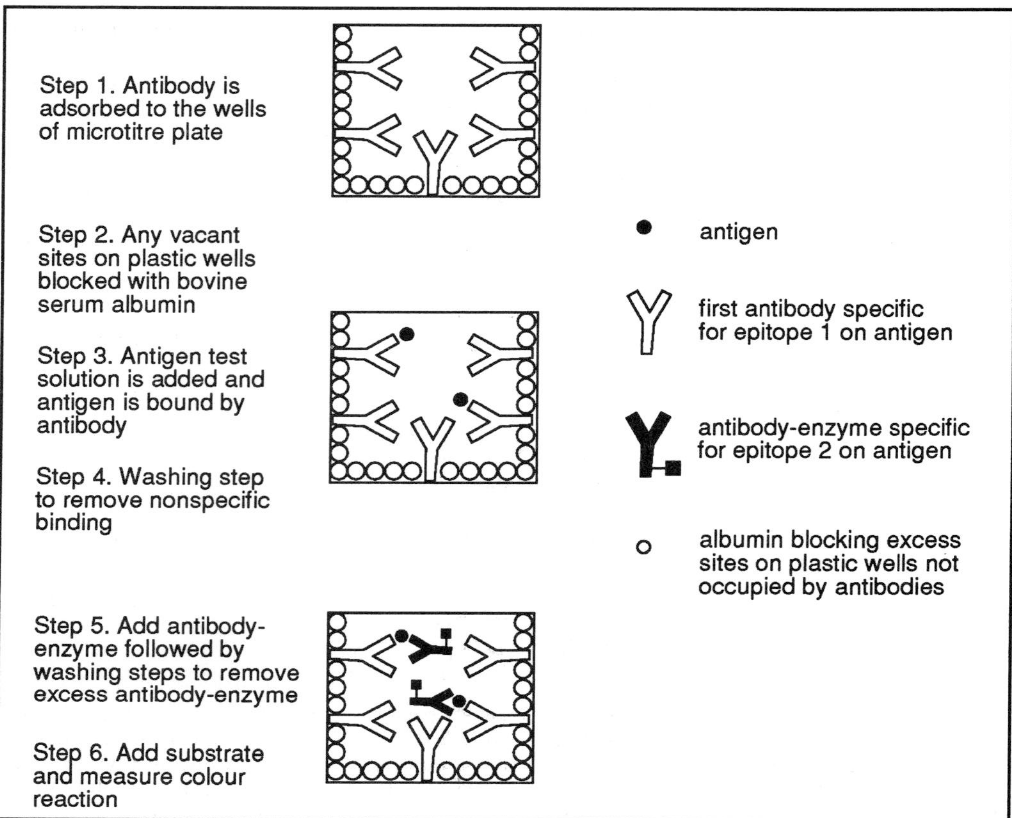

Figure 2.17 The principle of the two antibody sandwich assay. See text for details. Note that the diagram shows an excess of the first antibody over antigen. In practice, standardised solutions of different concentrations of antigen are used to determine the binding capacity of the antibodies in the wells. This is determined by measuring the colour intensity produced at Step 6. The test solutions of antigen are then suitably diluted so that the colour intensity (optical density) falls below the maximal values.

2.6.2 ELISA methods to measure antigens

two antibody sandwich assay

There are three principle methods to measure antigens by ELISA. The most useful method is the two antibody sandwich assay or the antigen capture 2 site assay. This method requires either two monoclonal antibodies with specificity for independent epitopes on the antigen or affinity purified polyclonal antibodies. One antibody is bound to the wells of the plastic microtitre plate and then the solution of antigen is added. The antigen is bound by the antibody. After washing steps to remove any unbound antigen, the second antibody with enzyme attached to it is used to quantitate how much antigen has become bound to the solid phase antibody.

Let us look in a little more detail at the practical procedures for the two antibody sandwich assay. These procedures are common to all variations of the ELISA. Some of these variations are schematically depicted in Figure 2.18. Firstly, the two antibody preparations will need to be purified and one of them has to be conjugated to the enzyme. The wells of a polyvinyl chloride microtitre plate are filled with a solution (concentration at least 4 μg ml^{-1}) of the unlabelled antibody and the plate is left at room temperature for a few hours for binding to occur. The plate is then washed to remove excess antibodies and incubated with a solution of bovine serum albumin (BSA).

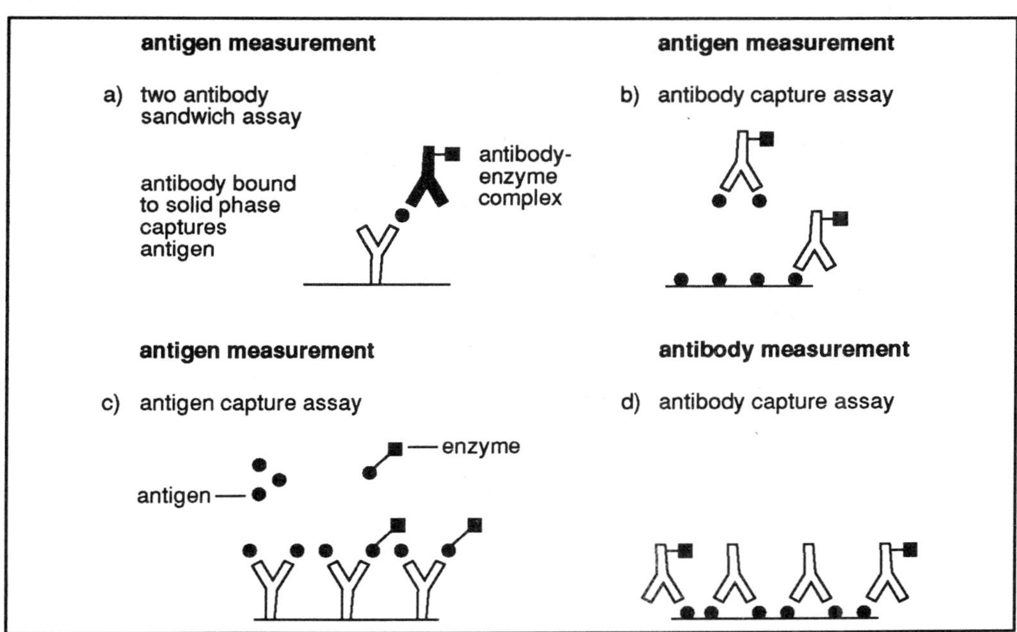

Figure 2.18 Examples of ELISA assays. a) Antibody attached to solid phase (eg well of a microtitre plate) binds antigen which is then quantified using a second antibody to which is attached the enzyme. b) A competitive assay in which free antigen in test (unknown) sample competes for antibody-enzyme conjugates with fixed antigen. The more free antigen present, the lower the colour development when chromogenic substrate added. c) In this assay, reference antigen is conjugated to enzyme and competes with free antigen in test (unknown); colour development as for b). d) To measure antibodies, we can use this antibody capture competitive assay. In this assay, antibody-enzyme competes for fixed antigen with test (unknown) antibodies. With a large excess of test antibodies, there would be no colour development.

Π Can you think what is the purpose of this treatment?

The BSA is used to block any remaining sites in the wells of the microtitre plate which have not bound antibodies. If these are not blocked, any protein subsequently added, whether it be antigen or antibodies, could nonspecifically bind to the plate rather than to the antigen or the antibody thus introducing errors in the quantitative assay.

A solution of antigen in blocking solution is now added and the plate kept at room temperature for at least 2 hours followed by 4 washes with saline. The second antibody-enzyme conjugate is now added and the plate is incubated for a further 2 hours. After further washing steps, the chromogenic substrate is added and the colour development is detected and measured in a spectrophotometer. We shall look at the most important enzymes and substrates later.

antibody capture assay

Alternative methods for measuring antigen are the competition assays. In the antibody capture assay, also called the antigen inhibition assay, the antigen is bound to the wells of the plate. The antigen specific antibody-enzyme conjugate is diluted out so that the amount bound per well gives an optical density of about 1.0 when you measure the substrate reaction resulting in a coloured product. In order to quantitate unknown amounts of antigen in test samples, a calibration curve is produced by mixing the antibody-enzyme conjugate with various dilutions of a standardised antigen solution prior to adding it to the wells. Obviously, this antigen will bind to the antibody-enzyme conjugate so preventing binding to the antigen bound to the wells. You can now determine the concentrations of antigen in unknown samples by comparison of the colour development with that of the standard curve.

In another competition assay, antibody is attached to the wells and antigen-enzyme is titrated on to the fixed antibody to determine the optimal dilution. As in the method above, various dilutions of a standardised antigen solution are used to compete with the antigen-enzyme conjugate for binding to the fixed antibody. Since the free antigen will compete for the antibody sites with the antigen-enzyme conjugate, there will be less of the latter binding resulting in lower intensities of the colour reaction on addition of substrate. In this way a standard curve is produced and concentrations of antigen in test solutions can be determined by reference to this curve.

2.6.3 ELISA methods to measure antibodies

By changing the individual components around you should be able to devise methods for measuring antibodies based on the methods used for measuring antigens.

Π Can you work out two antibody capture assays starting with antigen bound to the wells?

There are two approaches you could take to measure levels of antibodies in test solutions. In the first, you use a reference antibody-enzyme conjugate which will bind to the antigen. You then set up dilutions of a standardised solution of unlabelled antibody which will compete with the conjugate for the antigen thus producing a calibration curve. You can then use this to determine the concentration of antibody in unknown antibody solutions. The other approach is to use indirect antibody ELISA which simply involves the extra step of adding an anti-globulin-enzyme conjugate which will quantitate the amount of antigen specific antibodies in the well. Notice that this latter assay is not competitive; you simply add your unknown antibody solution and then, after washing, add your antiglobulin-enzyme conjugate to quantitate the unknown antibody.

indirect antibody ELISA

Π What are the the main advantage of using the antiglobulin in ELISA assays?

Since the second antibody has specificity for the Fc end of the first antibody, it must have been produced in a different animal. For instance, if the first antibodies are IgG mouse monoclonal antibodies, then the second antibodies could have been produced in rabbits by injection of murine IgG. These antibodies can, therefore, bind to any murine IgG molecule irrespective of the specificity of the first antibody. You should, therefore, be able to see the usefulness of such a reagent with the enzyme attached. Instead of conjugating the enzyme to each and every antigen specific first antibody that you use in the laboratory, it is much more useful and economic and far less laborious, to have just one antibody-enzyme conjugate which can be used to quantitate many different antibodies which have bound to antigens.

Π Close your text and construct diagrams of 1) two antibody sandwich assay to measure antigen 2) antibody capture assay to measure antibody and 3) antigen capture competitive assay to measure antigen. Check your diagrams with those in Figure 2.18.

SAQ 2.11

Which one of the following statements is correct?

1) The two antibody sandwich assay involves antigen bound to the microtitre plate.

2) The antibody capture assay to measure antibody levels involves competition between free and bound antigen for antibody-enzyme conjugate.

3) In the indirect antigen ELISA, the antibody enzyme conjugate contains the antigen specific antibody.

4) The indirect antibody ELISA quantitates the amount of antigen specific antibodies present.

2.6.4 Antibody-enzyme conjugates and their substrates

antibody-enzyme conjugates

horse radish peroxidase, alkaline phosphatase and β-galactosidase

anti-globulin-enzyme conjugates

It is not that difficult to prepare your own antibody-enzyme conjugates and it turns out to be much cheaper to do this than to use commercial products. There is a large choice of enzymes for labelling the antibodies but the three most commonly used are horse radish peroxidase, alkaline phosphatase and β-galactosidase. These conjugates can be used for not only ELISA assays but for other techniques as well for instance for immunoblotting and immunocytochemical techniques because they are all based on the same principle. Each of the 3 enzymes mentioned can be used for all these techniques because there are a range of substrates which yield soluble products for ELISA assays which are then quantitated using a spectrophotometer and others that yield insoluble products which are ideal for immunoblotting as well as immunocytochemical techniques. For most purposes, it is sufficient to prepare anti-globulin-enzyme conjugates since, as we mentioned earlier, they are the most useful since they can be used for many tests. However, if you need to prepare conjugates using monoclonal antibodies for use in the capture assays, then the methods are very similar.

anti-mouse IgG antibody

For example, if you prepare a lot of different murine IgG monoclonal antibodies, then an anti-mouse IgG antibody can be easily prepared by immunisation of a rabbit with pooled mouse IgG. This can then be purified by ammonium sulphate precipitation followed by ion exchange chromatography and the resulting antibody fraction is suitable for covalently linking with the enzyme. We shall be dealing with antibody purification in more detail later.

Π Notice that this a polyclonal antibody. Are there any advantages of this over a monoclonal antibody.

Using a polyclonal antibody probably increases the signal because it is likely that there will be a group of antibodies in the antiserum which will be specific for distinct epitopes on the first antibody. A monoclonal reagent will only bind to one epitope giving a ratio of only 1:1 for conjugate: first antibody. Since in the former case there are more enzyme molecules, the sensitivity of the reaction will be higher.

We have put the three enzymes, the substrates and the colour of the reaction products in Table 2.1. You will notice that the substrates are different for immunoblotting and immunocytochemistry because you need an insoluble product whereas the ELISA needs a soluble product for analysis.

Enzyme	Substrate	Colour of product	Solubility of product	Applications
Horse radish[2] peroxidase	tetramethyl benzidine (TMB)	yellow (450nm)	soluble	Enzyme immunoassays
	[1]chloronaphthol	blue-black	insoluble	Immunoblots and immunohistochemistry[3]
	diaminobenzidine	brown	insoluble	Immunoblots and immunohistochemistry[3]
Alkaline[2] phosphatase	nitrophenyl phosphate	yellow (405 nm)	soluble	Enzyme immunoassays
	bromochloroindolyl phosphatenitro blue tetrazolium (BCIP-NBT)	dark purple	insoluble	Immunoblots and immunohistochemistry
β-galactosidase	o-nitrophenyl-β-d-galactopyranoside (ONPG)	yellow (410nm)	soluble	Enzyme immunoassays
	5-bromo-4-chloro-3-indolyl-β-d-galactopyranoside (BCIG)	blue	insoluble	Immunoblots and immunohistochemistry

Table 2.1 Enzymes and substrates for immunoassays and related colour reactions. The absorbance peak is shown for the immunoassays. All initial substrates are colourless. Solubility is in water or alcohols. [1]relatively poor sensitivity compared to other substrates; [2]endogeonous activity in some cells and tissues which may interfere with assay; [3]includes immunocytochemistry where applicable.

The coupling procedure

glutaraldehyde method

The two most common methods to couple horse radish peroxidase (HRP) to antibodies is by the glutaraldehyde method or the periodate method. Alkaline phosphatase and β-galactosidase can be coupled using the glutaraldehyde method although there are many other coupling reagents which present certain advantages for some of the enzymes. We have already described one linking agent SPDP in Figure 2.14.

Glutaraldehyde, OHC-$(CH_2)_3$ -CHO, reacts with primary amino groups on proteins forming a bridge between the two proteins thus:

$$\text{protein 1} - \overset{\overset{\displaystyle H}{|}}{N} - \underset{\underset{\displaystyle O}{||}}{C} - (CH_2)_3 - \overset{\overset{\displaystyle O}{||}}{C} - \underset{\underset{\displaystyle H}{|}}{N} - \text{protein 2}$$

There is a very straightforward glutaraldehyde coupling in which the antibodies and the enzyme are mixed in phosphate buffer pH 6.8 and then 10% aqueous glutaraldehyde solution is added dropwise with stirring. The excess aldehyde groups left at the end of the reaction are then blocked with lysine. The antibody-enzyme conjugates are then separated from the non-coupled components and glutaraldehyde by passing the reaction mixture through a gel filtration column which separates the components by molecular weight.

The colour reaction

alkaline phosphatase

As we have said, once you have formed the antigen-antibody complex in the ELISA or the other methods mentioned above, the antibody-enzyme conjugate which binds to the antigen specific antibody somewhere on the Fc region is added. When the chromogenic substrate is added, the enzyme converts it to a coloured product. An example of such a reaction using alkaline phosphatase is given in Figure 2.19. This enzyme catalyses the cleavage of phosphate groups from substrate molecules and acts on the colourless substrate 4-nitrophenol phosphate converting it to 4-nitrophenol which is yellow. The increase in absorbance in the ELISA can be measured at 405nm at pH 10. The adsorbance reflects the number of enzyme molecules attached to antibodies in the microtitre well.

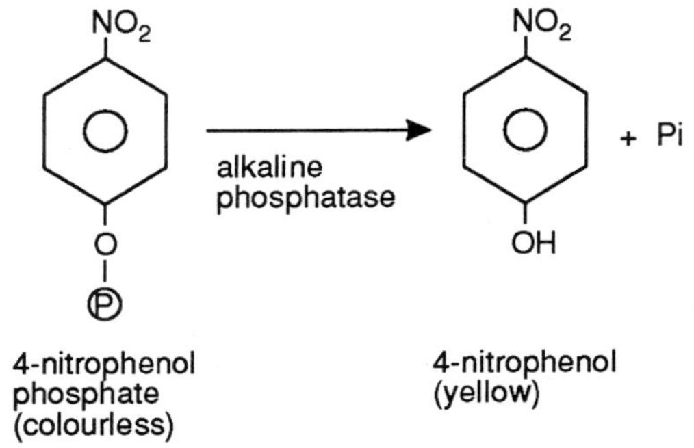

NO_2 NO_2

alkaline phosphatase → + Pi

O—Ⓟ OH

4-nitrophenol phosphate (colourless) 4-nitrophenol (yellow)

Figure 2.19 The colour reaction mediated by alkaline phosphatase.

2.6.5 The dot-ELISA assays

dot-blot assays

The microtitre plate ELISA can be adapted for qualitative or semi-quantitative assays such as 'dip-stick' or dot-blot assays. In dot-blot assays, the antigen is adsorbed as spots on to nitrocellulose or nylon membranes followed by antibody-enzyme conjugate and then substrate. Alternatively, you could use an indirect method using an anti-globulin-enzyme conjugate. By serially diluting out the specific antibodies onto the dots one can obtain a titre of the antibodies by looking for the end point when no colour remains in the dot. The dip-stick development allows such ELISA-type assays to be done outside specialised laboratories and one can obtain a result in 5 minutes. In some of these, for instance those for pregnancy testing, the membrane (stick) is coated with a monoclonal antibody to a hormone and then the colour is developed by dipping the membrane in various reagents in sequence.

pregnancy testing

Π What is one important difference between the plate ELISA and the dot-ELISA?

Since the conversion of substrate by antibody-enzyme is being done on a membrane, the coloured product has to be insoluble. One therefore could use alternative substrates as shown in Table 2.1.

A further important development we shall look at later (Section 2.9.1) which applies similar principles is Western blotting where electrophoresed proteins are identified using similar colour reactions to that used in ELISA's.

Π If you wished to detect antibodies secreted by B cells in microtitre plates, how could you adapt the ELISA to achieve this? Attempt to suggest a method before reading the next section.

2.6.6 The ELISA plaque (Elispot) assay

Elispot assay

The Elispot assay is yet another application of ELISA technology which allows us to detect antibody producing cells which are secreting a particular antibody. This could be used to detect antibody secreting clones in hybridoma technology (Chapter 4) or particular cytokines secreted by activated cells. In this technique, the cells are incubated in wells of the microtitre plate precoated with antibodies with specificity for the secreted product (in these examples, either antiglobulin or anti-cytokine antibodies). The cells are then removed and an antibody-enzyme conjugate is added followed by development of the colour with substrate.

2.6.7 Avidin-biotin assays

biotinylated antibodies

This is another advance which has much increased the sensitivity of the ELISA but has many other applications. Biotin is a small molecular weight co-enzyme (molecular weight 244 Daltons) which can easily be linked to proteins such as antibodies producing biotinylated antibodies. As biotin is small compared to the enzymes we have been discussing, it is likely to exert less risk of interfering with the antigen binding site so more biotin molecules can be attached. Four biotin molecules bind with extremely high affinity (Kd of 10^{-15} mol l^{-1}) to avidin which is a tetramer of 4 identical subunits of molecular weight 15 000 Daltons and the binding, as you can imagine, is stable under a wide range of conditions. If we couple biotin moieties to our antigen specific antibodies and then react these with the antigen, say in the wells of a microtitre plate, we can add avidin to which is linked the detection enzyme and this will then give a colour reaction with the substrate. The avidin-biotin system can be used in the ELISA and in Western blotting when enzymes are attached to avidin. It can also be used in radio-immunoassays when radio-isotopes such as ^{125}Iodine are coupled to avidin or a related compound, streptavidin or in immunocytochemistry when a fluorescent label or ferritin is attached to avidin. So you can see it is a very useful reagent.

There are 3 main methods of using the avidin-biotin system in enzyme immunoassays. Examine Figure 2.20 to understand the basic differences in these three assay systems. Notice that the biotin is attached in all cases to an anti-globulin so the same biotin-antibody complex can be used for many assays. In the first system (Figure 2.20a) the enzyme is attached to the avidin. In the second system (Figure 2.20b) the enzyme is attached to biotin. In this system, the avidin molecule simply bridges both antibody and enzyme. In the third system (Figure 2.20c) a large complex of enzyme-biotin-avidin is built up.

Π Can you place the three systems in order of sensitivity and explain the basis of your decision.

There is increasing sensitivity from a) through to c) in Figure 2.20 due to the increasing amounts of enzyme molecules that can be accommodated within the complex as an increased amount of enzyme will increase the amount of substrate converted to a coloured product.

To link biotin to the antibodies we can use commercially available N-hydroxysuccinimidobiotin which reacts with primary amines (usually the epsilon amino groups of lysine residues) on the antibody to form a peptide bond as shown in Figure 2.21.

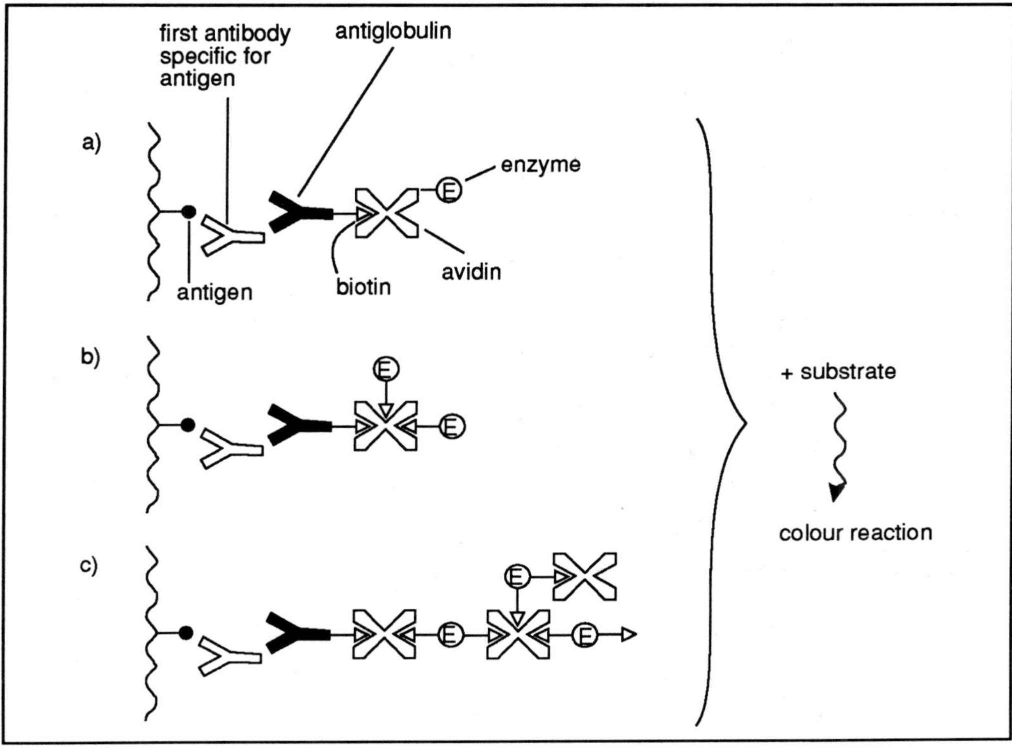

Figure 2.20 Avidin-biotin reactions. See text for details.

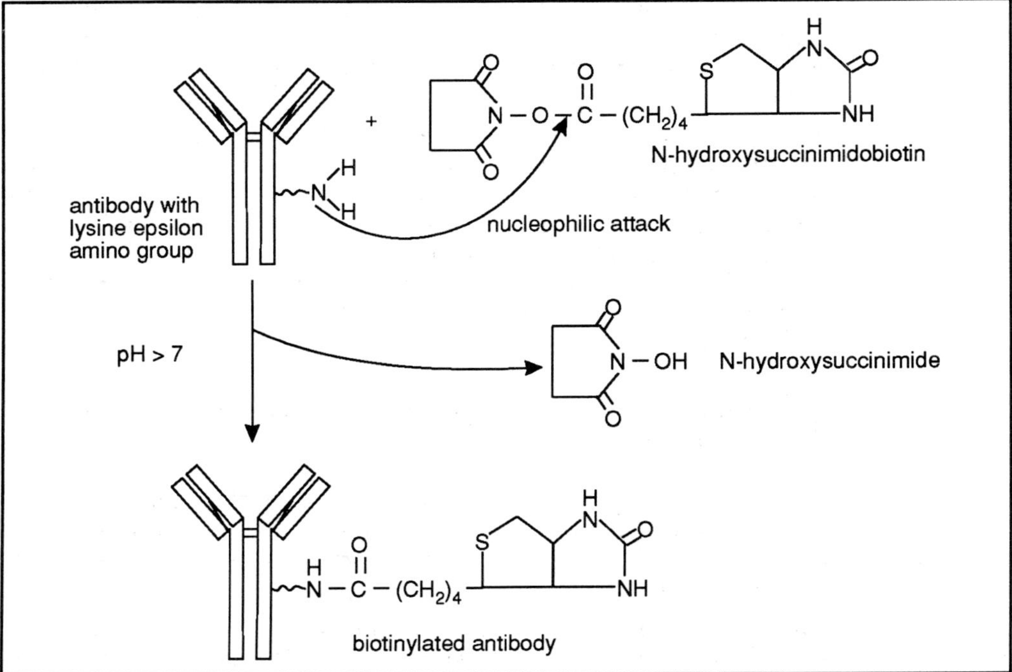

Figure 2.21 Biotinylation of antibodies.

Π Before you read on, try to think of any method you could adopt which avoids covalent linking of enzymes to the antibodies?

2.6.8 Bispecific antibodies for enzyme immunoassays

Bispecific antibodies can be made by a variety of methods but we shall leave that to later discussion. Imagine having an antibody in which one Fab is specific for mouse IgG and the other Fab binds specifically to horse radish peroxidase. This antibody could then be used in an ELISA assay for the quantitation step since you could add the bispecific antibody to the complex of antigen and first antibody. After washing steps, you could then add a solution of horse radish peroxidase and, after washing, the substrate. Another example would be the preparation of a bispecific antibody with specificities for the antigen and biotin. In all cases, such a biospecific antibody tool eliminates any possible effects of chemical modification derived from all the coupling procedures.

SAQ 2.12

Match the items in the left column with those in the right column using each item only once.

1) Dot-ELISA	a) Plasma cells
2) Plate ELISA	b) Solid phase antigen
3) Elispot	c) Fc region
4) Antibody capture	d) Insoluble product
5) Anti-globulin-enzyme	e) Soluble product

2.7 Radiolabel immunoassays

The extreme sensitivity of radio-isotopic procedures sometimes makes them the only choice for detection of minute quantities of antigens present in biological mixtures. They can measure antigens down to the picogram level (10^{-12}). They have been used to measure a wide range of molecules including IgE levels in serum, insulin, growth hormone, thyroid hormones, estrogens, bradykinin and morphine metabolites to name just a few.

The assays are very similar to the enzyme immunoassays we have discussed. The methodology can be divided into radioimmunoassays and immunoradiometric assays. Radioimmunoassays (RIA's) are competitive assays for antigen in the fluid phase and which can be extremely sensitive (down to femtogrammes ml^{-1}). In this assay the radiolabel is on the antigen. The slightly less sensitive immunoradiometric assays (nano- to picogrammes ml^{-1}) involve antibodies which are radiolabelled. Generally speaking, the major difference in methodology with enzyme immunoassays is that instead of conversion of a substrate to a coloured product, the reagents are labelled with an isotope and the amount of isotope detected in the antigen-antibody complex is counted and quantitated. The most popular isotope for antigen-antibody reactions is iodine (^{125}Iodine) which has a relatively short half-life of about 60 days and we shall concentrate our discussion on this isotope.

2.7.1 Radio-iodination procedures

chloramine T

There are a number of procedures for radiolabelling antigen or antibody. The most often used reagent is chloramine T and is used, alongside Iodogen, lactoperoxidase, Iodobeads and Enzymobeads, to radiolabel available tyrosine residues and to a much lesser extent histidine.

Another reagent, the Bolton-Hunter reagent, labels primary amino groups such as lysine residues by a very mild procedure. This is a particularly useful reagent when there are insufficient tyrosine residues in the protein to radiolabel. You should notice from Figure 2.22 that this reagent actually attaches radio-iodinated tyrosine residues to lysine residues in the protein. In so doing, of course, you lose one positive charge for each substitution and this should be remembered if you analyse the protein using 2-dimensional electrophoresis.

You could also avoid radiolabelling the prime reagents (antigens and antibodies) by using sandwich reagents such as ^{125}I-labelled protein A or G or ^{125}I-labelled avidin.

chloramine T method

Let us look briefly at some of these procedures for radiolabelling antigens and antibodies. The chloramine T method involves the oxidation of Na ^{125}I by the reagent leading to the iodination of tyrosine residues in the protein. In practice, a freshly made solution of chloramine T is mixed with ^{125}I and the protein. The reaction is very rapid and excess reagents are removed from the reaction mixture by adding a saturated solution of tyrosine. The iodinated proteins are then separated from the iodine by gel exclusion chromatography on either a Sephadex G10 or G25 column. Alternatively, you can use Iodobeads which are polystyrene beads with immobilised chloramine T. The reaction is terminated by removing the beads from the reaction mixture.

Iodogen

Another alternative iodination procedure is the method using Iodogen in which the iodinating reagent (See Figure 2.22) is coated onto the walls of a glass vial. The protein and ^{125}I are then added to the vial in buffer and at the end of the incubation the reaction mix is simply removed from the vial thus terminating the reaction.

lactoperoxidase

The lactoperoxidase procedure is possibly a more gentle procedure than the chloramine T procedure and involves iodination of tyrosine residues in the presence of hydrogen peroxide. In this method the lactoperoxidase, protein and ^{125}I are mixed in a polystyrene tube and then small aliquots of dilute hydrogen peroxide are added at intervals during the reaction. The reaction can be terminated by addition of a large excess of potassium iodide. The difficulty with this method is the concentration of hydrogen peroxide; too much poisons the enzyme, not enough provides too little substrate. An alternative is to endogenously generate the hydrogen peroxide using glucose oxidase. Enzymobeads are available commercially; they have both lactoperoxidase and glucose oxidase attached and they can be simply added to the reaction mixture of protein, glucose and ^{125}I. The reaction is controlled very easily by removing the beads.

In each iodination procedure, you need to separate the free radiolabelled iodine from the iodinated proteins as we described for the chloramine T method.

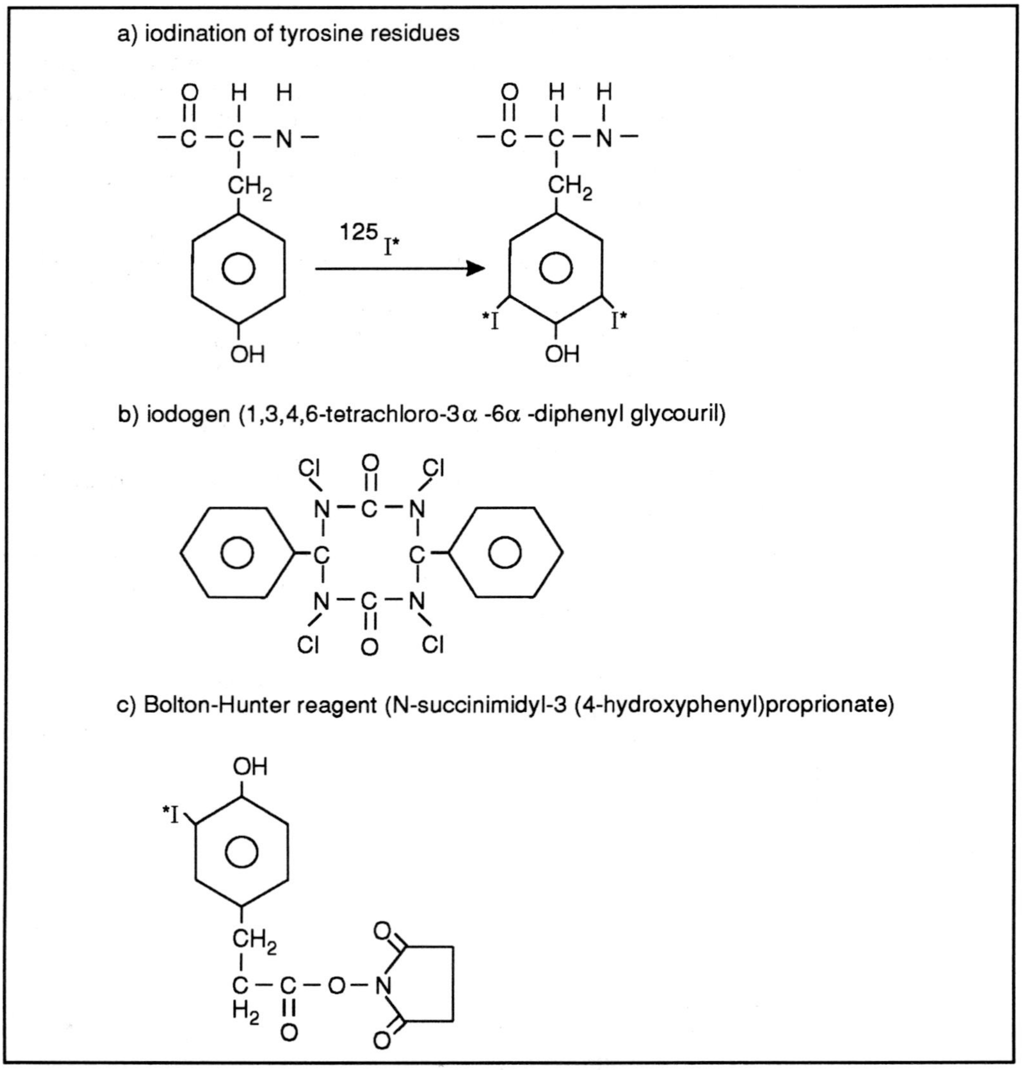

a) iodination of tyrosine residues

b) iodogen (1,3,4,6-tetrachloro-3α -6α -diphenyl glycouril)

c) Bolton-Hunter reagent (N-succinimidyl-3 (4-hydroxyphenyl)proprionate)

Figure 2.22 The iodination of tyrosine residues a) and two common iodination reagents b) and c).

Π If there are no available tyrosines in the protein and you do not wish to use the Bolton-Hunter reagent, what alternative approaches can you think of?

endogenously label the protein

Some proteins do not have available tyrosines. Examples of this are some MHC molecules on murine cells. If you have the cells in the laboratory which synthesize the protein or antibody then you can endogenously label the protein.

As an example, if you want radiolabelled monoclonal antibodies, then you can incubate the hybridoma cells in culture medium with either ^{35}S-methionine or ^{35}S-cysteine or ^{3}H labelled-lysine, phenylalanine, arginine or leucine or a mixture of these. The cells will then secrete the radiolabelled antibodies or proteins into the culture medium from which they can be purified.

2.7.2 The radioimmunoassay

radioimmuno-assay (RIA)

As we have said, the radioimmunoassay (RIA) involves the use of radiolabelled antigen which is captured by antibody (you can see the close similarity with the ELISA we have just discussed). It is a competitive assay that you can use to measure levels of unlabelled antigen in a sample. If unlabelled antigen is added to the system, it will compete with the radiolabelled antigen for the antibody resulting in a lower level of radiolabel in the antigen-antibody complex.

RIST (radioimmuno-sorbent test) test

An example of this assay is the RIST (radioimmunosorbent test) test which measures the total IgE in serum of patients suffering from allergies. Usually, in this test, anti-IgE antibodies are adsorbed onto the assay plate and increasing amounts of radiolabelled IgE are added to determine the capacity of the antibodies to bind the IgE. A standard curve is then produced by adding increasing amounts of unlabelled IgE to the radiolabelled IgE as shown in Figure 2.23. The concentration in IgE from a patient's serum can then be determined by mixing the serum with a standard amount of radiolabelled IgE and determining the radioactivity of the resulting complex. Obviously, the higher the amount of test IgE, the lower the amount of radioactivity will be detected. Notice that IgE is the 'antigen' in this test.

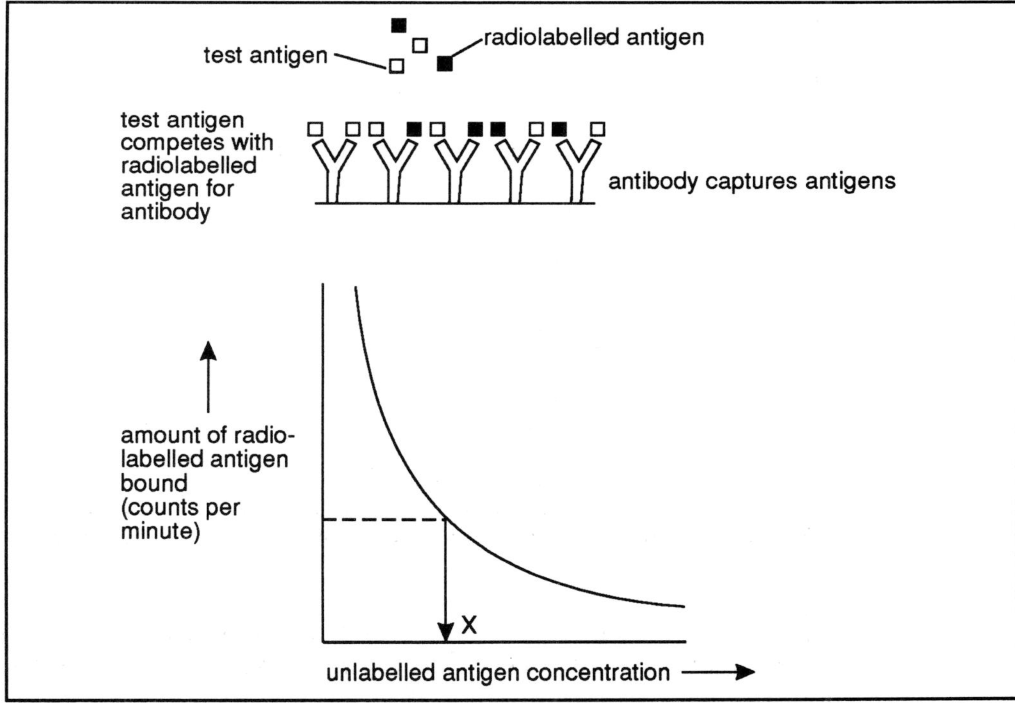

Figure 2.23 The competitive radioimmunoassay. Standard amounts of unlabelled antigen are mixed with radiolabelled antigen and antibody (this may be soluble or attached to a solid phase). The amounts of radiolabelled complex or radiolabel bound to the solid phase is measured and plotted against the concentration of unlabelled antigen. The curve can be used to determine the concentration of antigen in a test sample X by counting the bound radiolabel and reading the concentration from the standard curve.

2.7.3 Immunoradiometric assays (IRMA)

As we said previously, these assays use radiolabelled antibodies rather than radiolabelled antigen and are based on the same principles as the ELISA. They use a solid phase binding reaction which is monitored by determining the amount of radiolabelled antibody bound to the complex rather than by a colour reaction mediated by an enzyme conjugated antibody. The solid phase consists of plastic plates or tubes or antigen or antibody-coated beads. The sensitivity of IRMA and ELISA are about the same so the choice of method depends on the sort of criteria discussed in Section 2.6.1. The decision to use IRMA rather than the RIA may depend on whether you can radiolabel the antigen of choice. As with ELISA, it is more economic and convenient to radiolabel antiglobulins which could be used in many indirect IRMA's in the same way as the enzyme-labelled antiglobulins are used in indirect ELISA's. We will now briefly describe the major types of IRMA.

Π As you read through the next part of the text describing the major types of IRMA, construct diagrams illustrating the principle of each. Check your diagrams with those in Figure 2.24.

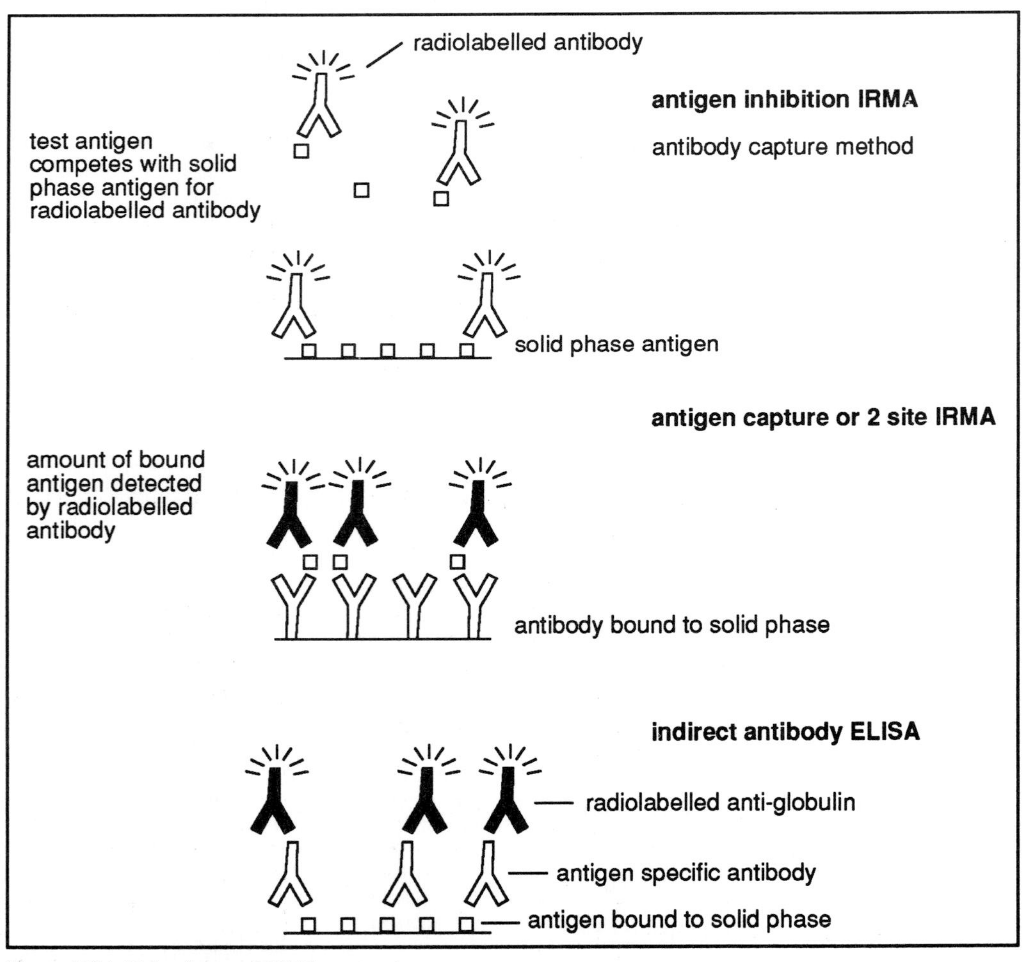

Figure 2.24 Major forms of IRMA.

Major forms of IRMA

antigen inhibition IRMA

The antigen inhibition IRMA is an antibody capture assay. The antigen is bound to the solid phase and a limiting amount of radiolabelled antibody is used which can be inhibited from binding to the solid phase antigen by the presence of test antigen in the solution.

antigen capture or 2 site IRMA

The antigen capture or 2 site IRMA uses an antibody bound to the solid phase which binds test antigen from the sample and a radiolabelled antibody specific for a different epitope on the antigen is used to quantitate the antigen.

indirect antibody IRMA

The indirect antibody IRMA uses antigen in the solid phase to which the test antibodies bind. The number of antibodies is then quantitated using a radiolabelled antiglobulin. By using radiolabelled anti-isotype antibodies it is possible to determine the class of antibodies at the same time.

SAQ 2.13

Which of the following statements is/are correct?

1) Immunoradiometric assays can use either radiolabelled antigens or antibodies.

2) Immunoradiometric assays use antibodies attached to the solid phase ie they are all antigen capture assays.

3) The radioimmunosorbent test (RIST) is a radioimmunoassay which involves capture of radiolabelled antigens by antibodies.

4) The lactoperoxidase procedure labels primary amino groups.

5) The Bolton-Hunter reagent increases the number of tyrosines in a protein.

2.8 Antigen-antibody reactions on cells and in tissues

immuno-cytochemistry

immuno-histochemistry

The detection of antigens, immune complexes and antibodies such as auto-antibodies in cells and tissues, called immunocytochemistry and immunohistochemistry respectively, is yet another example of antigen-antibody reactions. You should by now be able to see how you could apply the enzyme immunoassays and assays involving radiolabel to detect antigens in pathological samples such as tissue sections. However, there are other reagents which can be used and which are used mainly for immunocytochemistry and immunohistochemistry purposes. These include immunofluorescence, colloidal gold and immunoferritin.

2.8.1 Principals of immunofluorescence

fluorescein isothiocyanate (FITC)

Immunofluorescence assays are very similar in principle to the ELISA and radiolabelled assays, the main difference being in the label attached to the antibody. This technique, introduced by Coons in the 1940's. uses fluorescent dyes attached to antibodies which become themselves attached to antigens on cells or tissues. The position of the antigens is determined under a fluorescence microscope when UV light passes through the sample. The light is absorbed by the dye at one wavelength and emitted light at another wavelength. One such dye, fluorescein isothiocyanate (FITC) adsorbs light maximally at 490-495nm and emits a green coloured light at 517nm and is the popular choice for single colour immunofluorescence. The structure and spectral properties of some fluorochromes are shown in Figure 2.25 and the figure also indicates how the flüorochromes are covalently linked to antibodies.

Π Before reading on, see if you can develop two ways in which antigens in tissues can be identified by fluorescent antibodies?

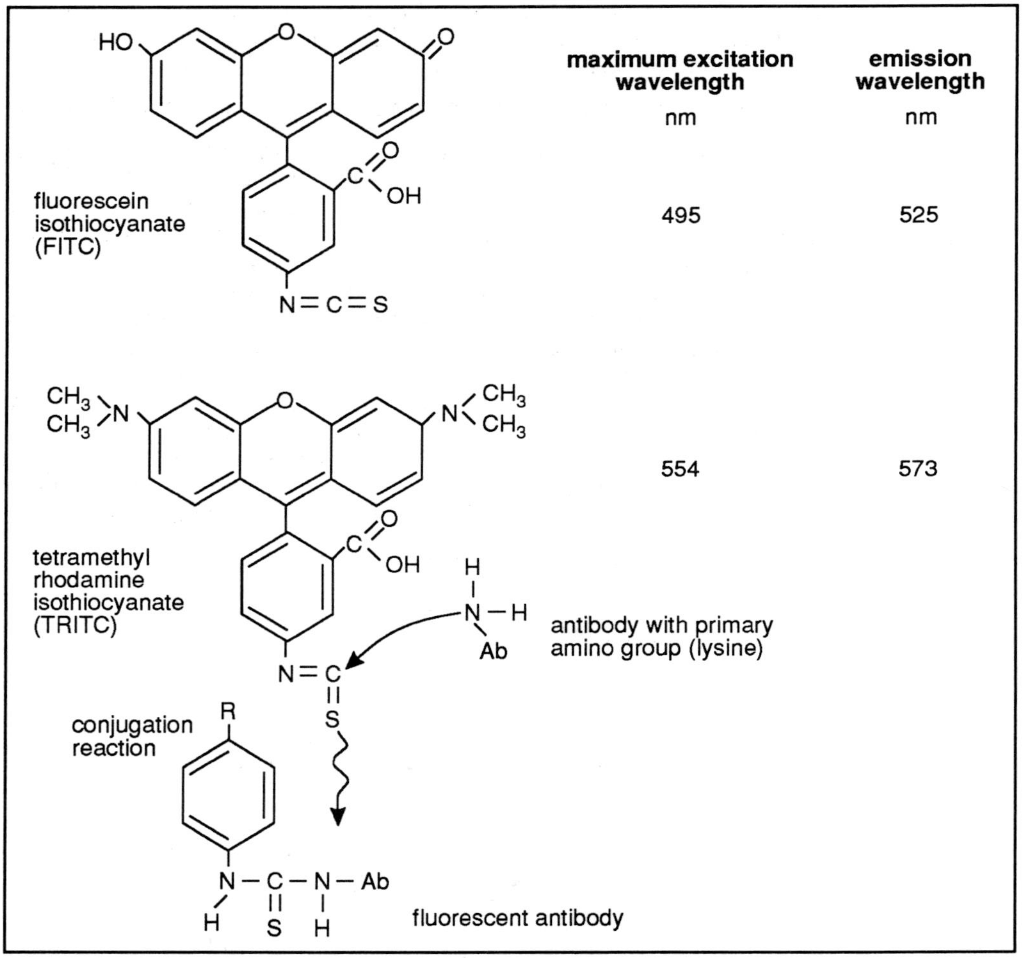

	maximum excitation wavelength nm	emission wavelength nm
fluorescein isothiocyanate (FITC)	495	525
tetramethyl rhodamine isothiocyanate (TRITC)	554	573

HO, O, O, C, O, OH, N=C=S

CH_3, CH_3, N, O, N, CH_3, CH_3, C, O, OH

H, N–H, Ab, antibody with primary amino group (lysine)

N=C, S

conjugation reaction

R

N–C–N–Ab, H, S, H

fluorescent antibody

Figure 2.25 The two most popular fluorochromes and their excitation and emission wavelengths. Also shown is the nucleophilic attack of the ← amino group of lysine on the isothiocyanate group resulting in a thiourea bond.

2.8.2 Direct and indirect immunofluorescence

direct and indirect immuno-fluorescence

Based on what we have previously discovered with the ELISA and assays involving radiolabels, the two approaches would be direct and indirect immunofluorescence. Indirect immunofluorescence, the antigen specific antibody also bears the fluorochrome whereas in indirect immunofluorescence, which is the more sensitive technique, an antiglobulin bears the fluorochrome.

Π How would you identify immune complexes in a pathological sample?

To identify immune complexes in a specimen, you could use a fluorescent labelled antiglobulin. This could also be used to identify auto-antibodies in a specimen.

Π Could you use an avidin-biotin system for fluorescence analysis?

Instead of attaching the fluorochrome to the antibody or antiglobulin you could biotinylate the antibody and attach the fluorescent dye to avidin. In this way you could increase the sensitivity of the assay as we saw in Figure 2.20c.

2.8.3 Flow cytometry and fluorescence activated cell sorting (FACS)

As we found out in Chapter 1, cells can be characterised by the possession of certain molecules on their surface. For instance T cells express CD2 and CD3; we call these T cell markers as they identify T cells. If you have antibodies to these markers, the cells can be analysed based on these markers and their sizes by flow cytometry. Hence you can determine the numbers of a particular cell type in a population of cells. The fluorescence-activated cell sorter (FACS) is, in addition, able to separate the cells which can then be used for further experiments. The essential principle of the FACS is that a cell suspension introduced into the instrument is forced through a nozzle as droplets containing single cells. These pass through a laser beam which excites the fluorescent dye and both fluorescence and light scatter induced by cell size are analysed. These are measured by detectors which trigger an electronic switch which creates an electronic field. Thus the charged cells are deflected. If however, droplets carrying unlabelled cells pass through the laser beam, the incorrect fluorescence signal is produced and the electronic field is not created and so the cells pass straight on.

Diagrammatically we can represent this by:

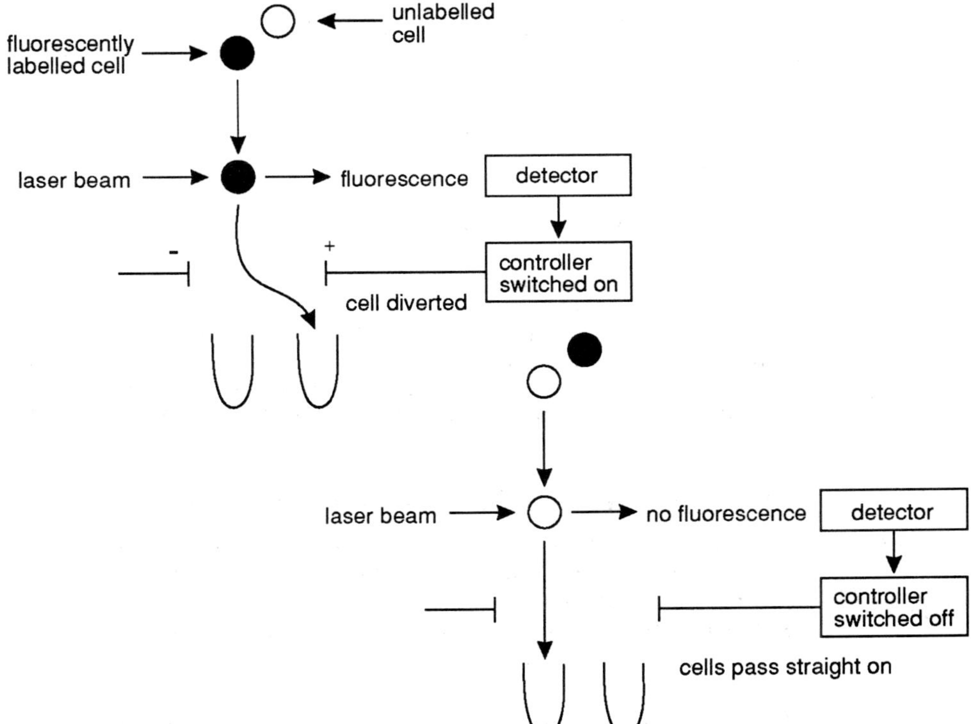

Thus if we have fluorescently labelled CD4 antibodies we could use these to help us separate CD4 cells (T_H cells) from, for example CD8 cells (T_C cells).

2.8.4 Other labelling reagents in immunohistochemistry

You will realise that the reagents used in ELISA can also be used to stain pathological tissues. In these methods, of course, you would choose substrates that produce an insoluble product. Since both peroxidase and alkaline phosphatase are found in many cells, β-galactosidase may be the enzyme of choice. This is a bacterial enzyme and is not found in mammalian cells. Examples of the substrates are indicated in Table 2.1.

colloidal gold

immuno-ferritin

Other labelling reagents include colloidal gold reagents and immuno-ferritin and these are used because of their high electron density to identify antigens in thin sections by electron microscopy. These reagents can be used in both direct and indirect assays and may be attached to avidin for the assay.

SAQ 2.14

Match the word in the left column with the most appropriate item in the right column.

1) Avidin — a) Radiolabelled antibodies
2) IRMA — b) Cell separation
3) FITC — c) Electron dense
4) Immunoferritin — d) Signal enhancement
5) FACS — e) Ultraviolet light

2.9 Western blotting

Western blotting

The last example we will cite of using antigen-antibody reactions to detect or characterise antigens or antibodies is Western blotting. This is a method by which proteins are separated by electrophoresis and these are then transferred to a membrane. Particular proteins are identified by labelled antibodies.

The antigen of interest may be a molecule on the surface of a cell or a soluble protein. In the former case, the cells would be lysed with a detergent prior to electrophoresis. The sample containing the protein of interest is electrophoresed in a polyacrylamide gel system after being boiled in a buffer containing sodium dodecyl sulphate which binds to the proteins making them all electronegative. The individual proteins migrate towards the anode based on their molecular weight ie small molecules move the farthest distance towards the anode, higher molecular weight molecules move at slower rates.

After electrophoresis, a nitrocellulose or nylon membrane is placed on the polyacrylamide gel containing the separated proteins and the two are sandwiched between filter papers and sponges wetted with transfer buffer. This is then placed in an electrophoretic tank containing transfer buffer or between electrodes in what is called a semi-dry apparatus with the membrane facing the anode. An electric current is passed through the apparatus and the proteins are transferred to the membrane where they become irreversibly bound. The membrane is then incubated in buffer containing a blocking agent such as serum albumin, casein (milk powder) or Tween 20 detergent which bind to all the remaining sites on the membrane not occupied by the protein bands.

Π What is the purpose of the blocking reagent:

This is done to prevent the antibody reagent which is added later, from nonspecifically binding to the membrane rather than specifically identifying the protein of interest.

The membrane is now incubated with the antibody in blocking buffer and the antibody binds to the protein on the membrane. After suitable washing steps, the membrane is incubated with an anti-globulin-enzyme conjugate which will bind to the antigen-antibody complexes on the membrane. After more washing steps, the substrate is added and colour development observed.

Since the colour reaction is on a membrane, the substrate must result in an insoluble product and Table 2.1 gives the list of suitable substrates for each enzyme. We have summarised the major features of Western blotting in Figure 2.26.

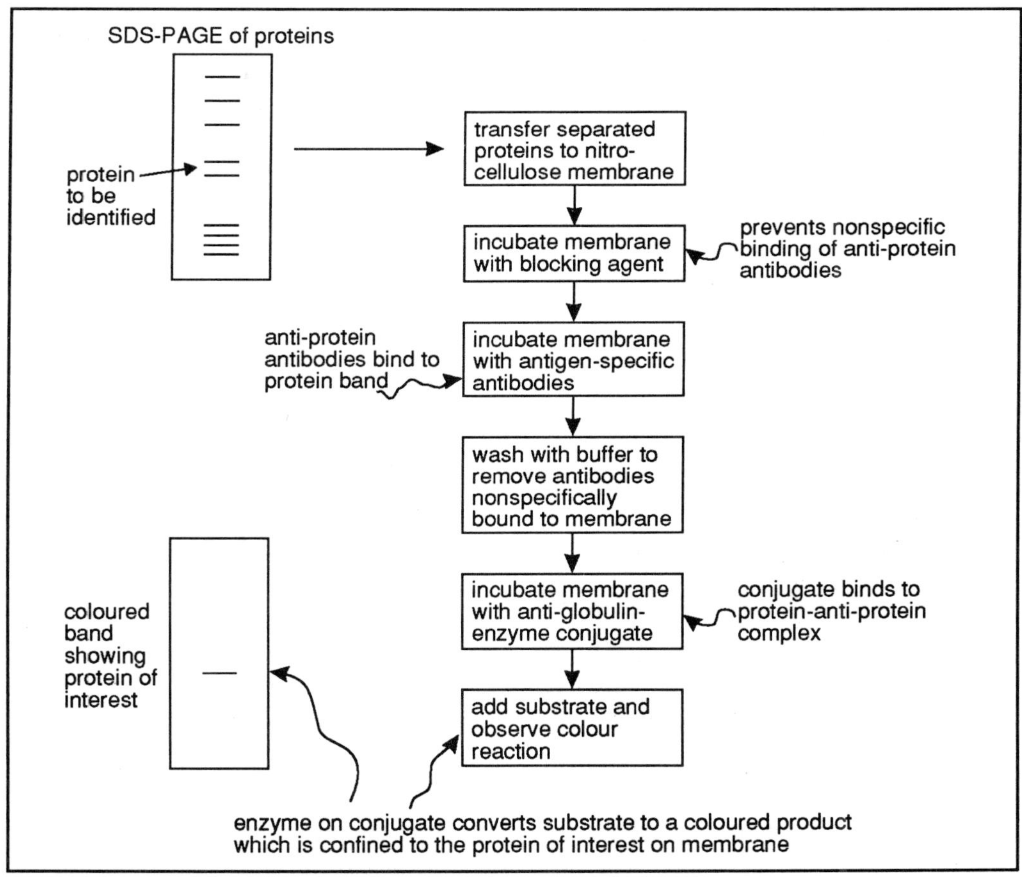

Figure 2.26 Western blotting. Note the boxes and straight arrows indicate the experimental steps and the wavy lines indicate what is happening at each stage.

SAQ 2.15

Which of the following could be used to identify antigen-antibody complexes in Western blotting?

1) Radiolabelled antiglobulins.
2) Radiolabelled protein A.
3) Horse radish peroxidase-labelled antibodies with tetramethyl benzidine substrate.
4) Colloidal gold antibody reagent.
5) Avidin-biotin system.

Summary and objectives

In this chapter, we have described the theoretical basis of antigen-antibody reactions and ways of measuring affinity. We then examined a variety of ways in which we can detect and measure antibodies and antigens using precipitation methods, agglutination methods, enzyme immunoassays, radiolabelled immunoassays. We have also looked at ways of detecting antibodies or antigens in cells and tissues and at specific detection of antigens in complex samples after electrophoresis (Western blotting).

Now you have completed this chapter you should be able to:

- derive equations for determining the affinity of an antibody, the valency of an antibody and the value of the average association constant using Scatchard analysis;
- calculate the amount of specific antibody in a mixture and the molar ratio of antibody:antigen;
- describe methods for determining affinities of antibodies;
- determine antibody concentration from the data in a Mancini assay;
- interpret experimental observations from immunoprecipitation reactions;
- list qualitative and quantitative precipitation methods;
- describe a range of techniques and applications based on agglutination and methods for coupling protein antigens to red cells;
- determine the total haemolytic complement of serum;
- decide on the most appropriate assay for the determination of antigen or antibody in a sample;
- explain the principals of the many enzyme immunoassays and radiolabelled immunoassays;
- describe what is meant by the terms dot-ELISA, Elispot, antibody capture, antigen capture, anti-globulin enzyme and solid phase;
- distinguish between radioimmunoassays and immunoradiometric assays;
- describe the methods for radiolabelling antibodies;
- explain the basis of immunofluorescence techniques;
- describe the procedure of Western blotting and list the labelled reagents which can be used to detect the antigen of interest.

Immunisation and generation of conventional antibodies

Immunisation and generation of conventional antibodies

3.1 Introduction

In this chapter we aim to introduce you to the production of conventional antibodies in Man and other animals resulting in specific antisera. This is obviously distinct from monoclonal antibodies; these are produced in the laboratory by hybridoma techniques and other forms of antibody production which result from manipulations using molecular genetics techniques. In spite of this, there are features which are common to the production of both antisera and monoclonal antibodies using hybridoma technology so those will be discussed. We shall also briefly examine passive immunisation and the use of vaccines to complete our studies of antisera production. In Chapter 4, we shall examine, in detail, how monoclonal antibodies are synthesised using hybridomas and in Chapter 5 we will discuss cloned antibody genes.

Π Before you read on, write down the main difference between polyclonal and monoclonal antibodies.

3.1.1 Polyclonal and monoclonal antibodies

As the name suggests, polyclonal antibodies are a mixture of antibodies from many B cell clones whereas monoclonal antibodies are all derived from a single clone of B cells and hence are all of identical structure.

3.1.2 The B cell response in a mouse

immune response in a mouse

Let us briefly examine the immune response in a mouse following injection of a pure protein antigen, A. As we saw in Chapter 1, within a few days plasma cells could be found in the spleen producing initially relatively poor quality antibodies of the IgM class. Under the influence of T helper cells, some of these cells will switch to IgG production and some cells will improve the quality of the product by undergoing somatic hypermutation. This maturation of the response improves with further booster injections leading to the production of anti-A IgG antibodies expressing high affinity antigen binding sites by a high proportion of the plasma cells in the spleen of the hyperimmunised mouse.

hyperimmunised mouse

Π Before you read on, will the mouse be synthesising antibodies apart from those to antigen A?

polyclonal response

There are two important points to note regarding this response in the murine spleen. Firstly, it is a polyclonal response because there will be many plasma cells of different clones producing a variety of antibodies to the different epitopes carried on immunogen A. Secondly, there will be other plasma cells in the spleen producing antibodies to antigens (non-A antigens) unrelated to the immunogen administered to the animal. These responses include the normal everyday physiological immune responses to environmental antigens and will include antibody production by plasma cells which arose from B cells stimulated by antigens some time before the final dose of the immunogen A was administered. These responses we can call the background response of the spleen.

serum contains a wide spectrum of antibodies representing the immediate immune history of the mouse

It is also important to think about the nature of the product. If we now obtain serum from this mouse, it will contain antibodies to the immunogen A but also background antibodies being made by the spleen and other lymphatic tissues throughout the body. The serum may also contain IgG antibodies which remain in the serum even though plasma cell production has ceased.

You can understand the difficulty in separating the anti-A antibodies from the remaining anti-non-A antibodies although this is possible as we shall see later. The greater difficulty would be in separating the anti-A antibodies from each other, ie to separate the products of each clone. In fact, this would be impossible and you would be left with, at best, a polyclonal product.

Additionally, once you have used up all the serum you cannot obtain exactly the same antibodies by repeating the immunisation procedure in another mouse. In other words, you have a limited supply of antiserum containing a fixed spectrum of antibody specificities and affinities.

3.1.3 The advantages of monoclonal antibodies

The disadvantages observed with polyclonal antisera can be abrogated using hybridoma technology. With this technique, you can pick out or select a particular plasma cell producing high affinity antibodies from all the plasma cells in the mouse spleen, endow it with the ability to proliferate *'ad infinitum'* and to produce antibodies for ever more. The cells that are used for these purposes are called hybridomas. We will discuss the production of hybridomas and monoclonal antibodies in Chapter 4. The production of hybridomas each producing a single type of antibody has become common place. So, instead of having a polyclonal mix in an antiserum which cannot be separated, we can generate a large collection of individual monoclonal antibodies, each of which has an exquisite specificity for a particular epitope on the immunogen. If we need to, we can mix the monoclonals together in any way we choose and we have an unlimited supply. These ideas are summarised in Figure 3.1. As we shall see in Chapter 5, there are even more recent advances which may consign hybridomas to the history books!

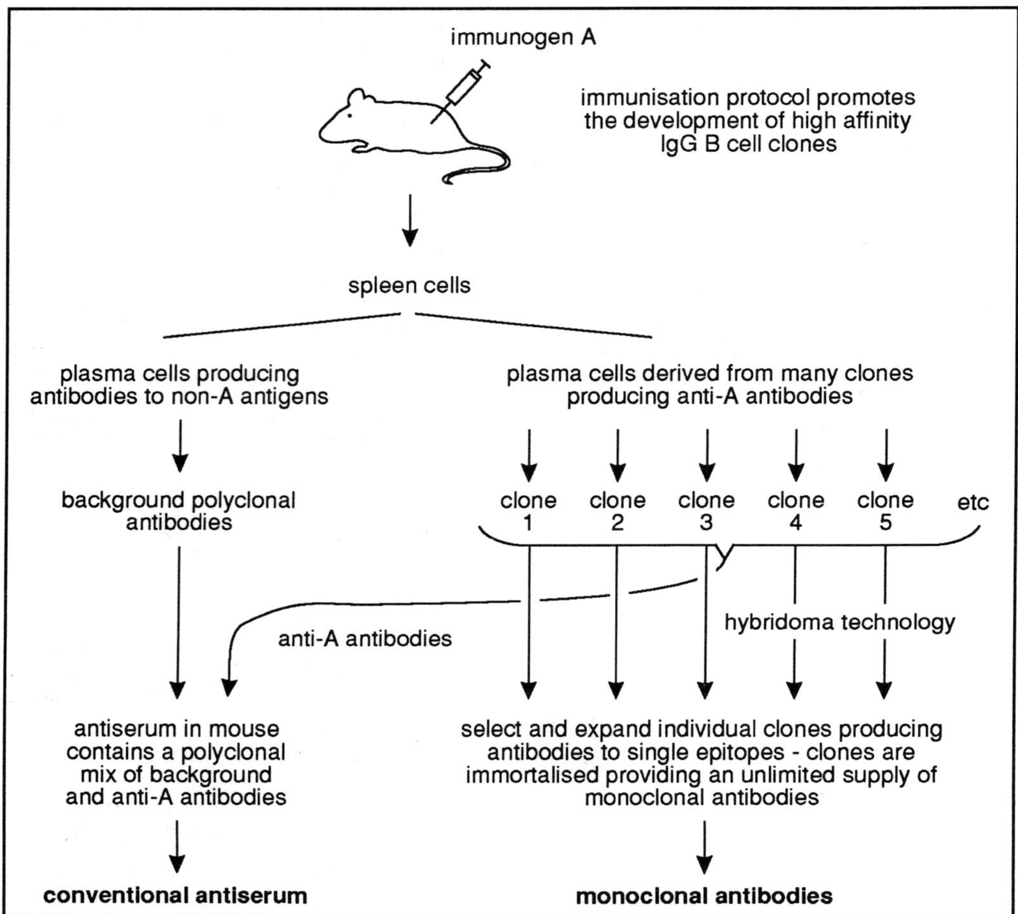

Figure 3.1 Monoclonal antibody technology using hybridomas. The selection of individual B cell clones from the spleen of an immunised mouse for the preparation of monoclonal antibodies.

3.1.4 The production of conventional and monoclonal antibodies

production of conventional antisera

The ideas expressed in Figure 3.1 are just as applicable to larger animals such as rabbits and sheep and because of the quantity of serum we can obtain these are the preferred animals for production of conventional antisera. So the methodology for antiserum production is relatively straightforward. The animals are given a priming injection followed by a number of booster injections resulting in hyperimmunised animals. Serum is collected throughout this process (test bleeds) to check on the quality of the antiserum and then a final collection of as large a volume of antiserum as possible is made. This is subsequently processed as described later. The production of antisera usually takes a minimum of 3 months.

monoclonal antibodies

In comparison, the method involving hybridomas for producing monoclonal antibodies (Figure 3.2) is much more complex and more costly. A few days after the final booster injection, the spleen is removed from the mouse and the cell suspension made from it is mixed with myeloma cells (cancerous B cells) which do not make their own antibodies. In the presence of polyethylene glycol, cell fusion occurs and a small proportion of the antibody producing spleen cells fuse with myeloma cells to form hybridomas. The reasons for fusing the spleen cells to myeloma cells is to produce a hybrid which will have a high proliferative capacity - a property of the myeloma partner.

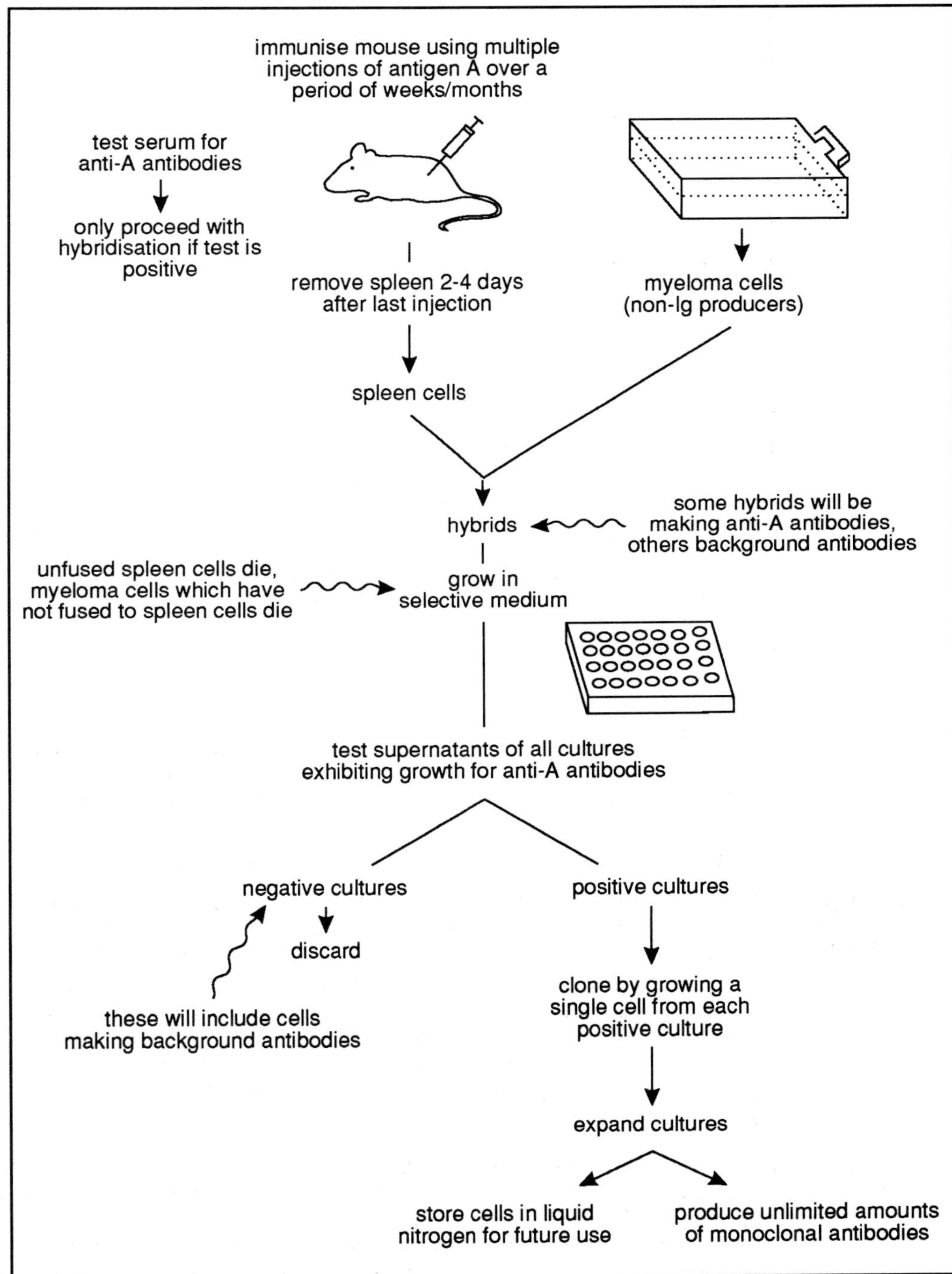

Figure 3.2 Production of monoclonal antibodies. See text for details.

The plasma cells, which do not fuse, die within a few days since they are terminally differentiated cells. The cells are cultured in selective medium which only allows the growth of the spleen-myeloma hybridomas and, after a short time in culture, the supernatants from the cells are tested for the presence of the antigen specific antibodies. Single cells from the positive cultures are then isolated and set up in fresh cultures to produce single clones producing monoclonal antibodies. These cells can then be expanded in bulk culture to produce large amounts of antibodies; they can also be stored in liquid nitrogen for future use. Thus the investigator is guaranteed a continuous supply of cloned cells and monoclonal antibodies.

We shall be looking in detail at the production of monoclonal antibodies and more recent alternative strategies using bioengineering in Chapters 4 and 5.

SAQ 3.1

Which one of the following statements is incorrect?

1) During the fusion procedure in preparation of monoclonal antibodies, unfused myeloma cells do not survive.

2) Most monoclonal antibodies are of the IgG class.

3) An antiserum does not have to be monospecific to be useful.

4) Inbred strains of mice produce the same antibodies.

3.2 Immunogens

3.2.1 Introduction

large chemically complex molecules are good immunogens

As you are already aware, large chemically complex molecules are good immunogens and most of these are proteins. Carbohydrates are often immunogenic especially when attached to proteins as glycoproteins. Native DNA is rarely immunogenic although it is possible to generate antibodies to polynucleotides. There are only a few reports on anti-lipid antibodies.

The general rule for the production of antisera to a single antigen is that the antigen has to be in a pure form or you will produce antibodies to the contaminants as well which may interfere in antibody assays. The stringency for preparing monoclonal antibodies is not as great since you can select clones of cells making antibodies against single epitopes.

Having said that, there are many examples of antisera being prepared against single antigens on cells by using the whole cells as immunogens. Additionally, we often prepare antibodies to immunogens linked to other molecules (see below). These approaches will obviously generate many antibodies of unwanted specificities and the investigator must have ways to remove these in order to produce a useful antiserum. We will now look at a few examples of how to prepare monospecific antisera following immunisation with 'impure' immunogens.

Π If you immunise a rabbit with a pure immunogen, is the antiserum produced monospecific?

When you collect this antiserum from the rabbit, you will, no doubt label it as anti-X (where X is the immunogen). However, as we saw in the initial discussion to this chapter, the rabbit is continuously producing various antibodies to many different antigens it meets throughout life and these will also be found in the serum which you collect. Therefore as well as anti-X, there will be many anti-bacterial and anti-viral antibodies among others. However, the antiserum is monospecific at an experimental level since if you use this antiserum in any of the tests described in Chapter 2 to assay for X it is unlikely that any of the non-X antibodies will bind unless there are some cross reacting epitopes. So we do often use antiserum without further purification if we can demonstrate reactivity to X and nothing else. However, we sometimes will have to take further steps to remove unwanted specificities and we will now describe some approaches for doing this.

3.2.2 Preparation of monospecific antisera using absorption techniques

Having said that, there are many examples of antisera being prepared against complex antigens such as lymphocytes which, after processing, resulted in monospecific antisera to a single surface molecule. One of the earliest examples was in the preparation of an antiserum to a membrane molecule Thy 1 on the surfaces of all mouse T cells. This was prepared by immunisation of rabbits with mouse thymocytes. The resulting antisera obtained after immunisation of the rabbit was then mixed in a step-wise procedure with many types of mouse cells in order to remove antibodies displaying reactivity towards common antigens present on both thymocytes and other cell types via a series of such step-wise absorptions. The absorbed antiserum finally obtained was monospecific for thymocytes but it showed some cross reactivity with mouse brain cells. You could obviously use this antiserum to detect thymocytes because brain cells were never a contaminant in the samples. Another example is shown in Figure 3.3.

absorptions

Π Having completed the absorption steps, does the antiserum contain only antibodies to thymocytes?

Having completed the last in-text activity, you should realise that even after the absorption step, the antiserum contains all the background antibodies we referred to previously and also all the other non-antibody serum proteins. These impurities may create high backgrounds for some of the tests described in Chapter 2. We shall look at ways of further purifying these antisera in Chapter 6. The purification procedures include affinity chromatography, a method by which you remove the specific antibodies from the serum rather than removing the unwanted antibodies. So with affinity chromatography you do end up with just specific antibodies.

You can often use complex antigens such as cells to produce monospecific antisera directly if the cells differ in the expression of a single antigen. Lymphocytes from one mouse strain when injected into another strain of mouse differing only in one MHC haplotype will induce antibodies only to the MHC antigens expressed on the allogeneic lymphocytes. If you wish to prepare anti-A blood group antibodies you can inject a rabbit with A positive red cells and absorb the antiserum with type O red cells. The rabbit will have produced antibodies to a host of different red cell antigens as well as against A but all these will be absorbed out by an excess of type O red cells leaving the antiserum monospecific for the A antigen. But do not forget, there are still background antibodies!

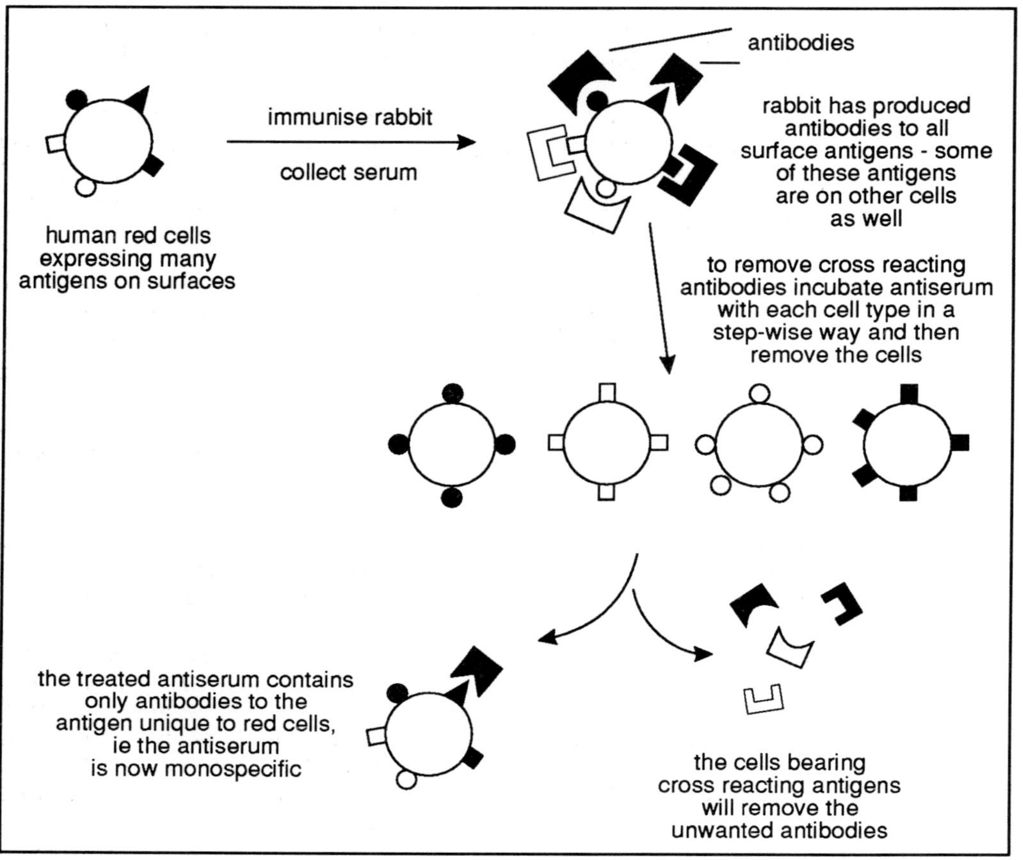

Figure 3.3 Absorption method to produce monospecific antisera.

Π Before you read on, can you think of a method of making a monospecific antiserum against thyroglobulin using sensitised red cells.

coupling of antigens to red cells

In Chapter 2 we discussed the coupling of antigens to red cells for passive haemagglutination tests. These cell conjugates can also be used to generate antisera against the antigen. The antiserum obtained after immunisation is simply incubated with an excess of red cells in order to remove all anti-red cell antibodies and this will result in an antiserum specific for the antigen, for instance thyroglobulin.

Π We looked into various ways of coupling proteins to red cells in Chapter 2. It might be useful for you to refresh your memory on tanned red cells, the chromic chloride method and heterobifunctional reagents (Section 2.4.4).

3.2.3 Monospecific antibodies using affinity chromatography

If you possess the pure antigen then you can recover monospecific antibodies by passing the antiserum over an affinity column of the antigen linked to an inert carrier such as Sepharose beads. After washing the column with buffer to remove non-specific binding proteins, you can elute the antibodies by changing the pH. For instance some of the antibodies will be eluted with glycine-HCl buffer pH 4. However some high affinity antibodies are difficult to remove and you inevitably lose some of these.

If there are known minor contaminants, you can avoid losing high affinity by antibodies of the desired specificity by preparing an affinity column to which the contaminating proteins are attached. Let us say that you have prepared an antiserum to human IgG by immunising a rabbit with whole IgG. The antiserum produced will cross react with other antibody classes due to the presence of anti-light chain antibodies.

You will remember from Chapter 1 that all classes of antibody carry the same light chains. This reactivity can be removed by passing the antiserum down an affinity column of light chains attached to Sepharose beads. In this case, you can recover all the anti-gamma chain antibodies in the eluate.

Π Are these monospecific antibodies the same as monoclonal antibodies?

Monospecific antibodies as described here are antibodies with specificity for one immunogen. Since this immunogen may have different epitopes, these antibodies are not monoclonal but form a collection of monoclonal antibodies.

We should say at this stage that most antibody reagents used to identify cell surface markers are now monoclonal. Using the hybridoma methodology you can avoid the problems of contaminating antibodies. However, antisera produced against pure antigens are extremely useful reagents and are still generated. These antisera against pure antigens may soon be rendered obsolete with the advent of the bioengineering techniques we shall discuss in Chapter 5. The one distinct advantage of the new technologies is that experimental animals may no longer be required.

We shall be looking at various ways of purifying antibodies in Chapter 6.

3.2.4 Modification of poor immunogens

Sometimes, the anticipated immune response after 2 or 3 injections of the antigen does not materialise suggesting that the antigen is a poor immunogen. Possible reasons for this and possible solutions are suggested in Figure 3.4. In such cases it may be advantageous to slightly modify the antigen to make it more immunogenic. One relatively simple approach is to denature the antigen by heating it or treating it with the detergent sodium dodecyl sulphate (SDS). If this is not effective, you can slightly alter the proteins by introducing dinitrophenol groups. Another approach is to make a complex of the immunogen with antibodies and inject this into the animal. The complex is thought to promote uptake of antigen and antigen processing. This procedure is obviously not possible unless you have previously obtained some anti-immunogen antibodies for some other purpose. You could also link your immunogen to red cells as we have just discussed. Otherwise you can link the protein to an immunogenic carrier as we will describe now.

denature the antigen

SAQ 3.2

You prepare an antiserum in a rabbit to sheep IgG and examine its specificity using immunoelectrophoresis with human and sheep whole serum as the antigens. You detect precipitin lines to sheep IgG, sheep albumin and human albumin. Which of the following statements are correct?

1) There are common epitopes on sheep IgG and human and sheep albumin.

2) The IgG used for immunisation was impure.

3) Passage of the antiserum over a sheep albumin affinity column would make the antiserum specific for sheep IgG.

4) The rabbit made antibodies to human albumin.

5) Addition of human albumin to the antiserum would remove the specificities for sheep and human albumin.

Linkage to carriers

Molecules of a molecular weight less than 1000 D such as many drugs or hormones as well as experimental haptens such as the dinitrophenyl group need to be linked to immunogenic carriers. However, you can also use this method with larger molecules which have proved to be poorly immunogenic.

Π Immunisation with a poor immunogen linked to a protein which is a good immunogen may lead to production of high affinity antibodies to the poor immunogen. Can you explain this carrier effect.

carrier effect

It is likely that B cells do bind to the poorly immunogenic protein but there may be a lack of T cell help. This would prevent production of good antibodies. However the immunogen specific B cells will probably bind the immunogen carrier complex, internalise it and produce peptides by cleavage of the complex. It is, however, likely that the peptides derived from the immunogen did not activate T helper cells when expressed associated with MHC II on the B cell surface and hence no activation of B cells resulted (refer to Figure 3.4). However, peptides derived from the carrier do activate carrier specific T cells which then will produce cytokines which activate the B cells into proliferation, class switching and somatic hypermutation resulting in good quality antibodies.

If you were unable to answer this in-text activity, it might be worthwhile to go back to Chapter 1 and re-read the sections on the role of T cells in B cell activation.

Examine Figure 3.4 carefully as it summarises the many different ways in which poor immunogens may be improved.

There are a host of methods for linking haptens to the carrier, the choice of method being dependent on the nature of the hapten and whether there are functional groups on the hapten which can be used in the linkage to the carrier. There are many very immunogenic carriers including keyhole limpet haemocyanin (KLH), bovine serum albumin (BSA), egg albumin (ovalbumin) and fowl gammaglobulin.

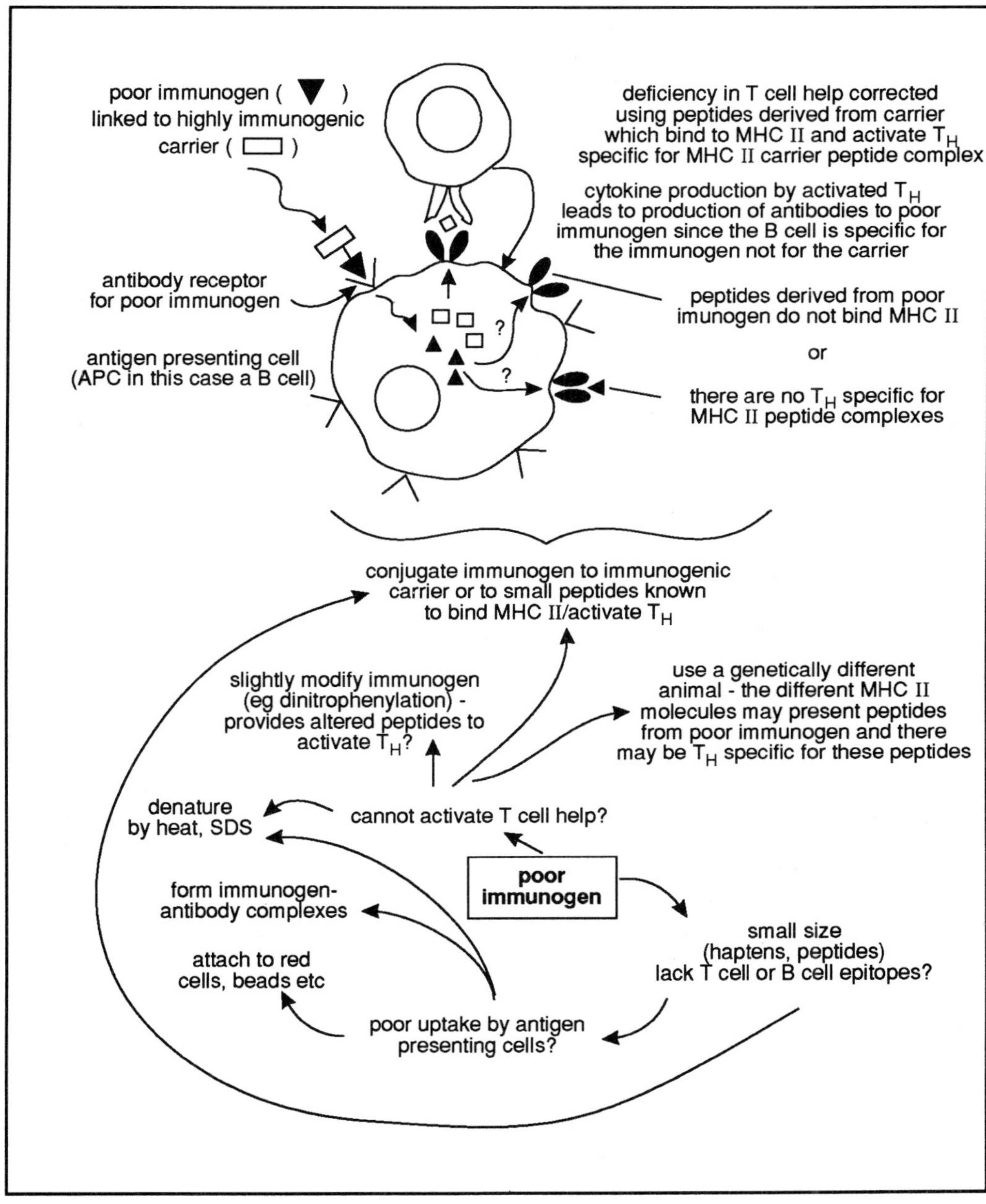

Figure 3.4 Approaches to counteract poor immunogenicity of antigens. Notice that attachment to red cells of haptens and peptides has the same effect as binding to a carrier since you provide carrier proteins from the red cells. In contrast, attachment to synthetic beads is primarily to promote uptake by antigen presenting cells such as macrophages. Denaturation of immunogens by heat or SDS could promote uptake by APC or activate T cells.

dinitro-phenylation

Dinitrophenylation of a protein carrier is relatively straightforward. In one method the protein carrier is dissolved in borate buffer and a solution of dinitrofluorobenzene is added. After stirring for one hour the solution is extensively dialysed against saline to remove the uncoupled DNFB leaving a solution of DNP carrier.

Π If you had prepared an antiserum against DNP-KLH in a rabbit and wished to purify the anti-DNP antibodies from those reactive with the carrier, which approach would you adopt?

making the antiserum specific for DNP

One approach based on our discussion thus far would be to couple DNP to a different carrier and then prepare an affinity column which would bind the anti-DNP antibodies. Such a conjugate could also be used in testing for the anti-DNP antibodies in precipitation reactions as obviously, antibodies and DNP do not precipitate. Another approach would be to prepare an affinity column of KLH which would remove the anti-KLH antibodies and the anti-DNP antibodies would be collected in the eluate from the column.

There is a wide choice of reagents which can be used to couple peptides or poorly immunogenic proteins to a carrier and many of these reagents are supplied with full instructions for linking from the commercial suppliers. We have already described a very popular one, SPDP (Section 2.4.4) in linking proteins to red cells but it can also be used to link peptides to carriers. In Figure 3.5 we illustrate the linking of polypeptide with carrier using another popular reagent, carbodiimide, which activates carboxyl groups on one of the reactants to link to amino groups on the other.

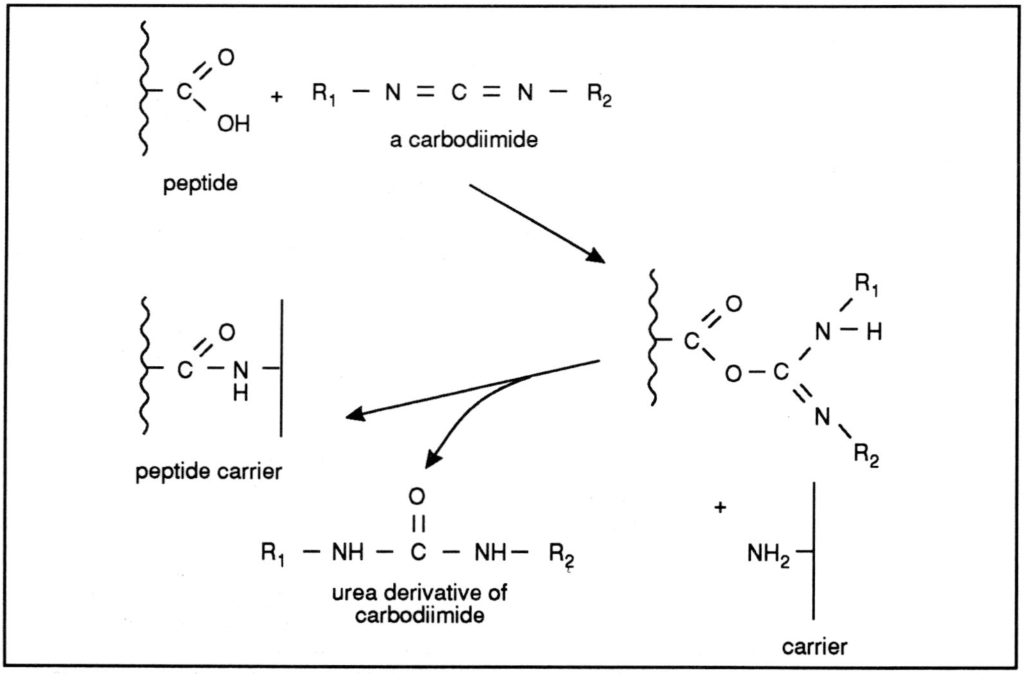

$R_1-N=C=N-R_2$

a carbodiimide

peptide

peptide carrier

$R_1-NH-C(=O)-NH-R_2$

urea derivative of carbodiimide

NH_2

carrier

Figure 3.5 The linkage of peptide and carrier using a carbodiimide.

Π Would you expect the anti-peptide antibodies to bind to the native protein?

This would depend to a large extent on the type of antibodies being generated. With polyclonal antisera, it is likely that some antibodies would bind the protein; with monoclonal antibodies, many would not. This is because many antibodies do not bind linear sequences and the conformation of the peptide and the equivalent sequence (epitope) in the native protein may be different. The idea behind the generation of peptides based on the sequences of immunogens is to predetermine the specificity of the antibodies for producing site specific antibodies since this would be useful in generating vaccines.

site specific antibodies

3.2.5 Processed antigens

Up to this point in the discussion we have described the use of complex antigens and various modified antigens and we emphasised earlier the need for a fairly high degree of purity of the immunogen, particularly for the production of conventional antisera. Even for the production of monoclonal antibodies, it is desirable to prepare as pure an antigen as possible so as not to generate too many non-specific clones which may hamper the detection of the ones producing specific antibodies. The principal methods for purifying antigens are polyacrylamide gel electrophoresis, isoelectric focusing or chromatofocusing, high performance liquid chromatography and affinity chromatography (Figure 3.6).

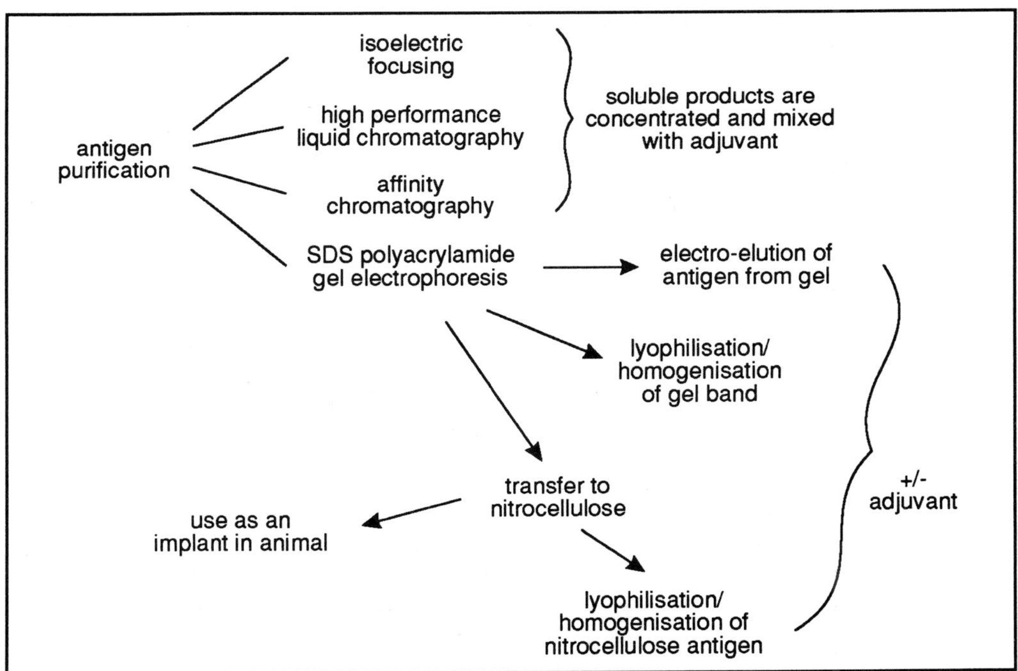

Figure 3.6 Major antigen purification methods and subsequent processing for immunisation.

In all except polyacrylamide gel electrophoresis (PAGE), the purified antigen is in the form of a soluble product which can be suitably concentrated and given to the animal in adjuvant (see below). However, antigens which have been separated on gels have to be treated differently for immunisation procedures. After PAGE the proteins may have been identified by staining with Coumassie Blue and after this identification step the stained band can be excised from the gel, lyophilised and crushed in a mortar and then it can be emulsified into Freund's adjuvant. In some instances, the Freund's adjuvant is not necessary because the gel itself can be acting as an adjuvant. Alternatively, proteins which have been transferred to a nitrocellulose membrane (as in Western blotting) can be used for immunisations. These membranes may be used as implants in animals or may be freeze dried (lyophilised) and homogenised and administered as a suspension. Alternatively, the protein can be electroeluted from the gel (Figure 3.6).

SAQ 3.3

You are presented with a mixture of 3 antisera consisting of antibodies to sheep red cells, DNP-KLH and human IgG. Propose a scheme to separate them.

3.3 Immunisation procedures

3.3.1 Choice of animal

Mice and rats are the only two species available for the preparation of monoclonal antibodies and they offer some advantages when studying cell interactions in conventional antibody production due to the availability of inbred strains. However, the volumes of antisera collected from such animals are extremely small and the animal of choice for most polyclonal or conventional antisera is the rabbit. Most rabbits will provide a minimum of 100 ml of antiserum. However, because they are outbred animals, individual rabbits respond differently and it is always wise to include about 3 rabbits in the immunisation protocol. The least costly method for producing very large quantities of antisera is to immunise sheep or goats. The sheep, in particular, is a very placid animal, easy to handle and bleed and it is possible to obtain 5 litres of antiserum from one animal. In addition, following production of the antisera, the sheep can be returned to the breeding stock with no adverse effects. Other animals used to a lesser extent for immunisation are guinea pigs, hamsters and chickens.

Sheep and rabbits make very good antibodies to all the human antibodies and to rat and mouse antibody classes as well as to each other so they are good sources of anti-globulin antibodies which we referred to in Chapter 2.

legal responsibilities

There are legal responsibilities of the investigator as to experimentation in animals which include the preparation of antisera and the production of monoclonal antibodies. The investigator needs both a personal licence covering the various techniques he/she wishes to use and a project licence which is specific for the particular research project. In the UK, for example, these licences are issued and supervised by the Home Office. Investigators in other countries should inquire into their legal responsibilities before attempting any animal research. For more details concerning the use of animals for experimental purposes we recommended the BIOTOL text 'A Compendium of Good Practices in Biotechnology'.

3.3.2 The use of adjuvants

particulate antigens

In general, particulate antigens are good immunogens and can often be injected alone into the animal. These include mammalian cells including red blood cells, viruses and bacteria, polyacrylamide gels and antigens attached to sepharose beads or similar particulate material.

adjuvants

For soluble materials such as proteins and carbohydrates and smaller molecules such as peptides, a much improved response can be obtained by the use of adjuvants. These are non-specific stimulators of the immune response and have a two fold action. Firstly, they form a local deposit of antigen which is slowly released to the immune system and secondly, they stimulate action by antigen presenting cells and cause an inflammatory reaction at the site of injection.

Freund's adjuvant

The local deposit of immunogen is formed by injecting along with the immunogen mineral oils or aluminium hydroxide precipitates. The most commonly used adjuvant, Freund's adjuvant, is an example of the former being a water-in-oil emulsion containing the immunogen whereas in the latter method the immunogen is adsorbed onto an aluminium salt such as aluminium hydroxide during precipitation. The non-specific stimulation of the immune system may be achieved by including in the injection heat killed bacteria such as *Bordetella pertussis* or *Mycobacterium tuberculosis* which stimulate the production of different cytokines which promote the inflammatory reaction and the stimulation of antigen presenting cells. It is now known that the prime stimulatory component of *M. tuberculosis* is muramyl dipeptide (MDP) and this is now available in several different forms to use as adjuvants.

Freund's adjuvant is available in two forms. The first, Complete Freund's Adjuvant (CFA) consists of both the water in oil emulsion and the bacteria whereas Incomplete Freund's Adjuvant (IFA) does not include the bacteria. CFA should only be used, if absolutely necessary, as a primary injection. Repeated use will cause painful necrotic lesions, hence only IFA is used for repeat injections. Some investigators also include bacteria when using alum precipitated immunogens.

immuno-stimulatory complex (ISCOM)

A more recent approach is to use liposomes for immunogen delivery. These are prepared from lecithin, stearylamine and cholesterol and it is possible to incorporate stimulants such as MDP as well as the immunogen. It is thought that these liposomes prevent general dispersion and loss of antigen and, by fusion with antigen presenting cells, effectively deliver the antigen to the immune system. Yet another recent innovation is in the use of saponin which, together with cholesterol and phosphatidyl choline, can form a stable complex with some immunogens to form a so called immunostimulatory complex (ISCOM).

3.3.3 Dose and route of immunogen

The dose range varies considerably depending on the immunogenicity of the antigen but the general rule is between 50 - 1000 μg in a rabbit and about 10 times less than this in a mouse or a rat. If cells are being used as the immunogen, about 1 million cells is a reasonable dose in a mouse and 5-10 times that amount in a rabbit .

subcutaneous route

In rabbits and larger animals, slow release of immunogen in adjuvant is promoted by intramuscular or intradermal injection whereas large volumes of immunogen are best given subcutaneously. The subcutaneous route, in fact, is that favoured by the Home Office in the UK for most immunogens in all species. The immunogen is delivered fairly rapidly to the local lymph nodes as we saw in Chapter 1 where the immune response takes place. A common route for injecting large volumes in mice, especially suspensions of cells, is the intraperitoneal route. Although it is sometimes considered advantageous to use the intravenous route for booster injections, primary injections, particularly of soluble antigens, may induce tolerance.

immunisation protocols

For most immunisation protocols, a number of injections spaced several weeks or sometimes months apart results in good quality high affinity antibodies in the serum and in high affinity plasma cells in rat and mouse spleens being prepared for hybridoma technology.

Π Can you give a plausible reason for this observation?

As we observed in Chapter 1, repeated responses to antigen results first in class switching and then in affinity maturation by mutational steps. B cells resulting from this process expressing high affinity receptors are selected by antigen whereas cells expressing lower affinity receptors do not survive. Hence the quality of the antisera after each response improves.

3.3.4 Bleeding animals for antibody assays

After each injection, the sera of the animals have to be tested for the presence of the antibodies and for assessment of the antibody titres. In mice, the least traumatic method is to bleed from the tail vein either by syringing or chopping off the end of the tail and recovering the few drops of blood which exude from the cut. An alternative method which results in about 0.25 ml of blood is to bleed from the retro-orbital sinus in the eye under anaesthetic. This can only be done by a skilled operator. Rabbits are bled from the marginal ear vein, sheep from the jugular vein and chickens from the brachial vein. Rats are usually bled from the heart under anaesthesia.

exsanguination

For the final bleed on sacrificing the animal, rats and rabbits can be bled from the heart under terminal anaesthesia, a process known as exsanguination. Larger animals such as sheep or goats are bled from the jugular vein and are not sacrificed as indicated earlier.

Π When should the first bleed be taken from the animal?

Animals should be bled before being given the first injection to obtain a control serum for subsequent tests.

SAQ 3.4

Pair the items in the left column with the most appropriate item in the right column. Use each item only once.

1) MDP	a) saponin and antigen
2) marginal ear vein	b) cytokines
3) mineral oils	c) hapten and carrier
4) ISCOM	d) test bleed
5) hyperimmunisation	e) antigen depot
6) conjugate	f) multiple injections

3.3.5 Recovery of antisera

decomplement the antiserum

Once the blood has been collected from the animal, it is allowed to stand at room temperature to clot. The clot is then detached from the sides of the vessel and the vessel is kept in the refrigerator overnight. During this period the clot retracts exuding the serum. The pale yellow liquid is then separated from the clot by centrifugation. It is normal to decomplement the antiserum by heating at 56°C for 30 minutes. The antiserum is then filter sterilised or 0.1% (w/w) sodium azide or methiolate (1 in 10 000 v/v) is added to prevent bacterial contamination. It is then aliquoted into tubes for storage at either 4°C (up to 6 months) or in the deep freeze (long term storage). Serum stored for long periods often acquires a precipitate. This has no detrimental effect on the antibodies and can be easily removed (cleared) by centrifugation.

collecting plasma

During harvesting of antiserum, varying degrees of haemolysis (damage to red cells) occurs. Some investigators avoid this by collecting plasma initially by bleeding into heparinised tubes or those containing citrate or EDTA (ethylene diamine tetra-acetic acid) to prevent clotting. After the cells have been removed by centrifugation, clot formation can be initiated by adding calcium chloride. The antiserum is then processed as previously described.

We shall be looking at the further processing, testing and purification of antisera in Chapter 6.

We have summarised the major points on the immunisation of animals in Figure 3.7. You will notice that, apart from the animals chosen for the procedure, the protocol is similar for production of conventional antisera and spleen cells used for the production of monoclonal antibodies. The difference is in the final step. For conventional antisera, you collect the serum, for monoclonal antibodies you collect the spleen cells.

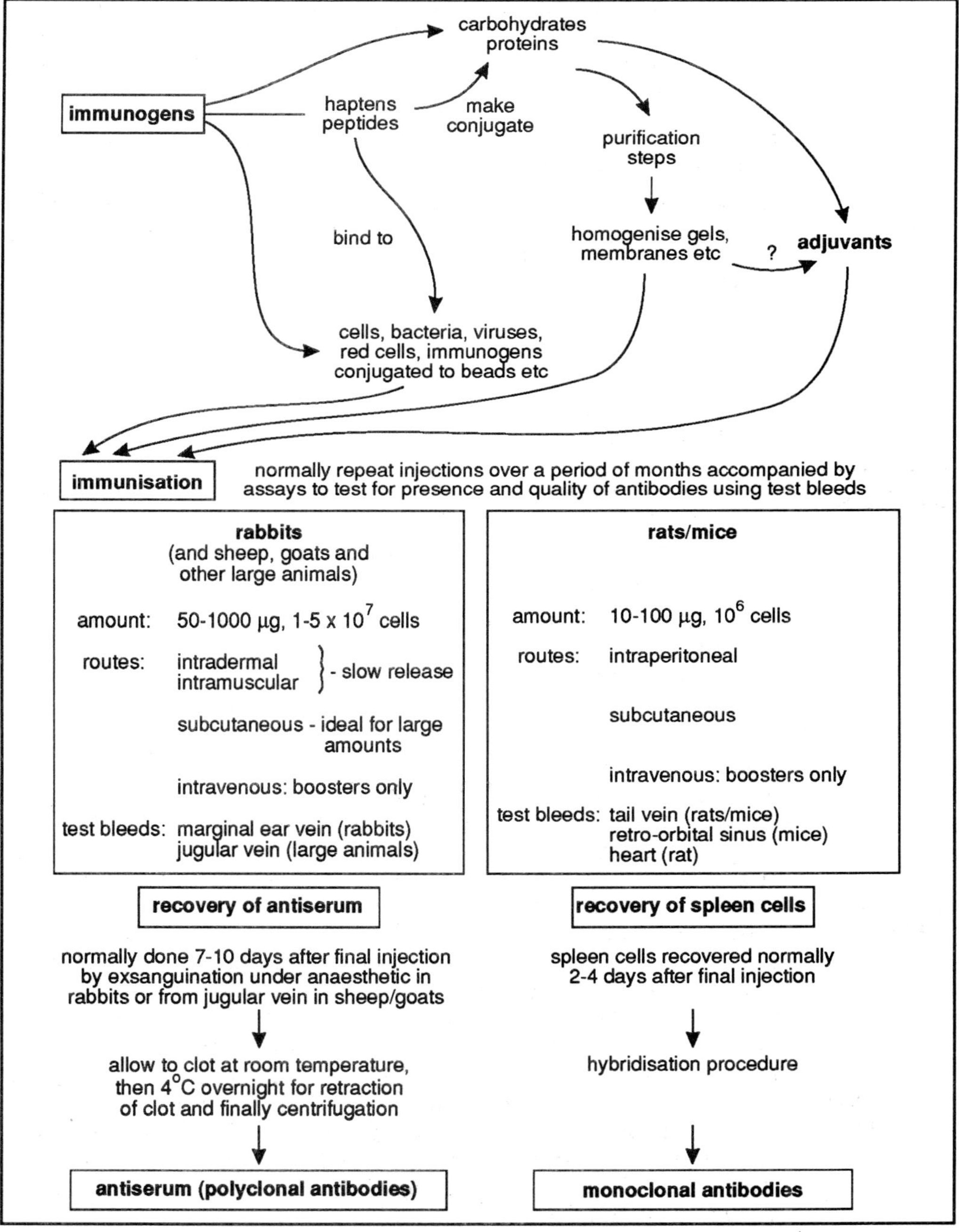

Figure 3.7 Immunisation protocols for the production of antisera and spleen cells for the production of monoclonal antibodies.

SAQ 3.5

Which of the following statement(s) is/are correct?

1) Absorption of unwanted specificities in an antiserum is less harsh than using affinity chromatography.

2) The presence of antibodies in the serum of an immunised mouse is a positive indicator for production of a monoclonal antibody from spleen cells of the mouse.

3) In preparing an immuno-affinity column to purify antigen the use of covalently linked monoclonal antibodies will yield a column of greater binding capacity than the use of covalently linked globulin fraction of high titre antisera.

4) Generally, antisera are better at agglutinating cells than monoclonal antibodies.

5) Plasma and serum contain the same proteins.

3.4 Passive immunisation

passive immunisation

We have just discussed the induction of antibodies by deliberate administration of antigens in various forms in animals. Passive immunisation is a method by which we can confer temporary protection on an individual or animal by supplying preformed antibodies specific for the pathogen against which the protection is required. Some examples are shown in Figure 3.8. Passive immunity may be acquired by:

- maternal antibodies transported via the placenta or in mother's milk;

- by injection of specific antibodies prepared in another animal;

- by administration of pooled human antibodies.

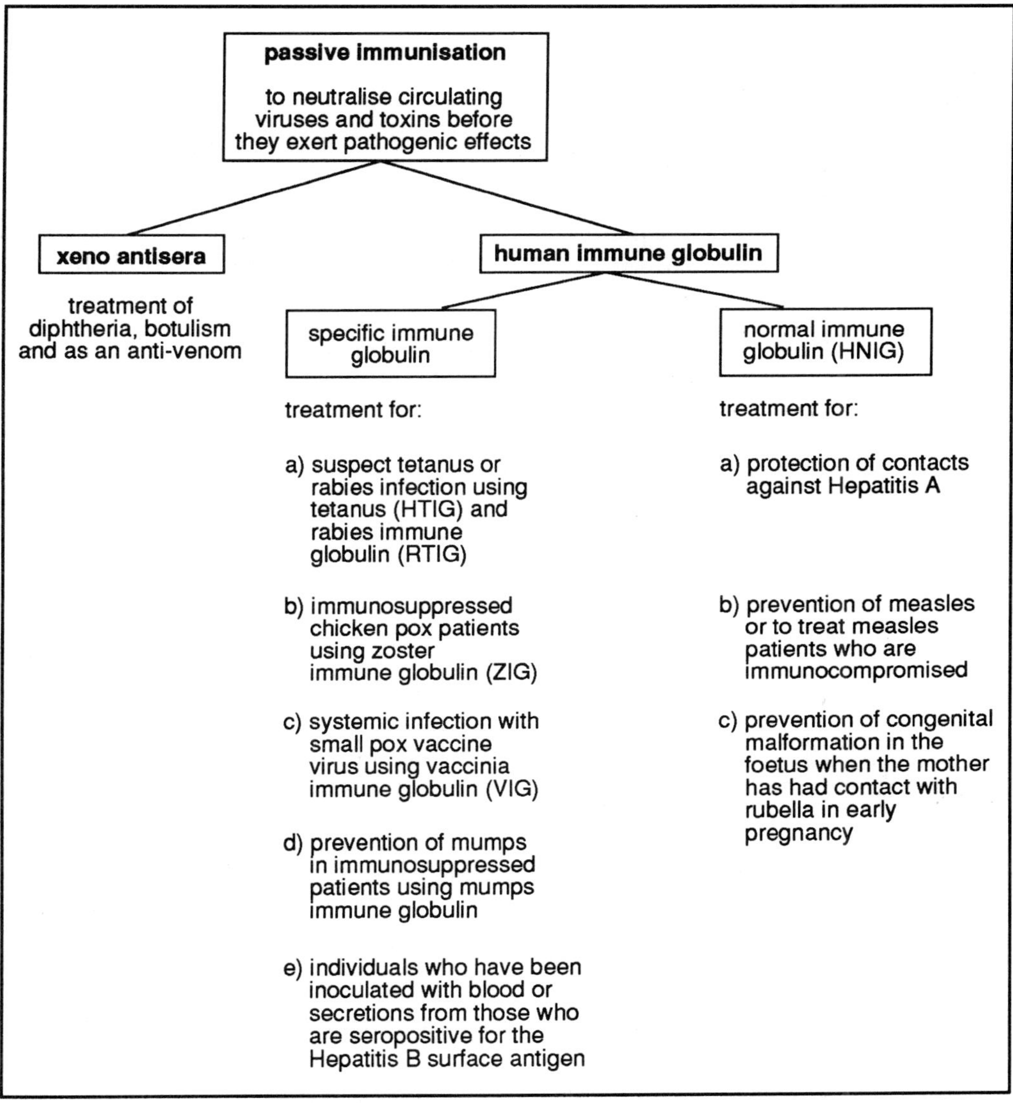

Figure 3.8 The use of animal antisera and human immune globulin in passive immunisation in the treatment of infectious diseases (see text for details).

3.4.1 Maternal antibodies

maternal IgG antibodies

During foetal life, the infant is supplied with maternal IgG antibodies which pass through the placenta and which will protect the infant from pathogens which the mother has recently encountered once the child is born. The foetus is normally protected from most pathogens while in the mother although there are some exceptions. The principal function of the maternally derived antibodies is to protect the infant as a newborn while the child's own immune system develops antibodies against the common pathogens which it will meet in early life. As you know from Chapter 1, these maternal IgG antibodies are protective antibodies due to their long half lives and high affinity.

secretory IgA antibodies

The other source of maternally derived antibodies is in the colostrum and milk containing secretory IgA antibodies with specificity for micro-organisms and antigens of the maternal gut. Although the half life of IgA is relatively short, the continuous supply from the mother for the first few months of life has an important role in protecting the infant while the secretory immune system of the child is being stimulated to produce its own antibodies.

3.4.2 Xeno antisera

hyperimmune sera from animals, diphtheria, botulism, snake bites

In the past it was quite common to obtain hyperimmune sera from animals such as the horse for treatment of a variety of life threatening diseases in Man. These proved useful for treatment of diphtheria, botulism and snake bites and are occasionally still used for this purpose.

Π Can you think of two reasons why the use of human immunoglobulin is preferred over horse antiserum?

serum sickness

Xeno antisera was obviously a useful source of antibodies since you could deliberately immunise the horse with a particular pathogen or antigen and obtain a good titre antiserum for use in Man. However, xeno antisera can induce various adverse reactions such as serum sickness which results from complex formation between the foreign horse proteins and antibodies made by the recipient. The second reason is that, even if you do not observe such an adverse reaction, the recipient may have become sensitised to any future contact with products from the same species which may cause problems in later life. Protection with xeno antisera is very short lived due to the short half lives of the protective antibodies.

3.4.3 Human immunoglobulins

human normal immune globulin

Hepatitis A

specific immunoglobulin

These are variously described as immunoglobulins, immune globulins or gamma globulins and generally mean an IgG rich fraction from serum. There are pooled human normal immune globulin designated HNIG and specific immune globulins. The former is extracted from blood donated from the general population and is mainly used to protect susceptible contacts of Hepatitis A, for instance unprotected people travelling in endemic areas of the world, or measles. It is sometimes used to protect newborn children in neonatal units against possible coxsackie virus infection. Specific immunoglobulin is prepared from the sera of people who have been recently immunised or are recovering from an infection from pathogens such as *Clostridium tetani* (causing tetanus), Hepatitis B virus, rabies virus or the varicella zoster virus (causing chicken pox).

Π When do you think someone needs the tetanus immune globulin?

tetanus immune globulin

Hepatitis B

rabies

varicella zoster immune globulin

Tetanus immune globulin is given as a precaution to individuals with a wound susceptible to tetanus infection who have no history of immunisation. The Hepatitis B immune globulin will prevent infection in individuals who have come into contact with Hepatitis B positive individuals and neonates whose mothers are positive. Rabies immune globulin is given immediately after an individual is bitten by a dog suspected to have rabies. The varicella zoster immune globulin is administered to people who have been in contact with chicken pox and are not able to produce an adequate immune response. These include patients who are immunodeficient and those who are undergoing immunosuppressive treatment for some other disease.

In the future, it is likely that monoclonal antibodies which have been humanised (Chapter 5) with well defined specificities against these infectious agents will become available and will replace immune globulin therapy.

3.4.4 Haemolytic disease of the newborn

anti-D (Rhesus) blood group antibodies

To close our discussion on passive immunisation we draw your attention to this interesting disease where prophylactic treatment with human immune globulin has saved many infants from death or debilitating disease. The most severe form of this disease is due to production of anti-D (Rhesus) blood group antibodies by RhD^- pregnant mothers carrying an RhD^+ foetus. The sensitisation of the mother occurs at the time of delivery of the first born when foetal red cells expressing the D antigen enter the maternal circulation. In about 70% of D^- mothers given sufficient foetal red cells, anti-D antibodies of the IgG class are formed. When the mother becomes pregnant again with an RhD^+ infant, these antibodies enter the foetal circulation and attack the red cells. The foetal liver degrades haemoglobin to bilirubin which causes brain damage (Kernicterus). This development can be prevented by administering anti-D human immunoglobulin.

Π When do you think this serum should be administered?

Obviously, the purpose of this treatment is to prevent the mother from becoming sensitised to the foetal red cells so the mother is injected with anti-D at each delivery. These antibodies will then bind to the foetal red cells in the maternal blood and they will be removed by the phagocytic cells of the reticuloendothelial system. The mother will therefore not respond to the D antigen and will not produce anti-D antibodies (Figure 3.9).

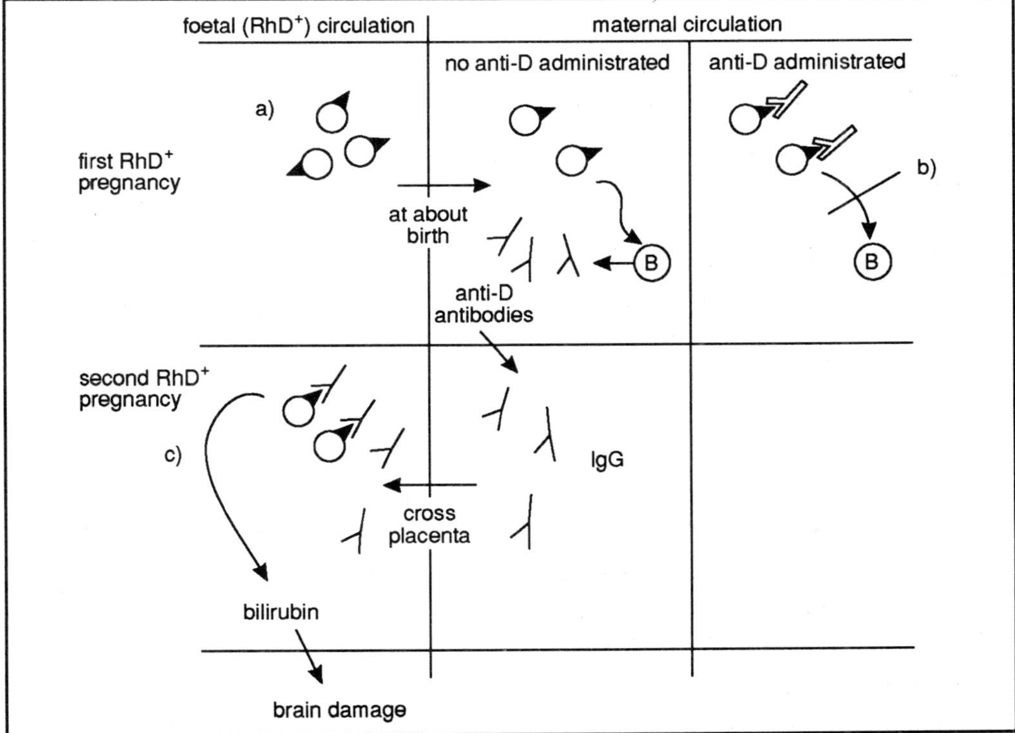

Figure 3.9 An example of the prophylactic effects of passive immunisation. The prevention of haemolytic disease of the newborn using anti-D antibodies administered to the mother at birth of each RhD^+ infant. a) foetal red cells express D^+. b) mother immunised with anti-D at delivery - foetal red cells destroyed and do not activate anti-D B cells. c) anti-D mediated damage to red cells.

SAQ 3.6

Which one of the following statements is correct?

1) Passive immunisation gives life long protection.
2) Passive immunisation with xeno-antisera gives longer protection than with human immunoglobulins.
3) Everyone travelling to high risk Hepatitis A areas should be given HNIG.
4) IgA antibodies in human milk are protective due to their long half lives.
5) HNIG is obtained from normal blood donors.

3.5 Active immunisation - vaccines

3.5.1 Introduction

Interest in vaccines to control infectious diseases have a long history beginning with the Chinese attempts to control smallpox over 1000 years ago. However, vaccination as a procedure started with the classical work of Jenner in 1798 when he used vaccinia (cowpox) virus to induce immunity to variola (smallpox) virus. You can see the origins of the term vaccination. About 100 years later a rabies vaccine was developed and in the early 20th century, vaccines were available for diphtheria, tetanus and pertussis (whooping cough). Since that time we have been very successful in protecting the world population from other communicable diseases. The two outstanding cases are smallpox which has been pronounced eradicated and poliomyelitis which is controlled at extremely low levels in the Western world due to the work of Sabin and Salk.

vaccination

Many other diseases are also being controlled by rigorous vaccination programmes particularly in the Western world. Infants are immunised with the triple vaccine for diphtheria, tetanus and pertussis, the polio vaccine and the measles vaccine (also rubella and mumps in the USA). Success from such programmes can be seen from the following data. In 1921 in the USA there were about 210 000 reported cases of diphtheria compared to 2 in 1985, in 1952 about 22 000 cases of polio compared to 5 in 1985 and lastly there were about 58 000 reported cases of rubella in 1969 compared to 604 in 1985

Present forms of vaccines and those under development are shown in Figure 3.10.

Π Make a list of what you consider would be essential features of a successful vaccine. You will find the answer in Section 3.5.2.

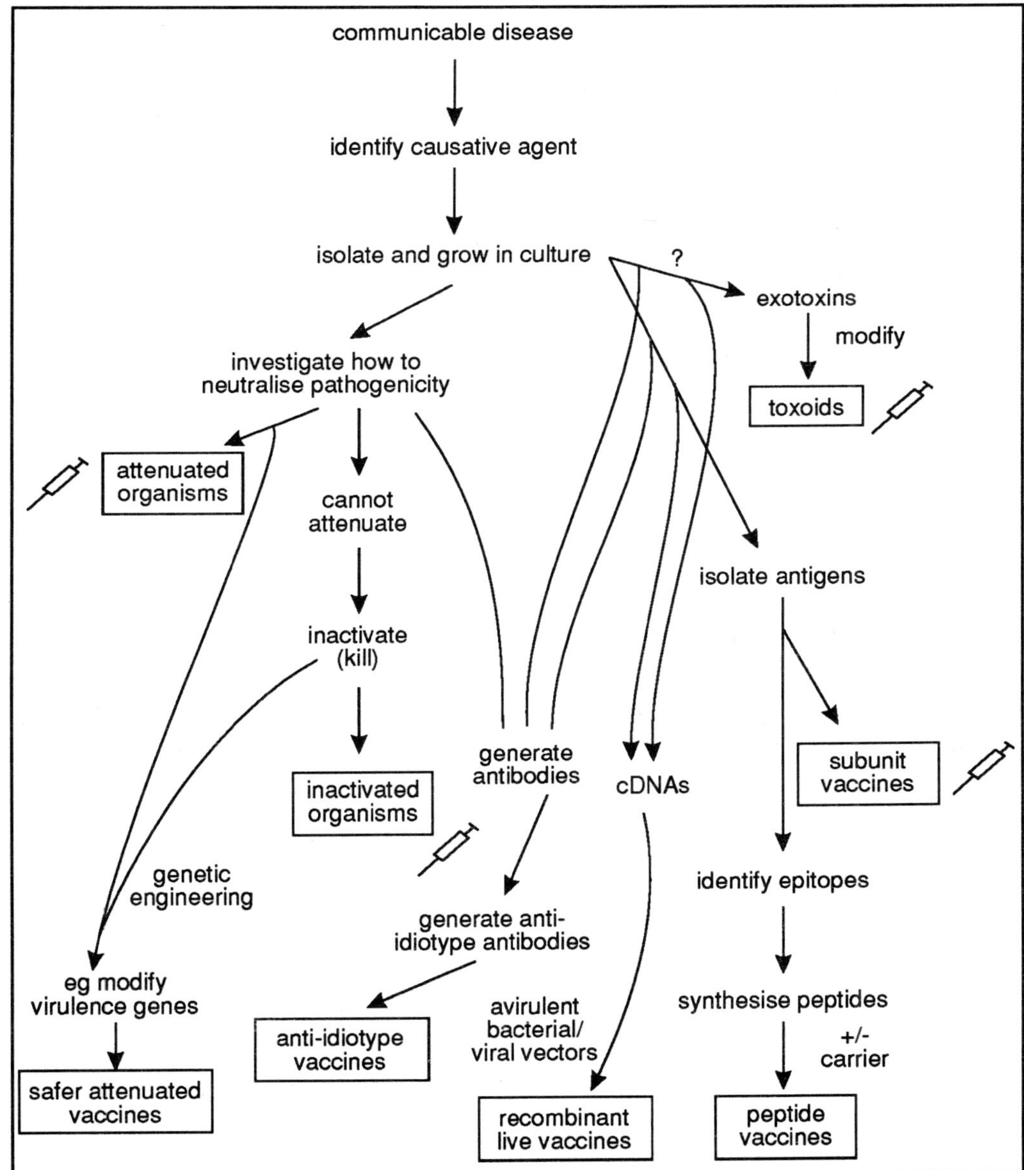

Figure 3.10 Types of vaccines presently in use or under development. Vaccines currently used are identified by a syringe (see text for details).

3.5.2 Essential requirements for the success of a vaccination programme

The idea of a vaccine is obviously to engender protective immunity to the infectious agent thus resulting in either high antibody titres and memory B cells and T helper cells or memory T cells which mediate cell mediated immune reactions, eg T_{DTH} or T_C. The route of administration of the vaccine is also important since this will influence the class of antibodies which will be made. For example, if you need systemic protection you will want to induce high affinity IgG antibodies but if you want protection on a mucosal surface you will want IgA.

requirements of a successful vaccine

You will also want to avoid the pathogenic effects of the organism while simultaneously inducing protective immunity to dominant antigens on the organism. With consideration for the Third World requirements, the vaccine must be cheap to make, cheap and safe to administer, must provide long lasting protection and it must be stable without requiring special storage facilities. In the Third World there are other difficulties such as assembling people for vaccination and ensuring that individuals actually obtain all the necessary booster injections to achieve long lasting immunity. Lastly, major international companies have little interest in vaccines which give them low returns on their investment so the Third World may have to make their own vaccines against organisms endemic in these areas. Did your list include all these features?

3.5.3 Vaccines using whole organisms

There are two forms of whole organisms which have been administered as vaccines and which induce immunity without the pathogenicity. The organisms are either killed or attenuated (altered). Details of some of these vaccines are shown in Figures 3.10 and 3.11.

killed organisms

Most bacterial vaccines are killed organisms and include typhoid and cholera vaccines. In contrast, most viral vaccines are live attenuated, the notable exceptions being the Salk polio vaccine, some rabies vaccines and the influenza vaccines. The idea behind inactivated vaccines is that the antigens which can be recognised on the live organisms in a subsequent infection are maintained thereby inducing an immune response but the ability to replicate and hence to be pathogenic has been destroyed. Due to the fact that these vaccines cannot replicate, the doses are large and full immunity generally needs multiple doses of the vaccine.

Π What do you think is the main advantage of attenuated vaccines over inactivated ones?

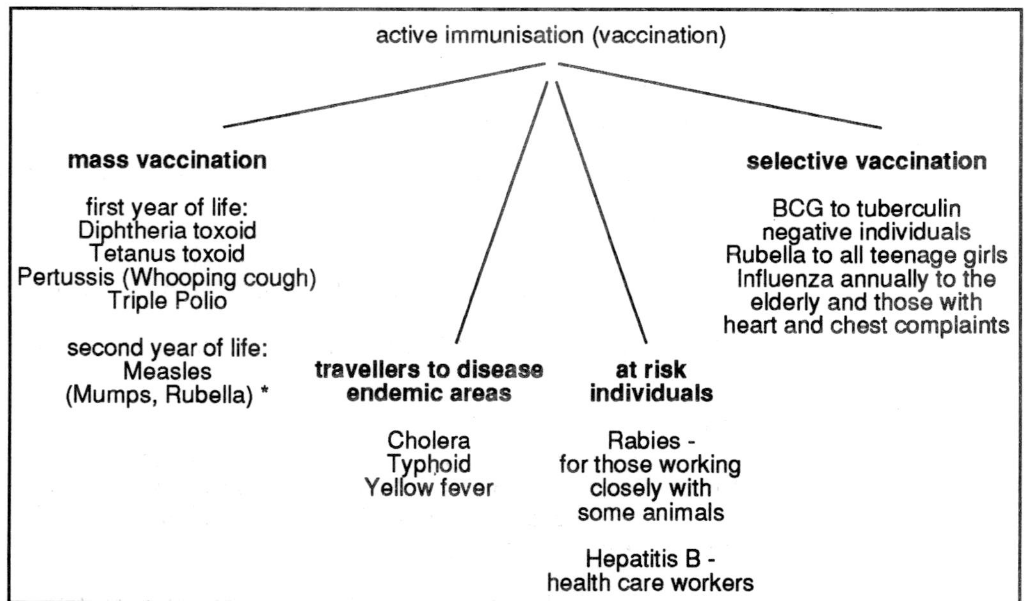

Figure 3.11 Major vaccination programmes in the developed world. * Mumps and rubella vaccines are given in some countries.

attenuated organisms

Attenuated organisms have one major advantage over the killed organisms since they can replicate in the host thus mimicking the natural organism but do not cause the disease. This means that a single small dose of vaccination is often sufficient to induce full protective immunity. The exception to this is the Salk triple vaccine which requires 3 doses. For anti-viral immunity, the advantage lies in the replication of the organisms in host cells resulting in processing and presentation to cytotoxic T cells. This activity is limited with killed virus.

Salk and Sabin polio vaccines

A comparison of the inactivated polio (Salk) and live attenuated polio (Sabin) vaccines illustrates some differences between the two whole organisms vaccines. Intranasal administration of the Salk vaccines induces a local IgA response for about 2 months whereas the Sabin vaccine results in long term production of IgA. The route of administration here is important to produce mucosal protection. However, the parenteral administration of the inactivated vaccine is effective since it induces circulating antibodies which block the spread of the virus from the entry point, the gut, to the central nervous system where the major viral damage is inflicted. The oral administration of the Sabin vaccine induces both secretory IgA and circulating IgG antibodies.

cholera

anthrax

BCG

Most attenuation methods involve altering the conditions in which the organisms are grown such as raising the temperature, changing the aerobicity or the supplements in the medium. Pasteur succeeded in producing attenuated cholera and anthrax vaccines using the first two modifications and the highly successful BCG (Bacille Calmette Guerin) vaccine against tuberculosis was made by growing a virulent strain of *Mycobacterium tuberculosis* in culture medium to which had been added bile.

The one danger, of course, with attenuated vaccines is that there may be some organisms present which have not been attenuated and are capable of inducing the disease. It is also possible that the attenuated strain might revert to virulence.

SAQ 3.7

Which of the following are correct procedures and/or can be expected to induce long term protective immunity?

1) Adminstration of the rabies vaccine 3 days after an individual reports being bitten by a rabid dog.

2) Intranasal and intramuscular administration of inactivated polio vaccine.

3) Mumps vaccine in an immunocompromised host.

4) Measles vaccine administered together with measles immune globulin.

5) A single dose of measles, mumps, rubella vaccine at 15 months of age.

3.5.4 Toxoids

exotoxins

Some bacteria produce exotoxins which produce major pathogenic effects in the infected individual. These include the exotoxins of *Corynebacterium diphtheriae* which causes myocardial, nervous system and renal/adrenal damage and that of *Clostridium tetani* which is a neurotoxin.

Π What immunisation strategy would be effective against these toxins?

Obviously you cannot immunise the individual with the active toxin. The approach for the production of a vaccine is to modify the toxins in such a way that they can no longer mediate pathogenic effects but still induce anti-toxin antibodies when used as an immunogen.

exotoxin is treated with formaldehyde

The exotoxin of *C. diphtheriae* is composed of an A and B chain joined together by a disulphide bridge. The B chain binds to oligosaccharide residues on the target cells and the whole exotoxin is then engulfed by the membrane and ends up in a vesicle in the cell. The A chain possesses enzyme activity which interferes with protein synthesis. The exotoxin is treated with formaldehyde which effectively detoxifies the molecule since the formaldehyde treated toxoid does not bind to host cells and looses its enzyme activity. Treatment of tetanus toxin with formaldehyde also produces a toxoid which is non-toxic but which retains the single epitope which induces protective antibodies in the immunised host.

3.5.5 Not all vaccines induce life long immunity

We have already implicated that a single dose of some vaccines is sufficient for life long immunity. In such cases, it is likely that the immunised individual is restimulated many times during life by the natural organism which activates the individual's memory B and T cells causing their clonal expansion and the production of more antibodies. Thus there are always sufficient levels of antibodies to protect the individual throughout life.

influenza virus

antigenic shift

In other cases, this is not so because the organism has the ability to change its antigenic epitopes and so re-infection of an immunised individual with the new strain can result in disease. Such is the case for influenza virus. The haemagglutinin (H) and neuraminidase (N) surface antigens change in major ways, (this process being called antigenic shift) or in minor ways, (called antigen drift). This means that antibodies against H and N antigens may be ineffective in preventing a subsequent infection. Persons at risk, such as those with heart and chest disease and the elderly are offered immunisation at the start of each influenza season and this approach is fairly effective.

polysaccharide vaccines

Killed whole organisms vaccines often produce relatively short term protection and so do polysaccharide vaccines.

Π Can you think of a reason why polysaccharide components on infectious agents do not induce long term protection?

polysaccharide toxin

Since T helper cells can only recognise linear peptide sequences associated with MHC II, polysaccharides alone will not promote the development of memory B cells which are T cell dependent. However, investigators have shown much improved protection if the polysaccharide is attached to one of the bacterial toxins.

Π Can you explain this phenomenon?

It is likely that B cells specifically bind the polysaccharide using its antibody receptors and the whole complex is internalised into an endosome. Peptides derived from the toxin then become attached to the MHC II molecules and some of these activate T helper cells specific for the toxin epitope-MHC II. The T cells then promote the development of memory cells to the polysaccharide as we saw in Chapter 1. This method, of course, involves a double vaccine since the individual would also become immune to the toxin.

Π Draw a diagram of these events and check your diagram against that in Figure 3.12.

3.5.6 New approaches to the development of vaccines

If you look back to Figure 3.10 you will observe that we have discussed the use of attenuated and inactivated whole organism vaccines and toxoids. These are the vaccines presently in general use. We shall now look at current trends towards new vaccines which include attenuation by genetic engineering (safer attenuated vaccines), subunit vaccines (isolated antigens), recombinant live vaccines, peptide vaccines and anti-idiotype vaccines.

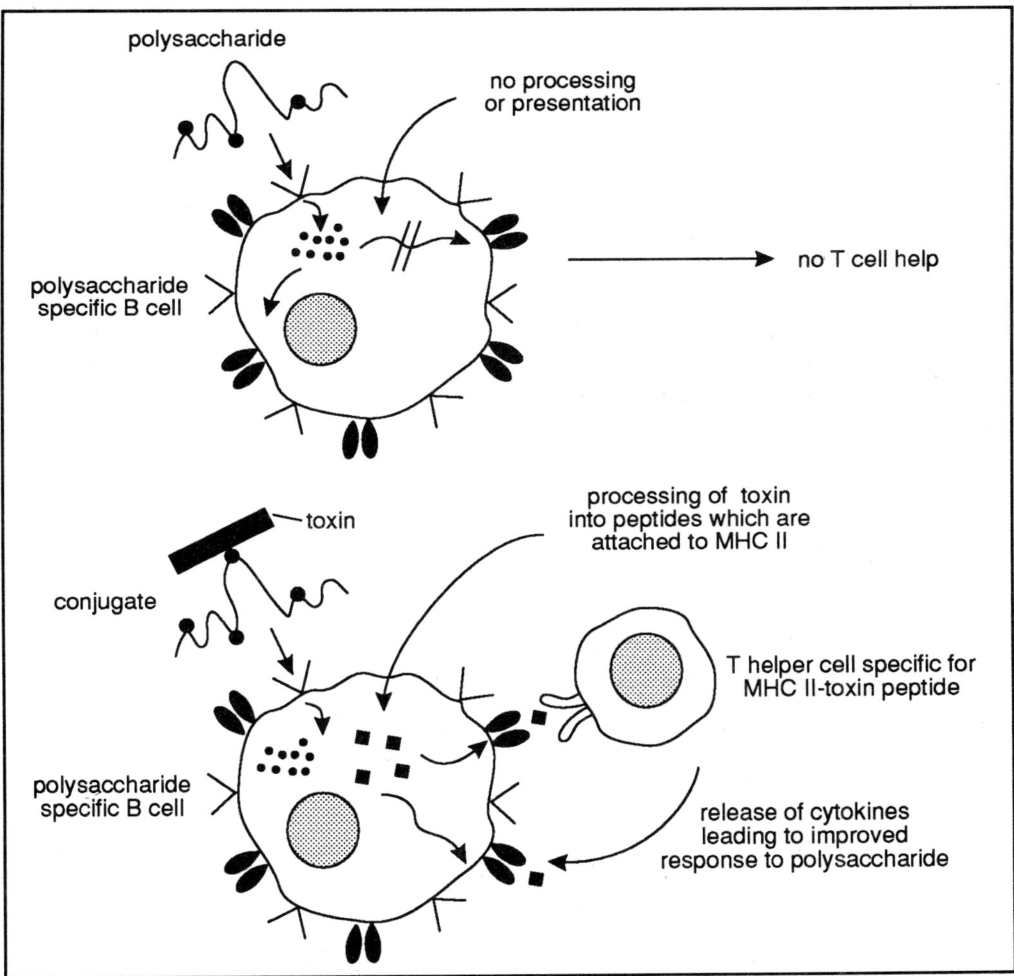

Figure 3.12 The enhancing effects of conjugating a polysaccharide to a toxin. Note that toxin specific B cells would also be activated and produce anti-toxin antibodies.

Attenuation by genetic engineering

As we have said there is a possibility of reversion to a virulent form when the classical methods of attenuation, as described above, are used. With the advent of genetic engineering, it should be possible to identify virulence genes in an infectious organism and to use genetic engineering such as site directed mutagenesis to either modify or inactivate them. A possible example of this is *Vibrio cholerae* which is responsible for cholera. Presently, the vaccine consists of killed organisms and the resulting protection is very short lived. This is because the vaccine induces protective circulating antibodies but no mucosal antibodies. If, however, the organism could be genetically engineered resulting in a live avirulent organism, this could be introduced into the gut to provide effective immunity. (A case study based on the production of attenuated vaccine by genetic engineering is given in the BIOTOL text 'Biotechnological Innovations in Health Care').

Subunit vaccines (isolated antigens)

Some whole organisms do not make good vaccines since it is possible that some of the epitopes may induce suppressor T cells in the host. We also have the problems of reversion to virulence. It would therefore be better and safer to select the immunodominant antigens to be used as vaccines, these are called subunit vaccines. Some of these are presently in use such as the diphtheria and tetanus toxoids and the split virion influenza virus vaccine. The bacterial polysaccharides of *Haemophilus influenzae* and *Pneumococcus spp.* provide short lived protection in high risk individuals but, as we mentioned earlier, there is no memory development. A single protein antigen from culture filtrates can protect individuals from anthrax and the Hepatitis B vaccine is a single protein. (A case study based on the production of the Hepatitis B vaccine is given in the BIOTOL text 'Biotechnological Innovations in Health Care').

subunit vaccines

Recombinant live vaccines

This is really a further advance on the subunit vaccines. The main problem with subunit vaccines being that they do not replicate in the host. The approach is to introduce genes for microbial antigens into non pathogenic viruses and bacteria and to infect the individuals with the recombinant live vaccines.

bacterial vectors

There has already been some success reported using both bacterial and viral vectors. *Salmonella typhi* has been shown to be a good choice for immunisation of the gut resulting in secretory IgA immunity against typhoid in Egypt and Chile. So *Salmonella spp.* could carry a plasmid containing a gene encoding an antigen from a micro-organism and hopefully induce mucosal immunity against the whole organisms from which the gene was derived. Another promising vehicle is BCG which is already a well established live vaccine in its own right. Additionally, mycobacteria are an essential component of Freund's adjuvant and should stimulate a good response to the antigens carried in BCG. Mice inoculated with BCG expressing a component of tetanus toxin, for example, produce reasonable levels of anti-toxin antibodies. There are many other bacterial species being investigated as vectors for live vaccines and we await further developments.

viral vectors

There are quite a few viruses being investigated as vectors for live vaccines as well. These include vaccinia, the avian specific avi-poxes and adenovirus. The most commonly used virus, however, is vaccinia which is an attenuated strain of the smallpox virus. This virus can accept up to about 30 kilobases of foreign DNA insert.

To produce a recombinant vaccinia virus the steps are as follows (use Figure 3.13 to help you follow the text). First, a bacterial recombination plasmid is constructed containing the foreign gene insert immediately downstream from a vaccinia promoter sequence inserted into the thymidine kinase site of vaccinia virus. This plasmid is used to transfect the host cells which are also infected with vaccinia virus. A process known as homologous recombination then occurs between the vaccinia virus and plasmid which results in the insertion of the plasmid sequence being incorporated into the vaccinia virus genome. The virus is then isolated and plated onto monolayers of thymidine

homologous recombination

kinase deficient cells and the TK recombinants are selected in the presence of bromodeoxyuridine which kills the cells which produce the enzyme. Cells which do not contain active thymidine kinase (TK) do not incorporate bromodeoxyuridine into DNA and thus these cells survive.

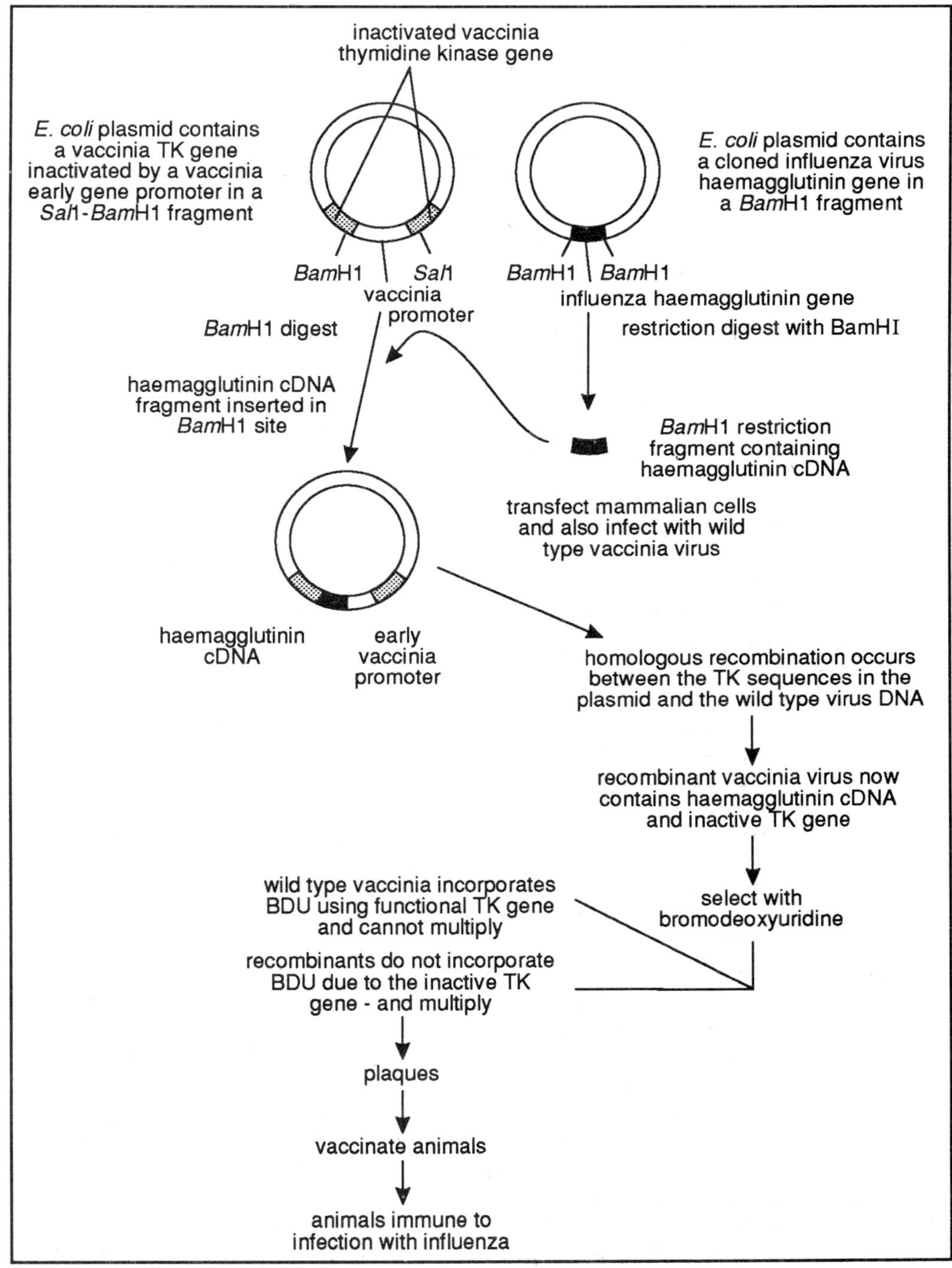

Figure 3.13 Construction of a live recombinant vaccine. TK = thymidine kinase.

Using this method, recombinant vaccinia viruses producing a whole variety of antigens including Hepatitis B surface antigen and influenza H (Figure 3.13) and N antigens have been generated and used to immunise various species of animals resulting in both antibody production and cell mediated immunity against the foreign antigens. There have been no human trials to date.

SAQ 3.8

Choose from the set of definitions or phrases in the left column, the most appropriate for each of the items from the right column (1-5):

a) an attenuated vector	1) toxoid
b) immunogenic sub component of a whole organism	2) antigenic drift
c) Inactivated neurotoxin	3) subunit vaccine
d) major changes in viral surface component	4) recombinant live vaccine
e) minor sequence differences	5) vaccinia
f) prophylactic antibodies	
g) viral immunogen	
h) inactivated whole organism	

Peptide vaccines

With the advent of genetic engineering it is now possible to isolate the mRNA and to make cDNA for many of the immunogenic proteins present in micro-organisms and thus to synthesise them. The advantages of such products is that they are non infectious and safe. The cDNAs can be inserted into a suitable vector as we have seen, or viral proteins can be synthesised in reasonable quantities as biologically active proteins in insect cells, yeast cells or mammalian cells and the products can then be incorporated into adjuvants for immunisation.

B cell epitopes are conformational

A recent approach has been to identify the actual epitopes on the proteins which induce antibodies and to synthesise peptides based on these sequences for use as immunogens. Many epitopes recognised by B cells are conformational, that is, they are formed from various parts of the protein which have come together due to folding and it is obviously difficult to imagine how one could construct a peptide which would mimic such a region. Even with linear epitopes it is difficult to substitute a short peptide for the sequence in the native protein due to random folding of the peptide with no support from the rest of the native protein. However, it has been suggested that peptides based on linear epitopes situated in regions where the protein folds are effective in generating antibodies which bind the native protein. It should also be noted that some of these linear epitopes may be recognising by T cells since this is the form of antigenic sequence more recognisable by T cells than B cells. We may, therefore, refer to these epitopes as T cell epitopes.

You can imagine, however, that this approach is attractive since peptides are relatively cheap to manufacture and if you can select the specificity of the antibodies by choosing the most immunogenic epitopes and eliminating the epitopes which induce suppression, then you may be able to engineer very precise vaccines.

Additionally, there is an emphasis now on T cell epitopes which were often neglected in the past with the result that very poorly protective vaccines were produced. Depending on the nature of the vaccine you might need helper T cell epitopes (epitopes recognised by helper T cells) and cytotoxic T cell epitopes (epitopes recognised by cytotoxic T cells). Another important point is whether you can generate a peptide which represents a T cell epitope which will be universally acceptable by both the MHC and T cells of individuals across the species. This problem is clearly apparent in experiments that have shown that cells from mice of different MHC haplotype respond to different sequences in the core protein of Hepatitis B virus.

Π How would you overcome this problem?

To prepare a vaccine which stimulates a T cell response (that is a T cell vaccine), you could use many peptides to cover MHC variation or clone the gene for the whole protein and use the protein product as the vaccine. This, however, might defeat the objective because it could introduce suppressor determinants or the whole protein might be toxic. Another approach would be to find motifs in sequences which determine binding to T cells and then attach these 'T cell' sequences to the B cell peptide. This sort of approach has been successful in animal experiments during development of a vaccine to foot and mouth disease virus (FMV). A peptide corresponding to amino acid positions 141-160 of the VP1 chain of FMV has been found to contain a major epitope recognised by B cells. Non responsiveness of some animals to this peptide has been overcome by attaching epitopes from non FMV proteins which are known to stimulate T helper cells.

Π In some instances, immunisation with a peptide alone has resulted in antibody production. Can you explain this phenomenon?

One would normally expect that such a low molecular weight peptide would not be immunogenic. However, we have to assume that the peptide was not only recognised by B cells but bound directly to MHC II on the B cell surface as well thus activating T helper cells.

We have already mentioned some of the successes in animals using synthetic peptides and some of the ideas are expressed in Figure 3.14. We will mention two more which demonstrate the potential of this approach.

streptococcal infections

A vaccine would be useful against *Streptococcus pyogenes*, the causative agent for meningitis. It has been found that only one surface component, the M protein, induces protective immunity. Past attempts to vaccinate Man against streptococcal infections have been hindered by the production of antibodies cross reacting with heart tissue and the high numbers of variation of the M protein. Using synthetic peptides, investigators have now determined which epitopes on the M protein induce protective immunity and which induce autoreactive antibodies. Thus there is now a multivalent vaccine consisting of peptides representing various M serotypes which induce protective antibodies without inducing auto-immune antibodies.

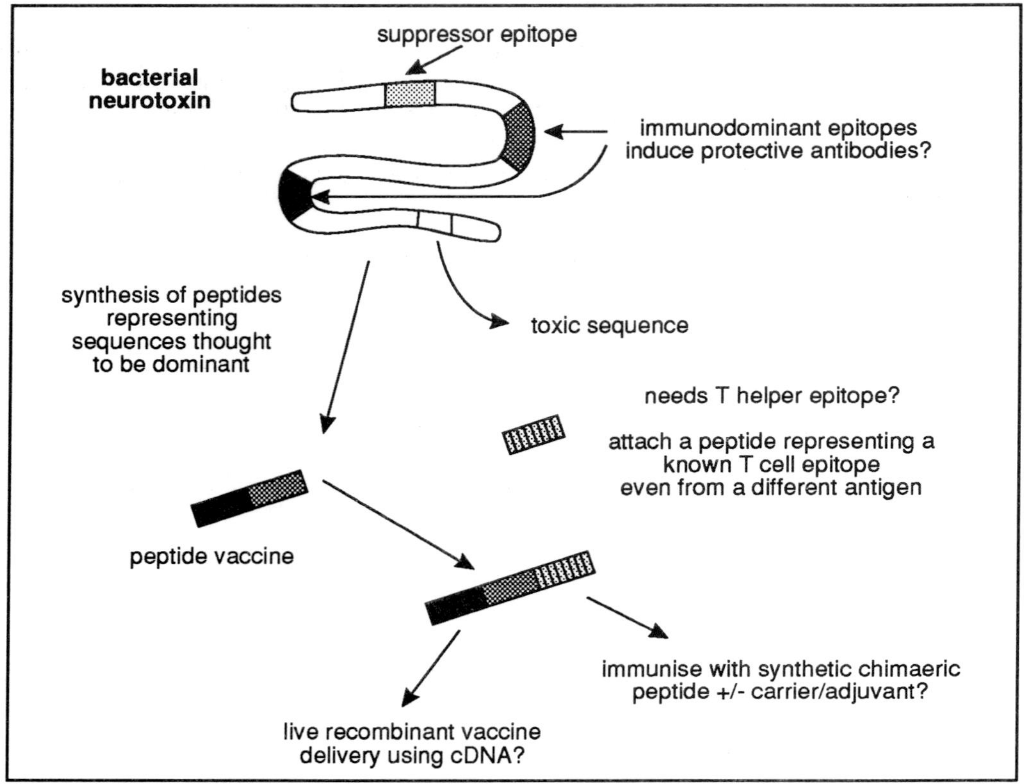

Figure 3.14 Synthetic peptide vaccines. The figure shows the ability to select particular sequences and ignore others such as suppressor or toxic sequences and prepare synthetic peptides representing immunodominant epitopes. However, the peptides selected may not activate T helper cells and genetic engineering enables insertion of a T cell epitope in the vaccine. Note that natural infection may not boost antibody levels due to the absence of a suitable T cell epitope in the natural sequence. In this case you may have to boost with the vaccine. Additionally, immunising with the peptides may not provoke protective antibodies due to conformational differences between the peptides and the sequences in the natural toxin. NB. T cell epitope = epitope recognised by T cells.

Some strains of *Escherichia coli* are the major cause of diarrhoea in developing countries due to the production of two antigenically distinct toxins. Investigators have now produced a 44 amino acid synthetic peptide which consists of a 26 amino acid sequence of one toxin and an 18 amino acid sequence of the other toxin. This peptide is totally non toxic and has been shown to be totally effective in inducing protective antibodies against both toxins.

Anti-idiotype vaccines

Π It might be useful to return to Chapter 1 to review what an anti-idiotype is.

This approach relies on the complementarity of the interaction between antigen and antibody. The antigen binding site of the antigen specific antibody (antibody 1) is the mirror image of the epitope and therefore some antibodies which are synthesised against this antibody (antibody 2) resemble the epitope and can be used to induce antibodies (antibody 3) which have the same specificity as antibody 1. Anti-idiotype vaccines have been tested in mice and have induced protection to some antigens. The immunogenic mechanisms may be non MHC restricted and therefore anti-idiotype vaccines may be useful immunogens in all members of the species. However, they are restricted in usefulness since one anti-idiotype represents only one epitope and protective immunity is generally raised to multiple epitopes.

SAQ 3.9

Which of the following statements describe the advantages of peptides as vaccines?

1) They are relatively cheap to produce and are safe.
2) They provoke good antibody responses.
3) They allow selection of immunodominant epitopes.
4) They activate T helper cells.
5) They induce antibodies which react with the native antigen.

Summary and objectives

In this chapter we have described ways to produce conventional antibodies and compared this methodology with that of monoclonal antibody production. We have looked at the different ways of obtaining monospecific antibodies from polyspecific sera and of the various types of immunogens which can be used to provoke the production of antibodies. We then examined the uses of passive immunisation using xeno-antisera and human immune globulin fractions. Finally we looked at the basis and methods of vaccination, its successes to date and the new approaches to the development of novel vaccines.

Now you have completed this chapter you should be able to:

- discuss the differences between monoclonal and polyclonal antibodies and their production;
- describe protocols for immunogen preparation, immunisation and antiserum production;
- devise procedures for separating antibodies from complex mixtures using absorption and affinity techniques;
- interpret experimental data involving precipitation reactions and cross reactivity;
- describe the use of xeno-antisera and human immune globulin in passive immunisation;
- discuss the benefits and limitations of present vaccination programmes using whole organisms and toxoids;
- discuss the potential benefits and limitations of recombinant live vaccines, subunit vaccines, peptide vaccines and anti-idiotype vaccines.

Hybridoma technology

Hybridoma technology

4.1 Introduction

4.1.1 The historical development of antibody production methods

We have depicted what we consider to be the major landmark advances in antibody production methods in Table 4.1. Following the discovery of antibodies in serum by Von Behring and Kitasato in 1890, it took 85 years before a major advance in antibody production methods was reported. This was, of course, the introduction of hybridoma technology for producing monoclonal antibodies by Kohler and Milstein in 1975. With the advent of molecular genetic techniques, events have moved rapidly with major advances in the past 10 years.

Kohler and Milstein

These advances have been motivated by a need for therapeutic antibodies in medicine as a major disadvantage of rodent monoclonal antibodies was their immunogenicity when they were injected into patients. This meant they had very limited use and could not be used for repeat treatments as the patients' anti-rodent antibodies generated following the first administration would rapidly promote their destruction before they could mediate any therapeutic effects.

need for therapeutic antibodies

The first advance was reported by a number of investigators beginning in 1984 who had synthesised humanised or chimaeric antibodies which consisted of mouse variable domains and human constant domains. A further advance was reported two years later by Winter's group in Cambridge. This group had further humanised the antibodies by grafting murine CDR regions onto human antibodies. The antibodies obtained were called reshaped antibodies.

humanised or chimaeric antibodies

reshaped antibodies

Π What is the difference between chimaeric and reshaped antibodies?

The difference between these antibodies is that the chimaeric antibodies possess whole V domains of both heavy and light chains from the mouse whereas in reshaped antibodies only the actual CDR regions of both heavy and light chains are derived from the mouse, ie the framework regions are human.

The remaining developments depicted in Table 4.1 suggest that the days of hybridoma technology are limited and monoclonal antibodies may be replaced by what has been termed coliclonal antibodies since they will be manufactured in prokaryotic systems. These advances include the ability to express functional antibody fragments in bacteria, the amplification of antibody genes using the polymerase chain reaction and a powerful screening system involving the display of antibody fragments on the surface of phage particles, so called phage antibodies. If the whole variable gene repertoire can be so expressed then selection of a required antibody can be made using antigen affinity techniques. These powerful techniques could exclude the requirement for immunisation of animals and the need for B cells in the near future.

coliclonal antibodies

phage antibodies

Date		Production Time
1890	Conventional antisera	Up to one year
1975	Hybridoma technology for monoclonal antibodies	Several months
1984	Chimaeric antibodies	2 - 4 weeks
1986	Reshaped antibodies	2 - 4 weeks
1989	Fab fragments expressed in bacteria	2 weeks
1989	Cloning antibody variable domains using the polymerase chain reaction	Represents a step towards producing antibodies using phage
1990	Phage antibodies	A few days

Table 4.1 The historical development of methods for production of antibodies. Note that for conventional antisera production, most of the production time is taken up with immunisation. Hybridoma technology does not take account of the immunisation time and is overall not likely to be quicker than conventional antisera production. The production time for chimaeric and reshaped antibodies assumed the availability of immune B cells. Production of antibodies in bacteria assuming availability of a cDNA library still takes about two weeks of production time due to the requirement for selecting the right antibodies by conventional cloning techniques. In contrast, phage antibodies, when a complete library is available, will only require a few days since the cloning procedure is eliminated as the phage antibodies are selected on an antigen column which takes little time. (See Chapter 5 for more details).

However, monoclonal antibody production using hybridoma technology is presently very much alive until these other techniques undergo further developments and this chapter is devoted to a description of the hybridoma technique as performed in rodents. We shall also briefly describe attempts to produce human monoclonal antibodies. We shall then complete our discussions on antibody production by describing the bioengineering of human antibodies chiefly for therapeutic purposes in Chapter 5.

SAQ 4.1

Match each item in the left column with one in the right column, using each item only once.

1) Chimaeric antibodies	a) Murine CDRs
2) Monoclonal antibodies	b) Antibody fragments
3) Reshaped antibodies	c) Antisera
4) Polyclonal antibodies	d) Hybridomas
5) Phage antibodies	e) Human and murine domains

4.2 Characteristics of the myeloma cell partners in hybridoma technology

4.2.1 HAT selection

We described the basis of monoclonal antibody production using hybridomas in the last chapter. The principal purpose of the fusion of the antibody secreting plasma cells to myeloma cells is to impart immortality to the plasma cell partners since unfused plasma cells cannot survive for more than a week in cell culture. This means that unfused plasma cells do not interfere in any way with the long term culture of the hybrids; but what about unfused myeloma cells? When you realise that the fusion frequency is of the order of 10^{-4} it should be apparent that you need a method by which you can promote the growth of the hybrids but not of the unfused myeloma cells which will be present in the culture in great excess. If these are not destroyed they will obviously overgrow your hybrid clones and you will end up with no antibodies at all.

So you can see that an essential requirement for the production of monoclonal antibodies using hybridoma technology is the ability to select for the antibody producing hybrids. This is done using a process known as HAT selection.

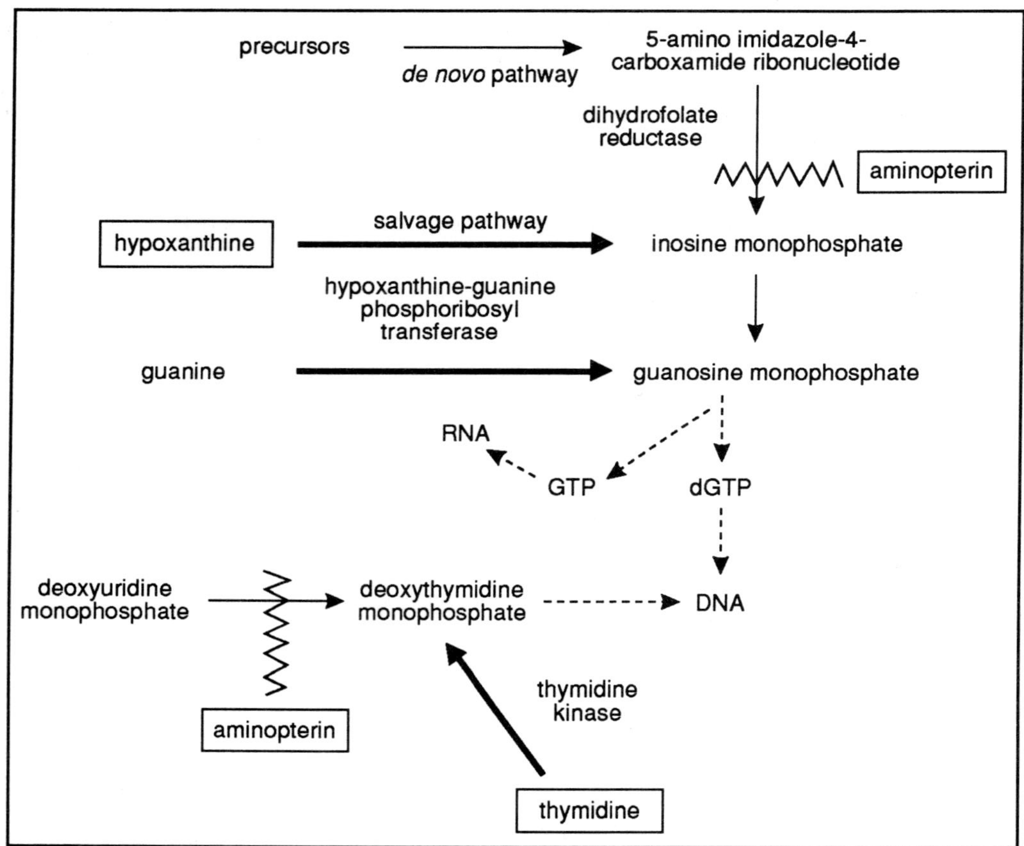

Figure 4.1 The basis of HAT selection. The salvage pathway is shown in heavy arrows, broken lines indicate multiple steps. The boxes indicate supplements in HAT medium; wavy lines show inhibition of enzyme activity. Aminopterin blocks the *de novo* pathway forcing cells to use the salvage pathway using the supplements hypoxanthine and thymidine added to the medium. Unfused myeloma cells cannot use this pathway as they are HGPRT$^-$ and do not survive. The myeloma-plasma cell hybrids survive by using the HGPRT supplied by the plasma cell partner.

de novo pathway

The selection process is based on the ability of mammalian cells to synthesise nucleotides by two pathways, the *de novo* pathway and the salvage pathway as shown in Figure 4.1. We can block the *de novo* pathway using drugs such as aminopterin, azaserine or methotrexate and so force the cells to use the salvage pathway as long as they have the necessary substrates hypoxanthine or guanine. It just happens that we can also produce myeloma cells which cannot use the salvage pathway thus providing a means of selecting for plasma cell-myeloma hybrids.

hypoxanthine-guanine phosphoribosyl transferase

Look at Figure 4.2 which depicts the preparation of HAT sensitive mutants. The enzyme which catalyses the most essential step in the salvage pathway is hypoxanthine-guanine phosphoribosyl transferase, designated HGPRT. The gene encoding this enzyme is on the X chromosome and is rendered inactive at a frequency of 1 in 10^7 by the normal rate of mutagenesis. If you culture, say 10^8 myeloma cells in the presence of excess 6-thioguanine, HGPRT will be induced in a majority of the cells and these will convert the purine analogue into the nucleotide monophosphates and these will finally be incorporated into RNA and DNA. Since these purine analogues are metabolic poisons, the cells will die.

HGPRT mutants

However, a few cells will have undergone mutations and will not be able to produce HGPRT, they are HGPRT mutants. These cells therefore cannot use the salvage pathway and will not take up the purine analogues but will use the *de novo* pathway for the synthesis of precursors of RNA and DNA. These cells will, therefore, survive and can be used as the partners for the antibody secreting plasma cells in the hybridisation technique when they are cultured in HAT medium.

We remind you that HAT medium contains hypoxanthine (H), aminopterin (A) and thymidine (T).

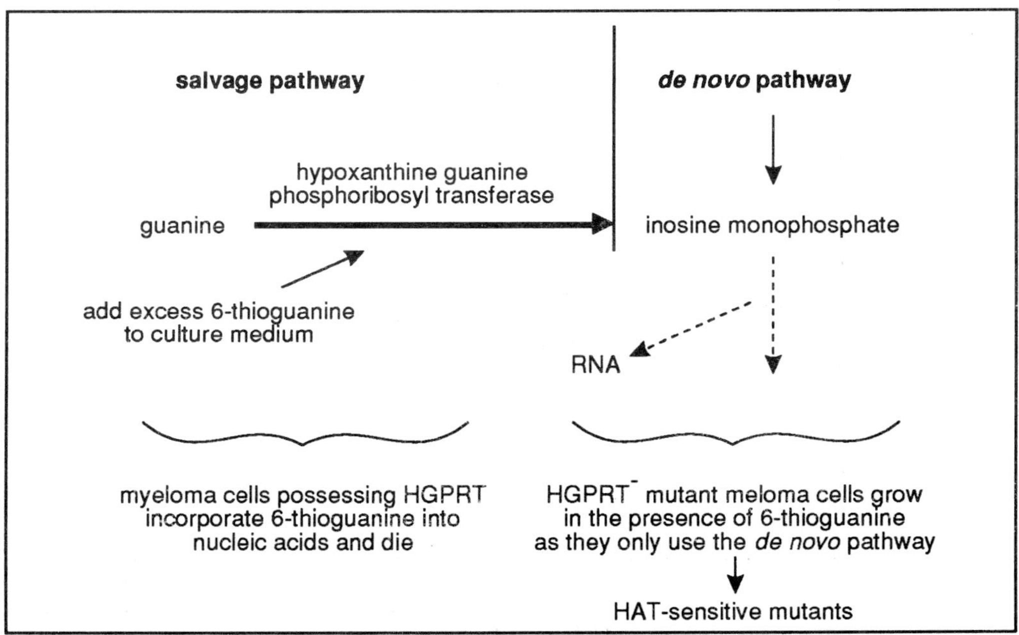

Figure 4.2 Selection of HAT sensitive mutants. The HAT sensitive mutants are ideal fusion partners for antibody secreting spleen cells as they can only survive in HAT as part of a hybrid (see text for further details).

Π Examine Figure 4.1. Can you explain why the hybrids (HGPRT$^-$ myeloma cells-plasma cells) survive in the culture when aminopterin is present?

When the hybrids are cultured in HAT medium which prevents use of the *de novo* pathway, only cells possessing HGPRT can survive. Thus myeloma cells which have not fused to plasma cells die whereas those which have successfully fused to plasma cells survive since the hybrid can use the salvage pathway due to the plasma cell partner providing the HGPRT. So you can see that within a few days of the actual fusion, the only cells still growing in the HAT medium are the hybrids. Examine Figure 4.3 which shows the main cell types and hybrids you would expect to find in the culture medium following fusion and explains why only the hybrids survive.

thymidylate synthetase

Aminopterin also blocks thymidylate synthetase which provides thymidine nucleotides. Thymidine therefore has to be added for bypassing this enzymatic step providing thymidine monophosphate using thymidine kinase. HAT medium, thus contains added hypoxanthine, aminopterin and thymidine.

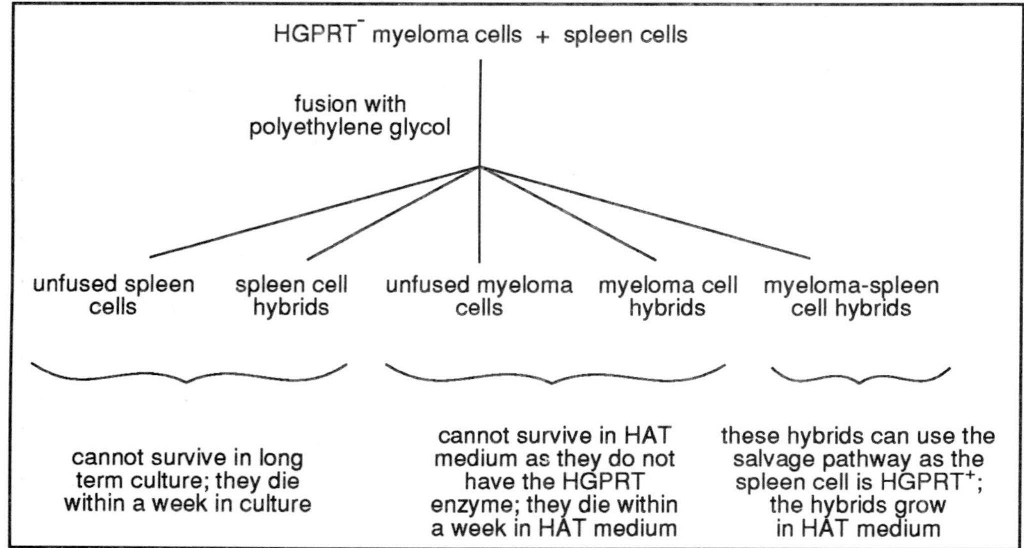

Figure 4.3 Cells and hybrids in the culture medium following fusion with · olyethylene glycol and the selective effects of HAT medium.

4.2.2 The source of myeloma cells

S· 2/0-Ag14

NS0/1

One method of inducing myelomas in mice is by injection of mineral oil into the peritoneal cavity. Potter in the USA isolated such myelomas from one strain of mouse called the BALB/c mouse and the cells were designed MOPC 21 (mineral oil plasma cytoma) from which was derived a cell line called P3K. HGPRT$^-$ mutants were derived from this cell line including the most commonly used myeloma cell lines Sp2/0-Ag14, NS0/1 and X63-Ag8.653. Since these myeloma cells are plasma cells they all have the capability of producing antibodies but these lines have been selected for non expression of Ig genes. There are also rat myeloma cell lines with similar characteristics. These include R210.RCY3 which secretes kappa chains and the non producer lines YB2/0 and IR 983F.

Π Can you think of a reason why failure to express Ig genes is an advantage in a myeloma cell to be used in the hybridisation technique?

Since these cells are to be used as partners for fusion to plasma cells secreting the antibody of choice, it would be a distinct disadvantage if the myeloma cells produced antibodies as well as the plasma cells since you would then have to go through the expensive process of affinity purifying your antibody.

freezing cells

The myeloma cell lines are easily obtainable from many research laboratories. They are maintained in liquid nitrogen. The process of freezing these cells is the same as for other mammalian cells such as the hybrids which will result from the fusion. The cells are washed in culture medium and centrifuged. The medium is removed and replaced by foetal calf serum. Dimethyl sulphoxide is then added (10% by volume) to prevent the formation of water crystals inside the cells during freezing. The cells are then gradually frozen in a -70°C deep freeze and then transferred to liquid nitrogen.

dimethyl sulphoxide

When the cells are required for fusion, they are removed from liquid nitrogen a few days before the planned fusion, rapidly defrosted at 37°C and then cultured in growth medium until they attain exponential growth. They are then ready to be used as fusion partners.

SAQ 4.2

Which one of the following statements is incorrect?

1) Hybridomas are immortal somatic cell hybrids that secrete monoclonal antibodies.

2) The antibody secreting cell partners in hybrids have undergone DNA rearrangement to produce functional antibody genes.

3) The myeloma cells possess genes which promote continuous growth in cell culture medium.

4) Drug selection of HGPRT$^-$ myeloma cells is performed immediately prior to each fusion.

5) HAT medium selects for cells which are able to use the salvage pathway.

4.3 Steps in hybridoma production

Now we have looked at the theory of HAT selection and the characteristics of the myeloma cells we are ready to proceed with the actual fusion, growth and cloning of selected hybridomas.

Π Before you read on, can you remember the main steps in monoclonal antibody production using this procedure from Chapter 3? To remind you, construct a flow diagram of the sequential steps and then check it with Figure 3.2.

4.3.1 Immunisation and testing of sera for antibodies

We have already dealt with immunisation procedures for the production of both conventional antiserum and monoclonal antibodies. In practice, a few BALB/c mice are selected for immunisation and the antigen is administered. In some laboratories a single injection has been found to be sufficient especially when immunising with cells but normally multiple injections are given over spaced intervals of time to promote the production of B cells producing high affinity antibodies.

Sera are collected by test bleeds from each animal a few days after immunisation and they are used to develop a suitable antibody screening procedure which can be applied to the supernatants of the hybridomas following fusion. It is also preferable that a rapid antibody test such as an automated ELISA assay which is able to deal with the many supernatants you have to assay (up to 1000!) is chosen. This is an important step since all hybrids will be rapidly growing in the wells of the culture plates and a good hybridoma may be lost if a quick method of detecting the desired hybridomis not available.

You would obviously not proceed with the fusion if you cannot detect antibodies in the sera of the immunised animals because hybridoma production is expensive in terms of materials.

The test you choose depends on the nature of the antigen. We described many tests in Chapter 2 and some of these are extremely rapid and are therefore suitable for testing of the hybridoma supernatants. These include passive haemagglutination assays and the ELISA-based assays for soluble antigens. For antibody production against cellular antigens a complement mediated lytic assay could be developed or you could assay the antibody for its ability to inhibit some function of the cell, ie a biological assay.

For instance, if you are generating an antibody against a hormone receptor on a cell ,the antibody may block the binding of the hormone thus blocking the biological activity which can be measured. However, you can imagine that such an assay takes considerable time to perform since it involves a period of cell culture.

The general rule is that if you are developing the antibody for a particular use then you should use a screening assay based on that use. For instance, if you want an antibody for use in ELISA or blot assays then you use that system to screen for the antibody during hybridoma development. Since the ELISA is such a popular assay for testing of monoclonal antibodies, you will find an SAQ on that topic later in the text.

Once you are satisfied about the way your assay will work, you administer a final injection of antigen to the mice and perform the fusion 3-4 days later.

SAQ 4.3

To emphasise the point of using an assay directly relevant to the proposed use of the antibody let us assume that you are trying to make an antibody which will block the binding of a hormone to its receptor. Interleukin 2 (IL-2) is a cytokine of about 15 000 D which promotes the growth of activated T cells. The growth signal is transmitted through receptors on the surfaces of the T cells. You wish to prepare monoclonal antibodies against IL-2 and perform a fusion after immunising mice with IL-2. You test the supernatants from 2 wells of one microculture plate and obtain the results shown in Figure 4.4.

Indicate by a +/- which of the following statements describe the activities of the two antibodies. If you cannot make a definitive conclusion indicate with a question mark.

	Antibody 1	Antibody 2
1) Inhibits the growth of T cells		
2) Blocks the effect of IL-2		
3) Specific for the IL-2 receptor		
4) Could be specific for the binding site on IL-2 for the receptor		
5) Is IL-2 specific		

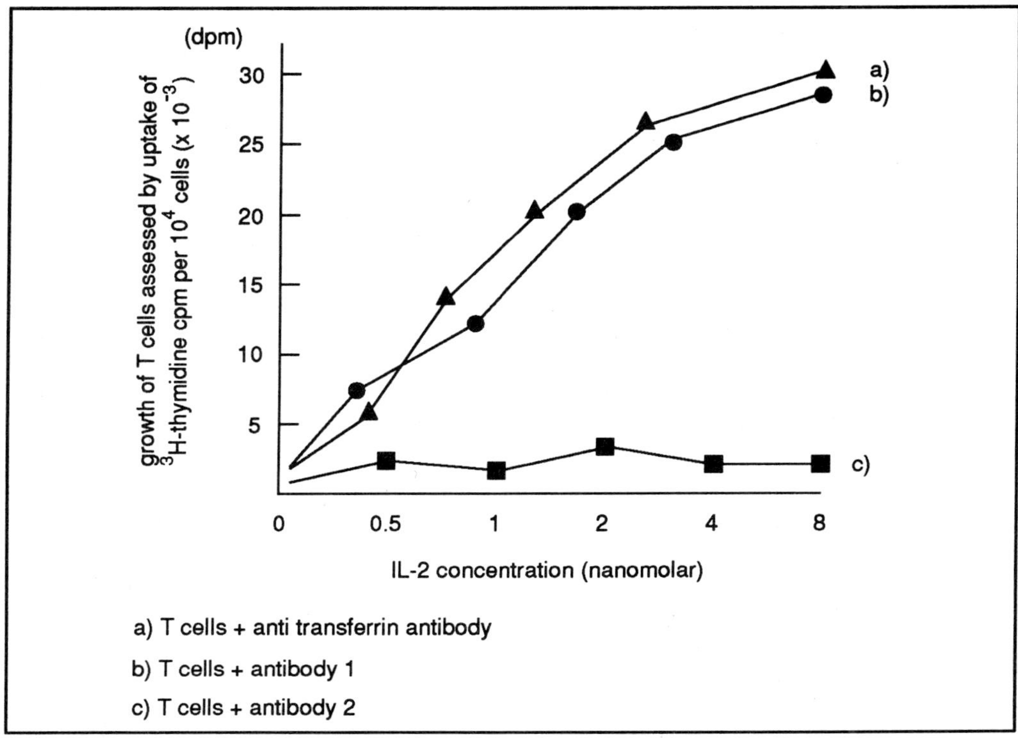

Figure 4.4 Supernatants from two wells of the microplate were examined for the presence of antibodies against IL-2. T cells were grown in medium supplemented with serial dilutions of IL-2 and the supernatants contained antibody 1 or antibody 2 or a control antibody, anti transferrin. The growth of the T cells was assessed 2 days later by addition of ^{3}H-thymidine followed by a further period of cell culture for 4 hours. The amount of radioactivity incorporated into DNA by the cells was measured, this indicating the growth activity of the cells.

4.3.2 Cell culture medium

Although many laboratories employ their own particular formulae for their cell culture media, it would be useful for you to know what are the principal ingredients required to support the growth of cells including hybridomas. The culture media contain various essential salts, glucose and amino acids and various other nutrients including vitamins and other cofactors. L-glutamine is usually added to the medium before use as it deteriorates with storage and sodium pyruvate is usually added as an additional substrate for metabolism. The medium is supplemented with a source of serum, usually foetal calf serum which supplies various essential nutrients such as insulin, transferrin and fatty acids as well as rarer nutrients such as selenium salts. Some investigators now use serum free media to prevent contamination of the monoclonal antibodies with serum proteins and to avoid the cost of the serum. Antibiotics such as penicillin and streptomycin are also included to prevent bacterial contamination of the culture.

L-glutamine

foetal calf serum

penicillin and streptomycin

pH of the culture medium

The pH of the culture medium is maintained at about 7.4 using bicarbonate-carbon dioxide buffering. Sodium bicarbonate is added to the medium and the cultures are maintained in an incubator containing 5-10% CO_2 in air and a water saturated atmosphere.

4.3.3 Counting spleen cells and myeloma cells

haemo-cytometer

Both spleen cells and myeloma cells are counted prior to the fusion to ensure the correct proportions of interacting cells. Counting is usually done using a haemocytometer which is a specifically made glass microscope slide with an imprinted grid made up of 1 mm squares each containing 16 equal sized smaller squares (Figure 4.5). With a cover slip in place the volume of liquid enclosed between the grid and the cover slip over each 1 mm square is 0.1 mm^3. A drop of cell suspension is introduced between the slide and coverslip and the total cells in each 1 mm square is counted using a microscope. It is normal practice to count not less than 100 cells to ensure accuracy of the cell count.

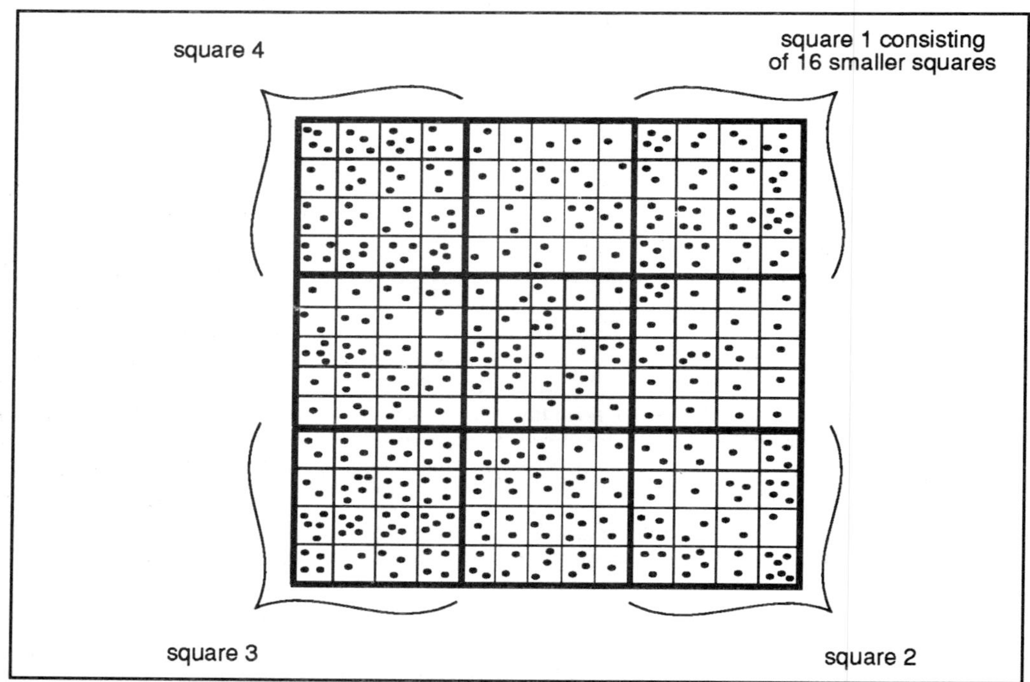

Figure 4.5 The haemocytometer. There is a central counting square and four squares at the corners of the grid each measuring 1 mm^2. Each of these 4 squares consists of 16 equal sized smaller squares. The depth of the counting chamber is 0.1 mm so the volume of cell suspension over each square is 0.1 mm^3. The cell suspension is suitably diluted and the exclusion dye added. The cell suspension is then introduced into the chamber and the non stained cells are counted in the outer squares until at least 100 cells have been counted. If you count the cells in square 1 you will find 47, in square 2, 43 and square 3, 60, making a total of 150. The average per square is 50 cells and this is contained within 0.1 mm^3. If you multiply this figure by 10 000 you convert the cell count number to cells per ml (cm^3), in this case, 0.5 x 10^6. In practice the actual cell concentration in your cell culture would be greater than this as you have to take into account any dilutions of the cells (for instance due to addition of dye) prior to counting.

Π Would you consider this method to be an accurate count of living cells?

If you do not use an exclusion dye, what the described method would do is to count total cells, dead as well as living. To determine the actual number of living cells you need to use a vital stain which stains dead cells but not living cells or vice versa. To do this you simply add one volume of trypan blue stain (0.25g 100 ml^{-1}) to one volume of cell suspension and count the unstained cells as previously described. In calculating the total count you would have to take into account the 1:1 dilution. Note tryptan blue is not taken up by viable cells, but it is taken up by dead cells. It is, for this reason, called an exclusion dye.

SAQ 4.4

To prepare for the fusion you take out a vial of myeloma cells from the liquid nitrogen storage facility and place them in 20 ml culture medium in the incubator. Four days later you remove a drop of cell suspension and count the cells. You obtain a figure of 200 cells in 4 squares. Determine the total number of cells in the culture flask. From this figure calculate the amount of cell suspension you will need to fuse 2 x 10^6 myeloma cells to 10^7 spleen cells. Choose the correct volume from those given below:

1) 0.5 ml 2) 4 ml 3) 12 ml 4) 2.4 ml 5) 10 ml

4.3.4 The fusion

polyethylene glycol

fresh HAT medium

A few days before the fusion some myeloma cells are removed from liquid nitrogen, rapidly defrosted and resuspended in cell culture medium in the incubator until they attain exponential growth. Typical doubling times are about 14 hours. On the day of the fusion, the mice are killed and the spleens are removed sterilely, are gently pulled apart in culture medium and are forced through a filter to remove spleen debris and to separate the cells. The cells are then resuspended in culture medium and counted. Typically, an adult mouse spleen will yield about 10^8 spleen cells. The spleen cells and myeloma cells (usually about 2 x 10^7) are then mixed in the same sterile tube and the fusing agent, polyethylene glycol 1700, is added to 30-50% concentration. After a short exposure to this agent, excess medium is slowly added to dilute out the polyethylene glycol and then the cells are resuspended in fresh HAT medium, aliquoted in 0.2 ml volumes into wells of sterile microculture plates and placed in the incubator. Clones should start appearing at about day 4 when examined by microscopy and can be seen by eye as small dots in the wells by the end of the first week.

4.3.5 Growth and screening of the hybridomas

slow recovery of the dihydrofolate reductase

HT medium

feeding of cultures

After the first week of culture in HAT medium, it can be assumed that only hybridomas have survived and the HAT medium can be changed for normal culture medium. However, due to the slow recovery of the dihydrofolate reductase enzyme which has been blocked by the aminopterin, it is necessary to maintain levels of hypoxanthine and thymidine to allow the cells to still use the salvage pathway. The HAT medium, therefore, is changed for HT medium and with feeding every 3 days, the aminopterin is gradually diluted out and the dihydrofolate reductase becomes active allowing the cells to use the *de novo* pathway for nucleotide synthesis. After another two weeks in HT medium it is safe to replace it with normal culture medium. The feeding of cultures is simple; using multichannel pipettes, 100 µl of medium is removed from the wells and replaced with fresh medium. This procedure does not normally disturb the cells as they are settled on the bottoms of the wells.

supernatants are tested for antibodies

In a similar fashion, 50-100 µl samples of supernatants from wells containing clones are removed and assayed for antibodies. Once positive wells are identified, the cells are removed for cloning. To remind you of a typical assay using the ELISA to detect positive wells examine Figure 4.6.

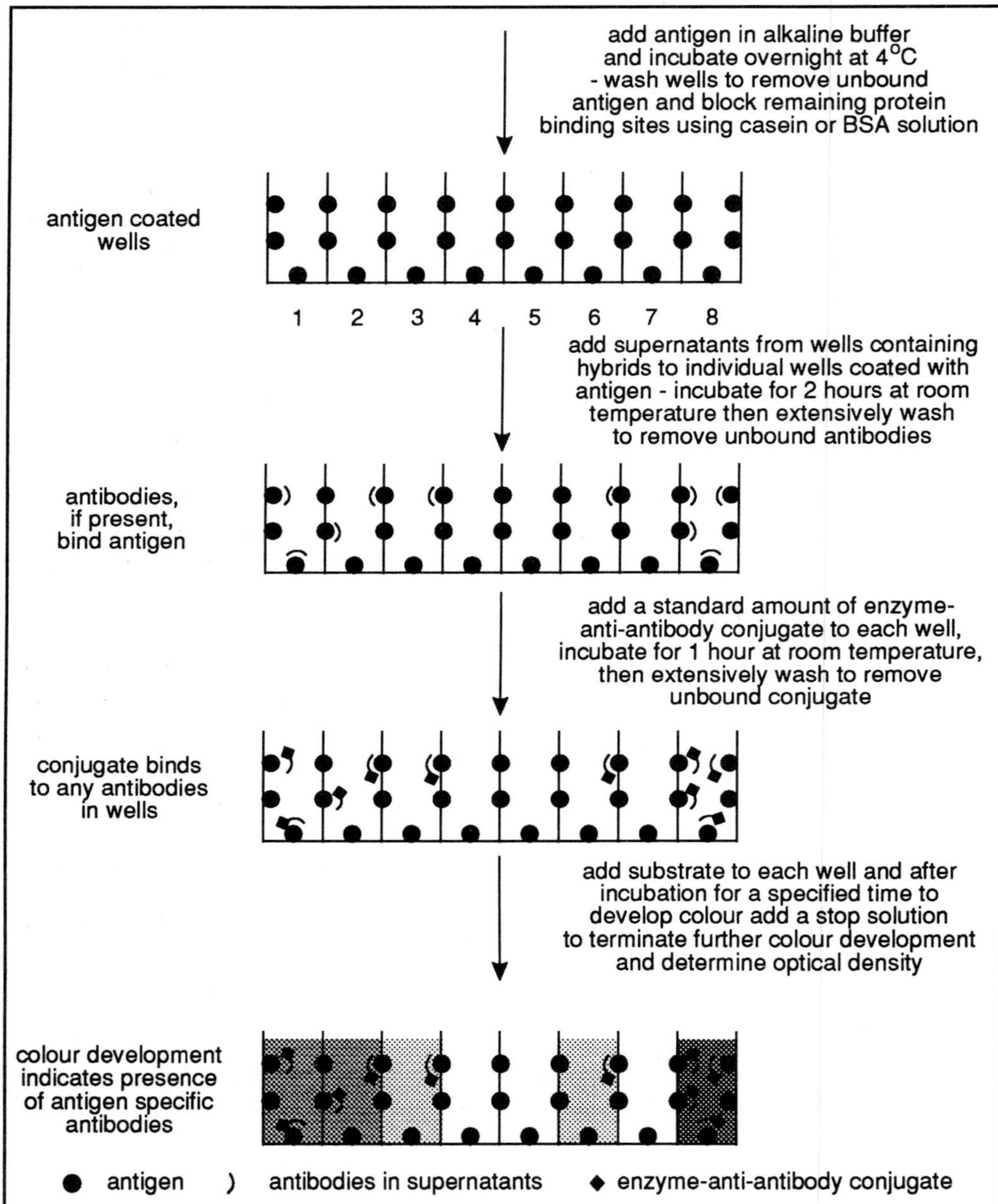

Figure 4.6 The ELISA assay to detect hybridoma antibodies. The test is done in wells of a microELISA plate of which only 8 wells are shown. In the stylised diagram, supernatants added to wells 1,2,3,6 and 8 contain antibodies specific for the antigen thus indicating that hybrids in the wells of the microculture plate from which these supernatants were derived are producing the required antibodies. These hybrids can now be selected for cloning. You will notice that wells 4,5 and 7 were negative. The supernatants in these wells contained antibodies to other antigens, for instance spleen background antibodies, or the hybrids were non producers.

SAQ 4.5

The ELISA assay - the wells of a 96 well microtitre plate are coated with an anti Interleukin 2 (IL-2) monoclonal antibody and any remaining sites are blocked with a solution of 5% bovine serum albumin. The following dilutions of IL-2 were then aliquoted into the wells to produce a standard curve. It will help you to look at Figure 4.7 to understand this protocol. The IL-2 (0.37 mg ml^{-1}) was diluted 1/50, 1/100, 1/500, 1/1000, 1/2 500, 1/5 000 and 1/10 000 and 100 µl were then added to the wells (a-g) in rows 2-9, ie the 1/50 dilution was added to row 2 (a-g), 1/100 dilution to row 3 (a-g) and so on. After an incubation period, the wells were washed and a second anti IL-2 IgG monoclonal antibody was added. Following another incubation period and washing steps, an anti IgG alkaline phosphatase conjugate was added to all the wells. After a further incubation period and washing steps, the substrate nitrophenyl phosphate was added and after colour (bright yellow) development, the optical densities were read at 405 nm.

A test sample X was also included in the assay. 100 µl of the test sample was incubation in an anti-Interleukin 2 monoclonal antibody coated well, the well was washed and incubated with a second anti-IL-2 IgG monoclonal antibody. After subsequently being washed, anti-IgG alkaline phosphatase conjugate was added to the well. After further incubation and washing, the amount of alkaline phosphatase remaining in the well was assayed using nitrophenol phosphate. The optical density at 405 nm was 1.5.

1) From the data in Figure 4.7 and above, determine the concentrations of IL-2 in the wells of each row in µg ml^{-1} and the mean optical density. Then plot a standard curve with OD on the y axis and log IL-2 concentration on the x axis. Then determine the concentration of the IL-2 in sample X. Choose your answer from those given below:

   a) 0.24 b) 2.1 c) 0.66 d) 0.34 e) 0.9 all in µg ml^{-1}

2) Given that the molecular weight of IL-2 is 15 000 Daltons and that the limit of sensitivity of the test is the 1/5000 dilution of IL-2, determine the test sensitivity in picomols IL-2 ml^{-1}. (Note picomol = 10^{-12} mol). Choose your answer from those below:

   a) 2.1 b) 9.4 c) 4.9 d) 25 e) 40

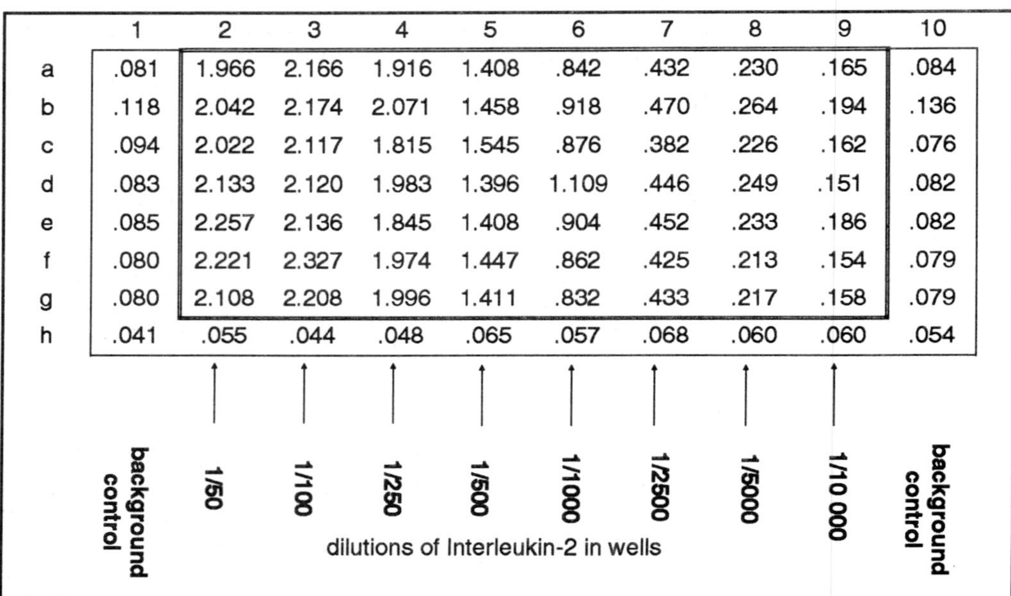

	1	2	3	4	5	6	7	8	9	10
a	.081	1.966	2.166	1.916	1.408	.842	.432	.230	.165	.084
b	.118	2.042	2.174	2.071	1.458	.918	.470	.264	.194	.136
c	.094	2.022	2.117	1.815	1.545	.876	.382	.226	.162	.076
d	.083	2.133	2.120	1.983	1.396	1.109	.446	.249	.151	.082
e	.085	2.257	2.136	1.845	1.408	.904	.452	.233	.186	.082
f	.080	2.221	2.327	1.974	1.447	.862	.425	.213	.154	.079
g	.080	2.108	2.208	1.996	1.411	.832	.433	.217	.158	.079
h	.041	.055	.044	.048	.065	.057	.068	.060	.060	.054

Figure 4.7 The ELISA assay. The figure shows vertical rows of wells in a micro-ELISA plate numbered 1-10 and horizontal rows a-h. Rows 1 (a-h) and 10 (a-h) as well as horizontal row H are various background controls and can be ignored. Interleukin 2 (0.37 mg cm^{-3}) was diluted as indicated and 100l μlreplicate samples dispensed into the wells of each vertical row (a-g) as shown. Hence the optical density (absorbance) readings given in vertical row 2 (a-g) represent a set of replicate results using the 1/50 dilution of IL-2. In the set problem you have to calculate the mean optical density value of these readings and similarly for each other vertical row of wells. (see SAQ 4.5).

4.3.6 The cloning procedure

Once the wells containing the best antibodies have been identified, the cells in the wells should be immediately cloned. This procedure is necessary to reduce the risk of overgrowth by non-producer cells and to demonstrate that the antibodies are truly monoclonal. Cloning may be done in soft agar or by limiting dilution and we shall concentrate on the latter as this is by far the most popular method. It is normal practice to clone the hybridoma lines twice to ensure monoclonality and to reduce the risk of chromosome loss which may lead to non producer clones. After two cloning procedures, chromosome loss is usually infrequent.

cloning by limiting dilution

The idea of cloning is that you dispense a single hybridoma cell into the well of the microculture plate and allow it to grow. Under these conditions all the progeny will be derived from the single cell thus representing a clone. There are various methods used for cloning by limiting dilution but the principle is the same. The method of limiting dilution involves greatly diluting out the cells from the positive well so that they are plated out in fresh medium at densities as low as 0.5 cells per well. You would expect about half the wells in any one group to exhibit growth if there are single clones in the wells.

picking out single cells

It is also possible to pick out single hybridomas using a capillary pipette from diluted cell cultures. This method requires some skill but does ensure that each clone is derived from a single cell.

feeder cells

The very small numbers of cells in each well (as small as a single cell) will not develop into clones unless supplied with various undetermined growth factors which can be supplied by other cells. It is, therefore, normal practice to provide what we call feeder cells in all the wells to supply these nutrients. These feeder cells can be rat or mouse thymocytes or splenocytes.

Π If you are trying to develop a clone, is it not bad practice to contaminate the B cells with feeder cells?

feeder supernatant

Since the cloning procedure takes at least two weeks, the feeder cells will all die within the first week leaving only the cloned B cells in each well. Some investigators, rather than using feeder cells, use feeder supernatant which is the culture medium derived from 2 day cultures of normal spleen cells or thymocytes.

4.3.7 Expansion of clones

Once the monoclonality of the cultures has been established, the clones are grown sequentially in ever increasing volumes of culture medium to develop a suitable stock of cells for storage in liquid nitrogen and to produce the antibodies. A typical procedure would be to place the cells from a microplate well (volume 200 µl) into 1 ml of medium and then when the cells attain 1 million cells ml^{-1} expand these to 5 ml of culture. When the cells have again grown to a concentration of 1 million cells ml^{-1} they are placed in 20 ml of culture medium, then to 100 ml, then to 1000 ml and so on.

antibody production

To obtain antibodies, the cells in a large volume of culture medium are allowed to grow beyond the normal cell concentration limit until the cells start dying thus maximising antibody production. Under these conditions the antibody levels may reach 50 µg ml^{-1}.

ascites fluid in mice

The antibodies may also be produced as ascites fluid in mice although this is not favoured by the Home Office in the UK. The advantage of ascites fluid is that the concentration of antibodies is about 1000 times higher than in supernatants, the antibody concentration ranging from 5-15 mg ml^{-1} and each mouse producing up to 5 ml of ascites fluid.

Mice are pretreated by injection of a stimulant such as pristane into the peritoneal cavity about 10 days prior to injection of about 10^6 to 10^7 hybrid cells into the same site. The pristane acts as an irritant inducing the migration of lymphoid cells and nutrients into the peritoneal cavity, these will provide support for the development of a tumour (Ascites tumour) from the injected hybridoma cells.

Π Based on what you learned about cell interactions in Chapter 1, what special condition is important regarding obtaining ascites fluid?

The mouse should have the same genetic background as the hybridoma, otherwise the hybridoma cells may be destroyed by the lymphoid cells of the injected mouse - essentially a transplant rejection. It is normal practice to use BALB/c mice as a source of the original spleen cell partners and to generate ascites in the same strain of mouse.

The ascites fluid may be tapped from the peritoneal cavity using a syringe about two weeks after injection of the cells and a similar injection may be made two weeks later. The serum can also be collected as it too will contain high levels of the monoclonal antibodies.

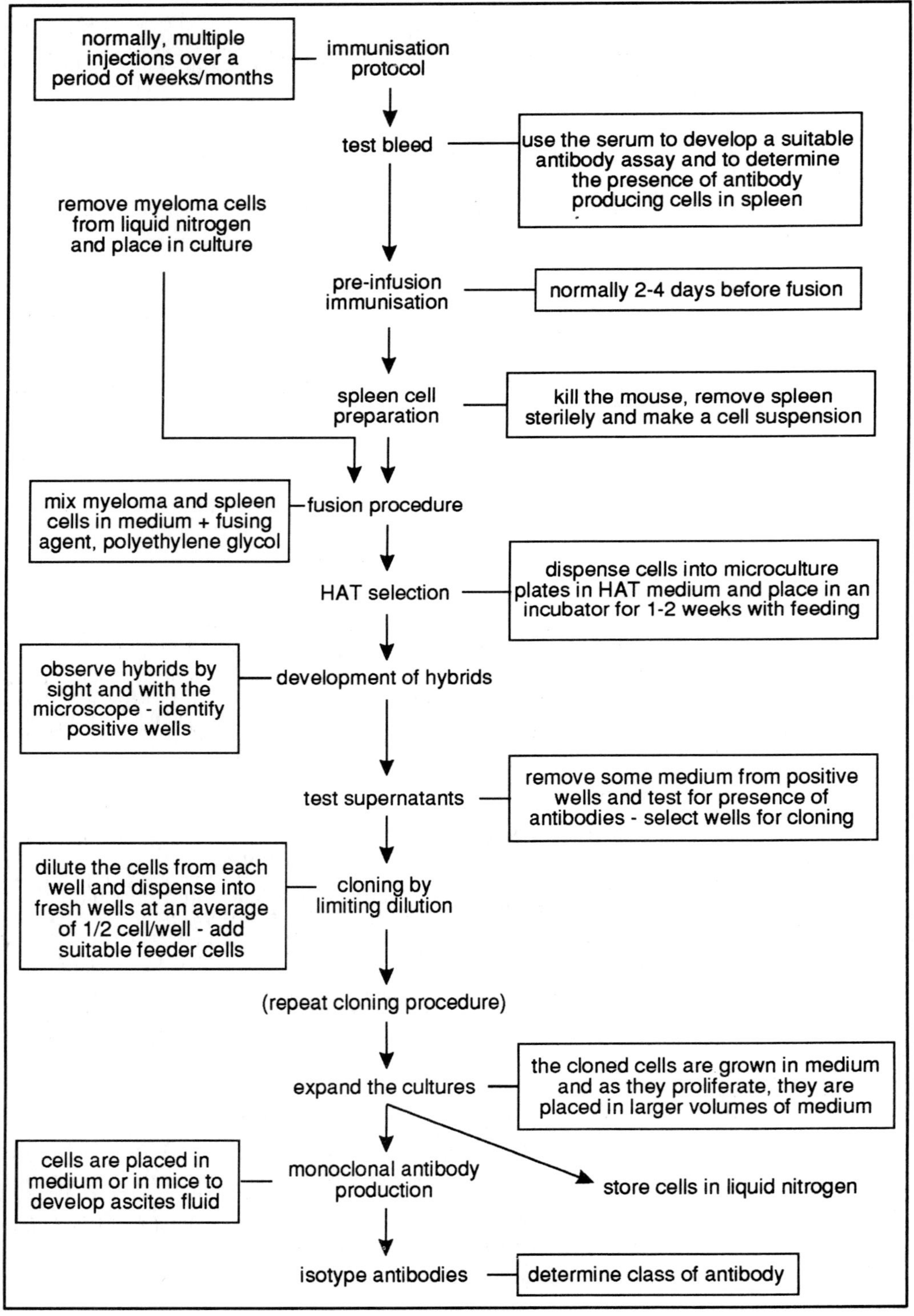

Figure 4.8 Major steps in the production of monoclonal antibodies using hybridoma technology.

After incubation at room temperature for one hour, the ascites fluid is kept at 4°C overnight after which time it is centrifuged to remove the cells. It is then stored in small aliquots in the deep freeze. We have summarised the major steps in the production of monoclonal antibodies in Figure 4.8.

SAQ 4.6

Which one of the following statements is correct?

1) Once a hybridoma is formed, it will produce antibodies forever.

2) The cloning procedure selects for high affinity B cells.

3) Ascites fluid and serum from an immunised mouse contain similar quantities of antigen specific antibodies.

4) The same mouse strain should be used as a source of immune B cells for fusion and for production of ascites fluid.

5) Feeder cells must be from the same mouse strain as the myeloma cells.

4.4 Isotyping the monoclonal antibodies

class/subclass of antibody

When you have finally produced a monoclonal antibody you will need to determine its class as this may be important regarding its use. For instance, mouse IgG1 adheres poorly to protein A so using protein A is not a good method for purification/concentration of this antibody class. In this case you would choose protein G to which mouse IgG1 binds much better. There are many methods for isotyping the antibodies and we will briefly look at two of these.

4.4.1 Ouchterlony assays

double diffusion method

You will remember the assays from Chapter 2. The monoclonal supernatant is placed in a central well in the agar and the class and subclass specific antisera are placed in the other wells (Figure 4.9). When the plate is placed in a diffusion chamber and left for a few hours, the antibodies in the supernatant diffuse towards the outer wells and the antibodies in these wells diffuse towards the central well. When the monoclonal antibodies meet in the agar with the class specific antiserum, a precipitin line will form indicating the class/subclass of the monoclonal antibody. If you look at Figure 4.9 as an example, we have shown a precipitin line between the central well containing the monoclonal and the well containing anti-mouse IgG2a thus indicating that the monoclonal is of the IgG2a class.

Π In many of these assays performed as indicated, you would find no precipitin line. See if you can explain why this is.

This test is not very sensitive and cannot detect the low levels of antibodies often present in monoclonal supernatants. If you use this test for this purpose you may have to concentrate the antibodies in the supernatant first.

4.4.2 Antibody capture assay - ELISA

Π You are now quite familiar with the ELISA technique. See if you can devise a suitable method of determining the class/subclass of a monoclonal using this method?

ELISA assays

You could coat a micro-ELISA plate with the antigen and then block the remaining sites with casein or albumin. Then you add your tissue culture supernatant to the cells and, after an incubation period, wash the wells with buffer. You then add your enzyme labelled class specific antibodies and finally substrate. The well exhibiting a colour reaction indicates the class/subclass of your monoclonal. Since this is an ELISA, it is much more sensitive than the double diffusion method and is a test of choice (Figure 4.9).

In some instances, a simpler approach would be to use the dot blot assay since all you are determining is the presence or absence of a particular antibody class/subclass.

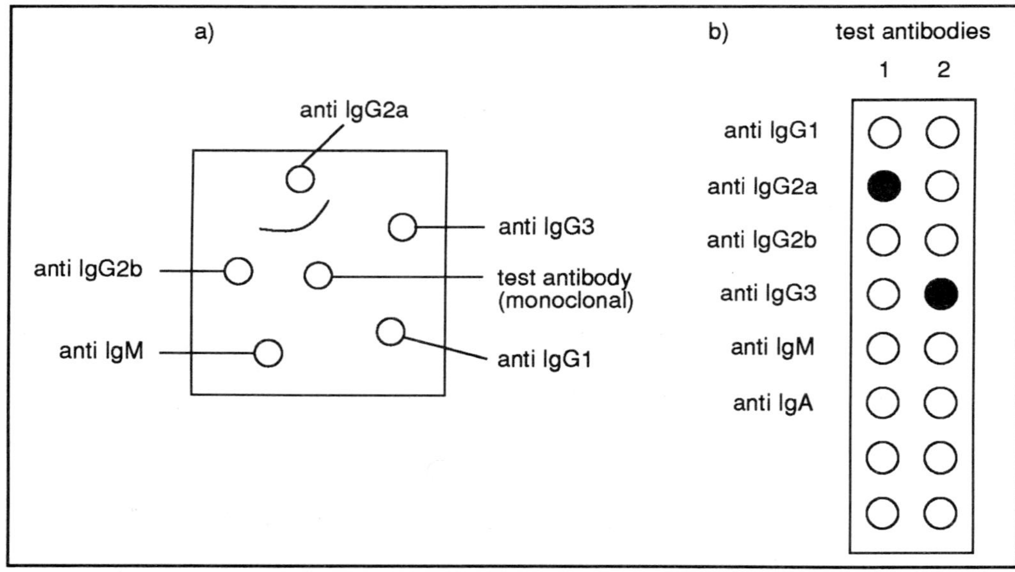

Figure 4.9 Determining the isotype of a monoclonal antibody. a) Double diffusion technique. After concentrating the supernatant, a sample of the monoclonal is placed in the centre well of an agar coated plate. The surrounding wells are filled with the various class and subclass specific antisera. The plate is then placed in a covered dish and left for a few hours. A precipitin line appears indicating the class/subclass of the antibody in the supernatant. In the illustration the test monoclonal antibody is of the IgG2a subclass. b) The ELISA technique. Wells of a micro-ELISA plate are coated with antigen and then supernatant containing the monoclonal antibody is added. The antibody is identified using enzyme labelled class/subclass specific antisera and substrate. In this figure, monoclonal supernatant 1 is added to the wells in row 1 and supernatant 2 in row 2, both monoclonals being specific for the same antigen. Subsequent development of the test indicates that one antibody is IgG2a, the other IgG3.

Π What are the two main differences between the dot blot and the ELISA described above?

The two main differences are firstly, that the antigen is attached to a nitrocellulose or nylon membrane as a dot and secondly, that the substrate used must be converted into a precipitate rather than into a soluble product.

Yet another approach is to coat the wells with the class specific antisera. When the supernatant is added the antibodies present will only be bound by the antibodies specific for that particular class/subclass. The complex is then detected using enzyme labelled anti-mouse immunoglobulin antibodies which bind all classes.

SAQ 4.7

We have summarised the major steps in murine monoclonal antibody production in Figure 4.8. Examine it closely, then close your text and arrange the following procedures for production of murine monoclonal antibodies in the correct order.

1) Fusion.

2) Selection in HAT medium.

3) Immunisation.

4) Expand the cultures.

5) Observe development of hybrids using microscope.

6) Isotype the antibody.

7) Cloning by limiting dilution.

8) Test serum for antibody.

9) Grow myeloma cells.

10) Test supernatants.

11) Pre-infusion immunisation.

4.5 Monoclonal antibodies produced by other species

4.5.1 Heterohybridomas

As we have indicated earlier, the only other animal for production of successful hybridomas is the rat. However, there have been many attempts, some partly successful, to produce cell hybrids using myelomas from mouse and antibody producing cells from other species. These have been called heterohybridomas or interspecies hybridomas. Cells from rat, hamster, rabbit, Man, sheep and cow have been used as fusion partners for mouse myelomas with the mouse-rat fusions being highly successful in producing stable hybrids on many occasions. With cell partners from the other species, there is often a loss of chromosome including those encoding immunoglobulin chains or antibody secretion only proceeds for a short time.

hetero-hybridomas

An alternative approach is to use a second order heterohybridoma which involves a second fusion. For instance, some investigators have produced mouse-human or mouse bovine hybridomas and these have been fused again with cells from the other species producing human-mouse-human or bovine-mouse-bovine heterohybridomas.

4.5.2 Human hybridomas

Major difficulties establishing human hybridomas

After all the successes with mouse hybrids, it was anticipated that it would not be long before we could produce human monoclonal antibodies which could then be used for a variety of clinical applications in a similar manner. However, this has proved difficult for two major reasons. Firstly, there is a paucity of suitable cell lines which can be grown easily in cell culture and which can act as fusion partners and support antibody production.

Secondly, it is difficult to obtain immune cells from Man since the best source of B cells is the spleen or lymph node. Obviously, it is unethical to firstly immunise a volunteer with a particular antigen and then to remove the spleen or lymph nodes to obtain the cells. The major source of cells in Man are peripheral blood cells (PBL) which do not provide a good supply of immune B cells even if you could immunise the volunteer. The cells in the blood appear to be much less reactive to antigen when stimulated *in vitro* than spleen or lymph node cells.

In many experiments, researchers have used cells from other tissue sources such as lymph nodes, spleen and tonsils from patients. One more recent approach is *in vitro* immunisation which has become possible because we now know more about the roles of cytokines in cell interactions and this has resulted in some successful responses. Another approach is to transfer human cells to immunodeficient mice, the so called SCID (severe combined immunodeficient) mice which lack both B and T cells and then to immunise the mice. The B cells could then be removed for immortalisation (see below). This work is presently in its infancy with no reported outstanding successes.

As we shall see in the next chapter, the relatively slow progress in this area may be superseded by the very rapid developments in producing human monoclonal antibodies using gene cloning techniques. Because of this likely outcome, we shall only briefly review the techniques for production of human hybridomas so that you are aware of the type of work that is proceeding.

Fusion cell partners

There have been three major sources for provision of fusion cell partners for establishing human hybridomas. These are human myelomas, lymphoblastoid cell lines and Epstein Barr virus immortalised B cells.

Human myelomas

human myeloma partners

There are a few examples of successful human myeloma partners. One is the SKO-007 plasmacytoma which was fused with lymphocytes from the spleen of a Hodgkin's lymphoma patient resulting in the production of IgG antibodies to the dinitrophenyl hapten. Another human myeloma, KR12, was used as a fusion partner with human lymphocytes and the resulting hybrid produced anti-tetanus toxoid IgM antibodies. However, presently all the fusion partners reported produced antibodies themselves and are not non-secretors like their murine equivalents.

Human lymphoblastoid cell lines (LCL)

The first LCL was derived from a B cell line from a patient with multiple myeloma which was fused with peripheral blood lymphocytes from a patient who had a very high titre of anti-measles antibody. This resulted in clones secreting IgM antibodies. A derivative of this B cell line was fused with cells from patients with a variety of auto-immune disorders and the hybrids produced auto-antibodies thus demonstrating the presence of auto-immune B cells in these patients. The disadvantage of using LCL is that they generally do not sustain high level antibody production. Some researchers have grown these hybridomas as solid tumours in immunodeficient mice (lacking T cells) and these tumours have produced high levels of antibodies in ascites fluid.

Epstein Barr virus immortalisation of B cells

Epstein Barr virus (EBV)

An alternative approach to finding a fusion partner as described above is to immortalise antigen specific B cells. Epstein Barr virus (EBV) is a herpes virus shown to be the infectious agent of infectious mononucleosis. In patients infected with EBV there is a high production of antibodies of a wide specificity suggesting polyclonal stimulation by the virus and this stimulation of a wide spectrum of B cells is known to be T cell independent. This transformation leads to permanent B cell growth, the B cells having the appearance of large lymphoblastoid cells. The disadvantage with this procedure is that, in many instances, the EBV transformed B cell lines lose their ability to produce antibodies and generally the amount of antibody produced is small and the antibody is usually IgM. A more recent approach to improve antibody production is to fuse the EBV transformed cells with a human fusion partner such as those discussed above. A few laboratories have produced established hybrids using this technique.

permanent B cell growth

SAQ 4.8

Which one of the following statements is correct?

1) Human hybrids generally produce monoclonal antibodies.

2) EBV transformed B cells are stable antibody producers

3) Most fusions involve class switched B cells.

4) Human hybrids can be grown in immunodeficient mice.

5) Heterohybridomas are so called as they produce more than one antibody.

Summary and objectives

In this chapter we have described in considerable detail, the production of monoclonal antibodies using hybridoma technology principally in mice. We examined the principle of HAT selection and have looked at all stages of hybridoma production from the immunisation of the mice to the isotyping of the resulting monoclonal antibodies. We also briefly examined the attempts to produce human monoclonal antibodies for therapeutic purposes.

Now you have completed this chapter you should be able to:

- list the major advances which have been made in antibody production methods;
- discuss the principle of HAT selection and describe the major characteristics of the myeloma cell partners;
- interpret data from biological assays of monoclonal supernatants;
- estimate the number of cells in a sample based on haemocytometer data;
- describe in detail the ELISA assay for detection of antibodies in hybrid supernatants;
- manipulate and interpret data from ELISA assays;
- describe all stages of monoclonal antibody production using hybridoma technology;
- discuss the difficulties met, and progress made, in the production of human monoclonal antibodies.

Bioengineered antibodies

Bioengineered antibodies

5.1 Introduction

5.1.1 Bioengineering involves molecular genetics

This chapter assumes that you have a basic knowledge of molecular genetic techniques such as the use of restriction endonucleases, cloning vectors and transformation of *E. coli* using antibiotic resistance markers. You should also know the background to the techniques of site directed mutagenesis and DNA sequencing (this knowledge can be gained from the BIOTOL texts 'Techniques for Engineering Genes' and 'Strategies for Engineering Organisms').

We have to warn you that the subject matter of this chapter is very complex and rapid progress is being reported both in the development of more sophisticated technologies for manipulating DNA and in the number and variety of newly engineered antibody based products. Many of these developments are beyond the scope of this text and will not be dealt with. Our intention is to present a more simple picture of the major approaches over the past few years so that you can appreciate the potential and implications of such developments especially with regard to how we shall be making antibodies of choice in the near future. Shall we have dispensed with the mouse? We will leave that for you to decide when you have completed this chapter.

5.1.2 The scope for antibody engineering

Look at Figure 5.1 which depicts the various antibodies and fragments which have been produced; all of them have the potential to replace hybridoma antibodies. We have already referred to some of these in Chapter 4 (Section 4.1.1). You will remember that the major drive behind these developments was to produce antibodies for immunotherapy which were not immunogenic in Man. You will therefore notice that the mouse component of the successive products has decreased until there is essentially no murine component in the antibody products at all.

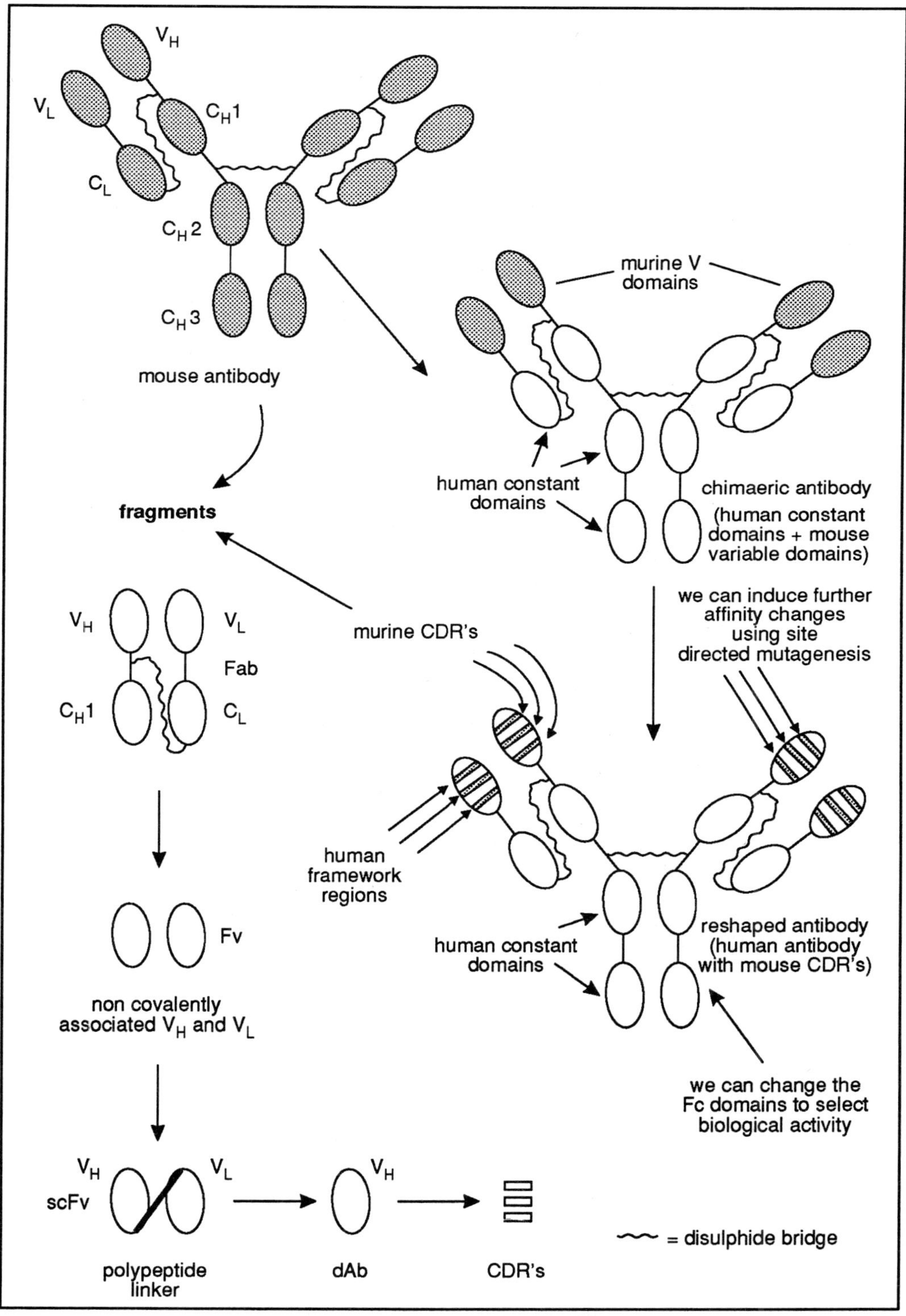

Figure 5.1 Bioengineered antibodies and fragments. The fragments have been expressed in bacteria, the whole antibodies in eukaryotic cells. The CDRs are synthesised in the laboratory. CDR = complementary determining region; dAB = antibody domain; Fv = variable fragment; scFv = single chain Fv fragment.

chimaeric antibodies

reshaped antibodies

antibody fragments

The first attempts to replace mouse antibodies with human antibody components was in the development of chimaeric antibodies which were mainly human antibodies but with mouse heavy and light chain V domains. This was rapidly succeeded by reshaped antibodies which were more human, the only murine components being the CDR regions. With the development of PCR and phage technology, the emphasis switched to antibody fragments, the most important being the single chain Fv fragment (scFv) which consists of the heavy and light chain variable domains covalently linked by a linker sequence. Even smaller fragments with antigen binding activity are dAB's (consisting of a single domain) which consist of heavy chain variable regions and smaller still, CDR peptides. In this chapter we shall restrict our discussions to the production of various antibodies and fragments which possess antigen binding activity and can be produced in large quantities ie replacements for hybridoma antibodies resulting from an initial immunisation of a mouse. In Chapter 7 we shall look at modifications of such fragments resulting in a variety of tools for use in medicine and other areas of science and technology.

SAQ 5.1

Match each item in the left column with one from the right using each item only once.

1) Fv	a) linker polypeptide
2) chimaeric antibody	b) single domain
3) scFv	c) murine CDR
4) dAb	d) non-covalent association
5) reshaped antibody	e) 2 parts human, 1 part mouse

5.2 Chimaeric antibodies

5.2.1 Selection of vector and host

whole antibody molecules have not been made in bacteria

periplasmic space

insoluble inclusion bodies

At the time of development of chimaeric antibodies and even to the present day, there are no reports of expression of whole antibody molecules in prokaryotic hosts. The difficulties with expressing such large multi-chain molecules in bacteria involves problems with assembly and secretion of the molecule into the periplasmic space and with the non-glycosylation of the antibodies by bacteria. In many instances, the expressed products are to be found in insoluble inclusion bodies in the bacteria and have to be extracted. Major problems can arise during the process by which the molecule is renatured or refolded which, of course, also involves formation of the correct disulphide bridges. This has been successful with antibody fragments but there have been no attempts to express whole antibodies in bacteria due to these difficulties.

Π What alternative host could you suggest which would avoid these complications?

myeloma cells

The alternative approach has been to use eukaryotic hosts which could both glycosylate the antibodies and secrete them in their natural form thus avoiding the problems with renaturation. The obvious early candidates for expression were the professional antibody secretors, the myeloma cells. There were some complications in this approach since the cDNA fragments encoding the heavy and light chains have to be cloned in bacteria but expressed in a different host and it was necessary to develop suitable vectors which would allow the cloning and transfer of the antibody chain cDNAs from bacteria to mammalian cells.

Π Based on your knowledge of molecular genetics, can you list the special requirements for a mammalian expression vector?

mammalian vector

The mammalian vector must have a prokaryotic origin for bacterial replication and an antibiotic resistance marker so that the recombinants can be selected and cloned before they are used to transfect the mammalian cells. The vector also needs eukaryotic elements which control transcription in mammalian cells, polyadenylation elements and, perhaps, some sequences which promote processing of transcripts. Some method of selection of the mammalian cell transformants will also be required.

pSV2 vector

simian virus 40 (SV40)

pSV2gpt

pSV2neo

A plasmid which was developed by Mulligan and Berg in the early 1980s and fulfilled all these requirements was the pSV2 vector. This possesses an origin of replication from the bacterial plasmid pBr322 and an antibiotic (ampicillin) resistance gene (selection marker). These are essential for the replication and cloning of the recombinants. pSV2 vectors also have promoters and polyadenylation elements for mammalian cell expression derived from simian virus 40 (SV40). Derivatives of pSV2, pSV2gpt and pSV2neo have resistance genes to the drugs mycophenolic acid/aminopterin and the drug G418 respectively. You should, of course, be aware that there are many other mammalian vectors which have since been constructed.

Π Before you read on, can you devise an assay to detect myeloma cells transfected with pSV2gpt?

resistance genes for aminopterin

transfectomas

You may have noticed that pSV2gpt vectors possess resistance genes for aminopterin and this vector, on transfection of a myeloma cell such as SP2/0, encodes for the enzyme hypoxanthine phosphoribosyl transferase, which, you will remember, these cells lack. If we, therefore, place these transfected cells (transfectomas) in medium containing HAT they will survive. Some of these transfected cells will contain pSV2gpt containing cDNA inserts for heavy chain, light chain or whole antibodies (see below) and these cells should secrete the antibodies. Similarly, cells containing pSV2neo vector, will survive in the presence of the drug G418 whereas non transfected cells will not. So you can see we now have the means to identify and clone the recombinants, transfer them into mammalian cells and get the antibodies expressed. The essential details of the pSV2 plasmid are shown in Figure 5.2.

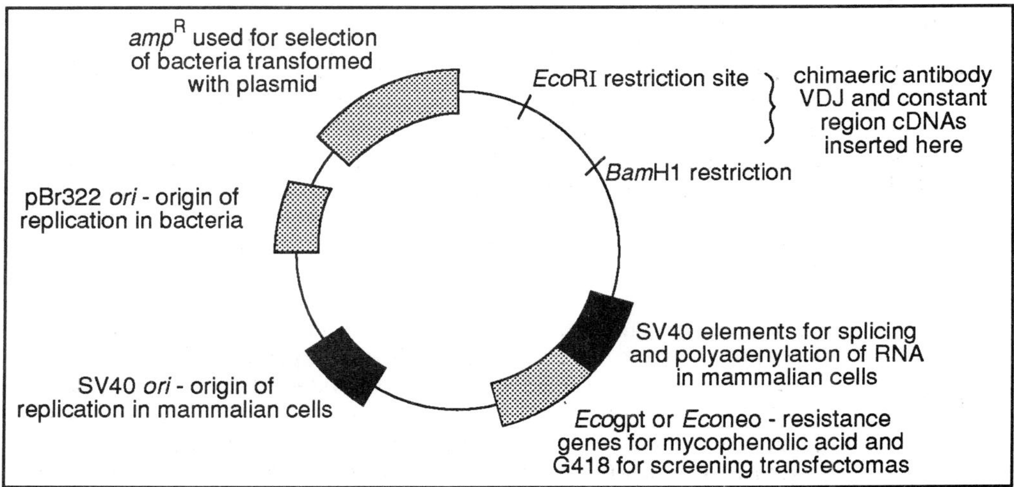

Figure 5.2 Major features of the pSV2 vector (not to scale) used for the production of chimaeric antibodies. See text for details.

5.2.2 Preparation of chimaeric antibodies

Examine Figure 5.3 while you read this section which depicts the major stages resulting in expression of antibodies in mammalian cells. As we said at the start of this chapter we are assuming that you are familiar with the basic techniques in cDNA cloning.

cloning of antibody cDNAs

In the early experiments to produce chimaeric antibodies, cDNAs were prepared from mRNAs extracted from antigen specific hybridoma cells using reverse transcriptase (first DNA strand synthesis) and DNA polymerase (produces double stranded cDNA) and then cohesive ends are attached so that it can be ligated into a vector. The fragments were then ligated into a plasmid vector and cloned in *E. coli*. The recombinants were then screened for murine heavy and light chain cDNA inserts and after bulk culture, plasmid DNA was prepared. The plasmids would be carrying restriction fragments for either antigen specific heavy or light chain cDNA from the hybridoma. Other plasmids could be constructed which carry cDNA inserts for the human heavy and light chain constant regions.

exon shuffling

In the simplified procedure shown in Figure 5.3, begin at the top left hand corner. We begin with a hybridoma cell which is producing antibodies of the desired specificity. From this we extract the mRNAs for the antibody and produce doubled stranded DNA copies (cDNAs) of these nucleotide sequences using reverse transcriptase.

We attach cohesive ends to these nucleotide sequences and ligate them into suitable bacterial plasmids and use these to transform bacteria. Bacteria carrying cDNAs derived from the hybridoma mRNA are identified by using DNA probes that will hybridise with the cDNA. Suitable recombinants are then grown up and their plasmid DNAs extracted. These plasmids carry DNA encoding for mouse heavy and light chains. In practice it is usually to use restriction enzyme fragments of the cDNA derived from hybridoma mRNAs. In this way it is possible to produce plasmids carrying selected fragments of the mouse light and heavy genes. The two plasmids shown in the centre of Figure 5.3 carry gene portions encoding for different parts of the mouse antibody.

Plasmids carrying portions of human immunoglobulin portions are produced in an analogous way. In these cases, mRNA from human myeloma cells are used. In Figure 5.3 we have shown two such plasmids. The one on the right carries the constant domain of κ light chains, the one on the left, the C_μ gene cluster.

The idea was to select the murine variable domain cDNAs of both heavy and light chains and ligate them into the pSV2 vectors along with the heavy and light chain human constant region fragments respectively. You will observe that the murine variable cDNAs of both heavy and light chain were contained within *Eco*RI-*Hin*dIII restriction fragments and the human constant region cDNAs in *Hin*dIII-*Bam*HI restriction fragments. Thus if we treat these plasmids with *Eco*RI and *Hin*dIII, we will produce fragments containing the appropriate gene portions. Using this approach it is then possible to ligate these into an appropriate plasmids. In practice, the variable and constant heavy chain restriction fragments were ligated into pSV2gpt vectors and the light chain restriction fragments together into pSV2neo vectors. The pSV2 vectors, then contained the whole sequences of either the chimaeric light or heavy chain cDNAs.

Π Why were the heavy and light chain restriction fragments ligated into the different vectors? (Think about this and then read on, you will then be able to check your ideas)

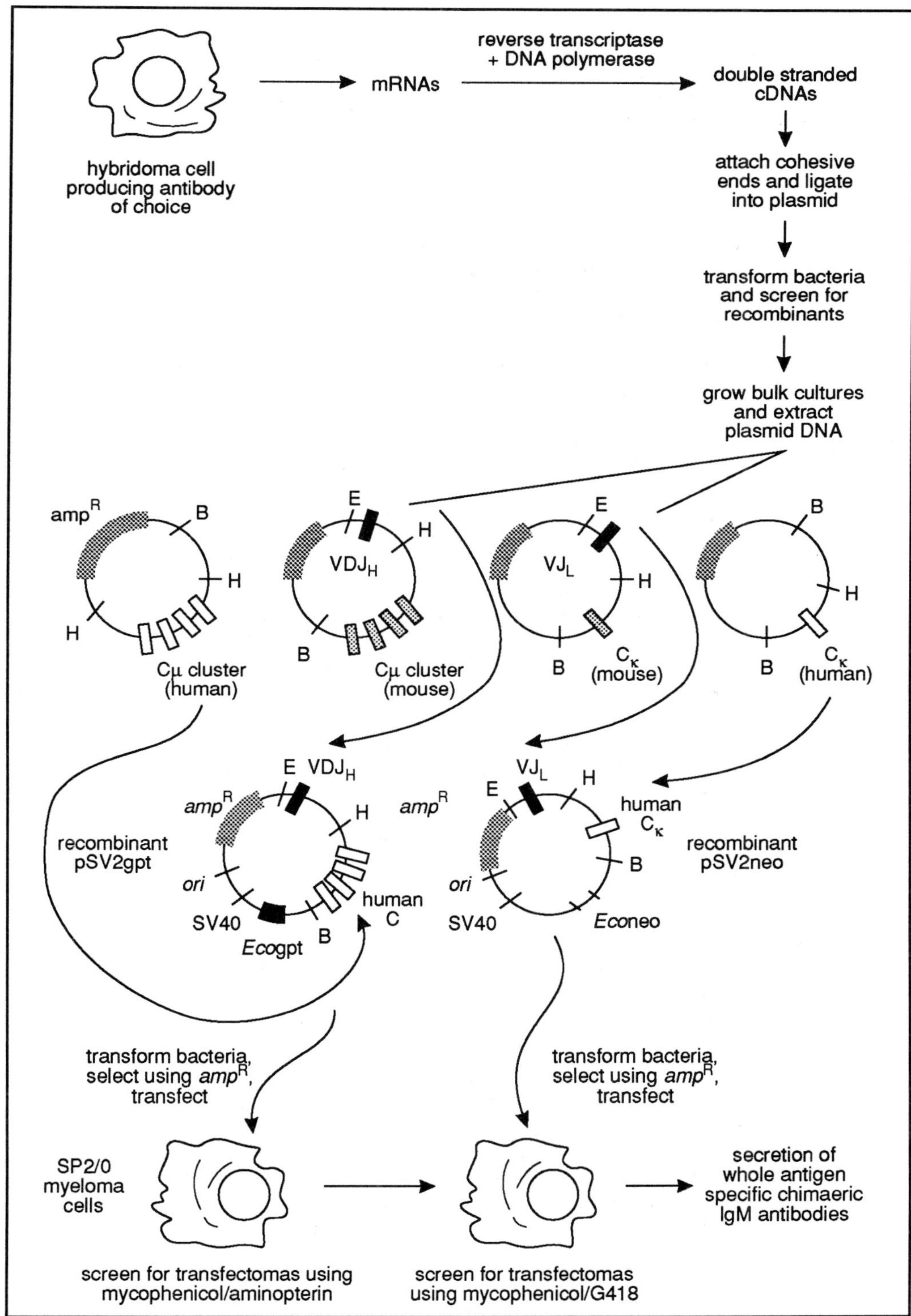

Figure 5.3 The production of chimaeric antibodies. See text for further details. Note that the vectors containing human inserts were prepared separately from human cells. Amp^R = ampicillin resistance gene, E = *Eco*RI restriction site, B = *Bam*HI restriction site, H = *Hin*dIII restriction site. For details on pSV2 see Figure 5.2.

preparation of transfectomas

The different recombinant plasmids were then cloned into *E. coli* and transformants identified by ampicillin resistance. The amplified recombinants were then used to transfect the myeloma cells using a variety of methods including protoplast fusion and electroporation (if you wish to find out more about these methods we refer you to the BIOTOL text 'Techniques for Engineering Genes'). The normal procedure was to transfect with pSV2gpt (heavy chain) first and then introduce the pSV2neo (light chain) into the transfectomas known to be producing heavy chain. However, the frequency of transfection was extremely low. The transfectomas carrying vectors bearing inserts for both heavy and light chains were then identified by their resistance to both mycophenolic acid (and/or aminopterin) and G418 respectively.

use of both pSV2gpt and pSV2neo

low frequency of transfection

Because of the low frequency of transfection, some of the transfectomas containing initially pSV2gpt (heavy chain) would not have been transfected with the pSV2neo (light chain). However, in answer to the previous in text activity, if only the one vector had been used for both heavy and light chain inserts, growth in mycophenolic acid would have identified transfectomas carrying either just the heavy chain pSV2gpt vectors or those transfected with heavy and light chain pSV2gpt vectors. You could then have only identified the cloned transfectomas by the secretion of whole antibody - quite an arduous task!

The experiment we have described to you is only one of a number of approaches to produce chimaeric antibodies but it should demonstrate the basic technology which developed in the early 1980s which then led to more dramatic discoveries as we shall see later.

residual immunogenicity

biological half life

Using methods such as this a few laboratories did show that transfectomas did produce chimaeric antibodies. In some clinical trials it was found that such chimaeric antibodies were less immunogenic than murine monoclonal antibodies but there was still residual immunogenicity provided by the murine variable regions. It was also interesting to note that the biological half life of chimaeric antibodies is intermediate between that of whole murine and human antibodies.

SAQ 5.2

With reference to Figure 5.3 decide which one of the following statements is incorrect?

1) pSV2 is a mammalian expression vector.
2) A restriction digest of the pSV2 plasmid using *Eco*RI and *Bam*HI would allow the ligation of both light and heavy chain restriction fragments into the plasmid.
3) Only a small proportion of the myeloma cells would be transfected with the recombinant pSV2 plasmids.
4) The production of chimaeric antibodies is mainly about shuffling exons.
5) SP2/0 myeloma cells can survive in the presence of mycophenicol and G418.

5.3 Reshaped antibodies

5.3.1 Contribution by murine CDRs

The advent of reshaped antibodies is mainly due to work by Winter's group in Cambridge, UK and the original purpose of such a manipulation was to further humanise mouse antibodies resulting in non-immunogenic therapeutic antibodies. As for chimaeric antibodies, the procedures for producing reshaped antibodies are extremely complex and we shall just outline them so that you can identify the main steps in the process.

Π Can you remember what is the special feature of reshaped antibodies?

rodent CDR regions

site directed mutagenesis

Reshaped antibodies are human antibodies with rodent CDR regions grafted in to the variable regions. The principle behind the procedure is to immunise mice with the antigen of choice, produce a hybridoma, clone the variable cDNAs of both light and heavy chains and determine the sequence of the CDRs of both variable regions. Oligonucleotides representing these sequences are then constructed and the CDRs are grafted into the human variable framework sequences by site directed mutagenesis. The contribution by the mouse, then, is solely to provide sequence information on the CDRs and there is no actual insertion of murine cDNA material into the human antibody cDNA sequences.

5.3.2 Production of reshaped antibodies

oligonucleotides

homology matching

Follow the protocol depicted in Figure 5.4 while you read the text. Mice were immunised with the antigen of choice and hybridomas produced from the spleen cells. The pool of messenger RNAs was extracted from the cells and single strand cDNAs produced using reverse transcriptase. The cDNAs were then amplified using either conventional cloning techniques or the polymerase chain reaction (see Figure 5.4). Note that the polymerase chain reaction is an *in vitro* technique for making multiple copies of nucleotide sequences. The technique is described in detail in the BIOTOL text 'Techniques for Engineering Genes'. We will elaborate on this technique in Section 5.4.2. The variable regions of both light and heavy chain DNA were then sequenced and oligonucleotides synthesised based on the murine sequences of the three CDR regions on each chain. It is normal practice to choose human V_H and V_L sequences in which the CDRs bear similarity to the murine CDRs to encourage base pairing (this is called homology matching). The sequence of each oligonucleotide also extends beyond the CDR thus promoting base pairing with the flanking framework regions.

CDR grafts

M13 based vectors

The human framework regions designed to accept the CDR grafts were derived from human myeloma cells producing antibodies of unknown specificity. Messenger RNA extracted from such cells was used as template for cDNA synthesis and then the normal cloning steps were performed. Restriction fragments containing V_H and V_L were then ligated into M13 based vectors and the single stranded DNA was subjected to site directed mutagenesis using the synthetic oligonucleotides (see Figure 5.5). In some laboratories this process was stepwise, changing one CDR at a time, in other laboratories all CDRs were mutated simultaneously. Once the oligonucleotide had annealed to the parent strand second strand synthesis was mediated by the Klenow fragment of DNA polymerase I and T4 ligase. Transformation into *E. coli*, cloning and selection resulted in plaques containing the mutant molecule and others containing the parent molecule. The plaques containing the mutant molecule were identified by hybridisation with radiolabelled oligonucleotides capable of annealing with the mutant molecules.

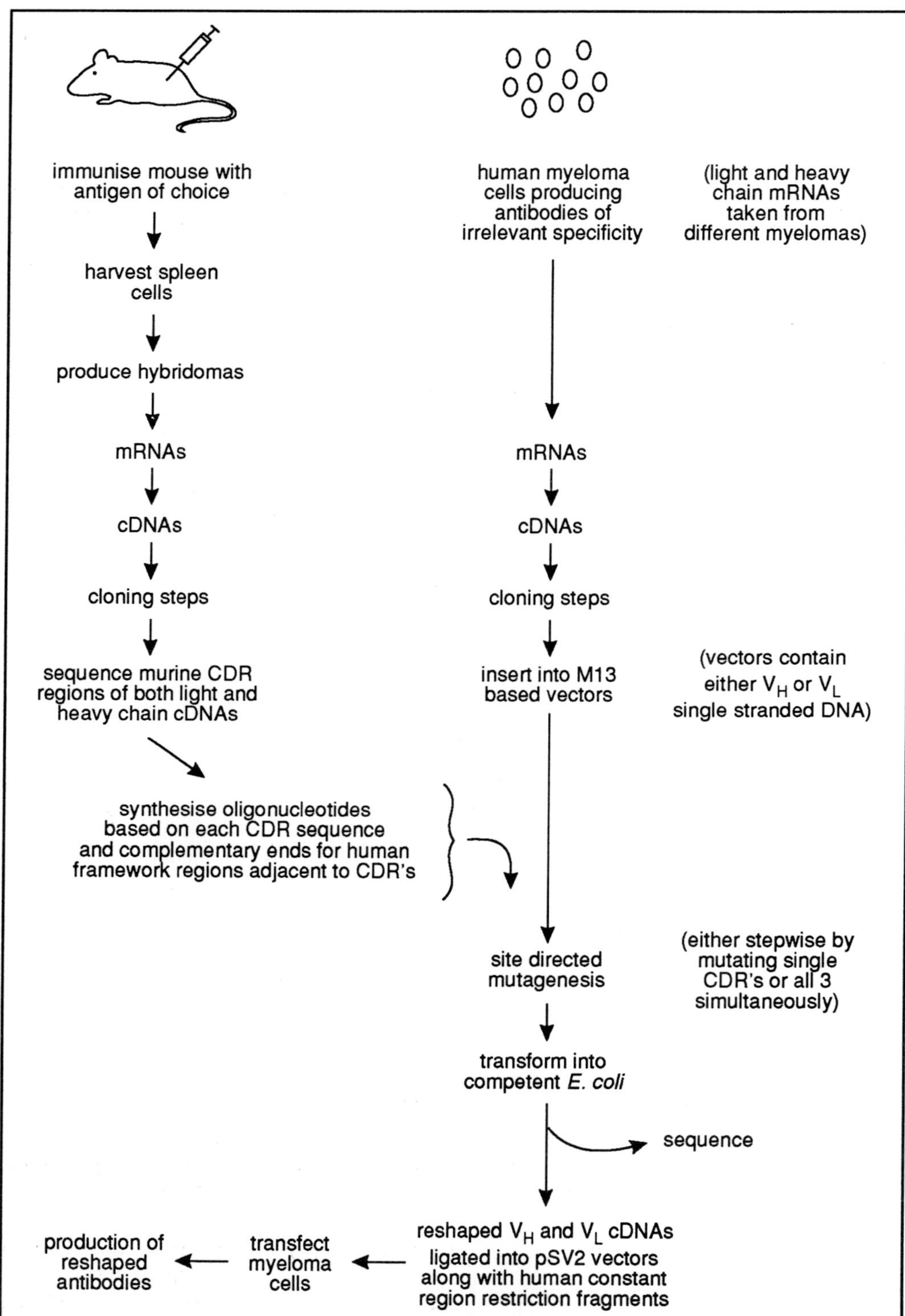

Figure 5.4 Production of reshaped antibodies. The protocol depicted hear shows the major steps involved. Many intermediate steps have not been shown for the sake of simplicity. Some laboratories used the polymerase chain reaction to amplify the cDNAs.

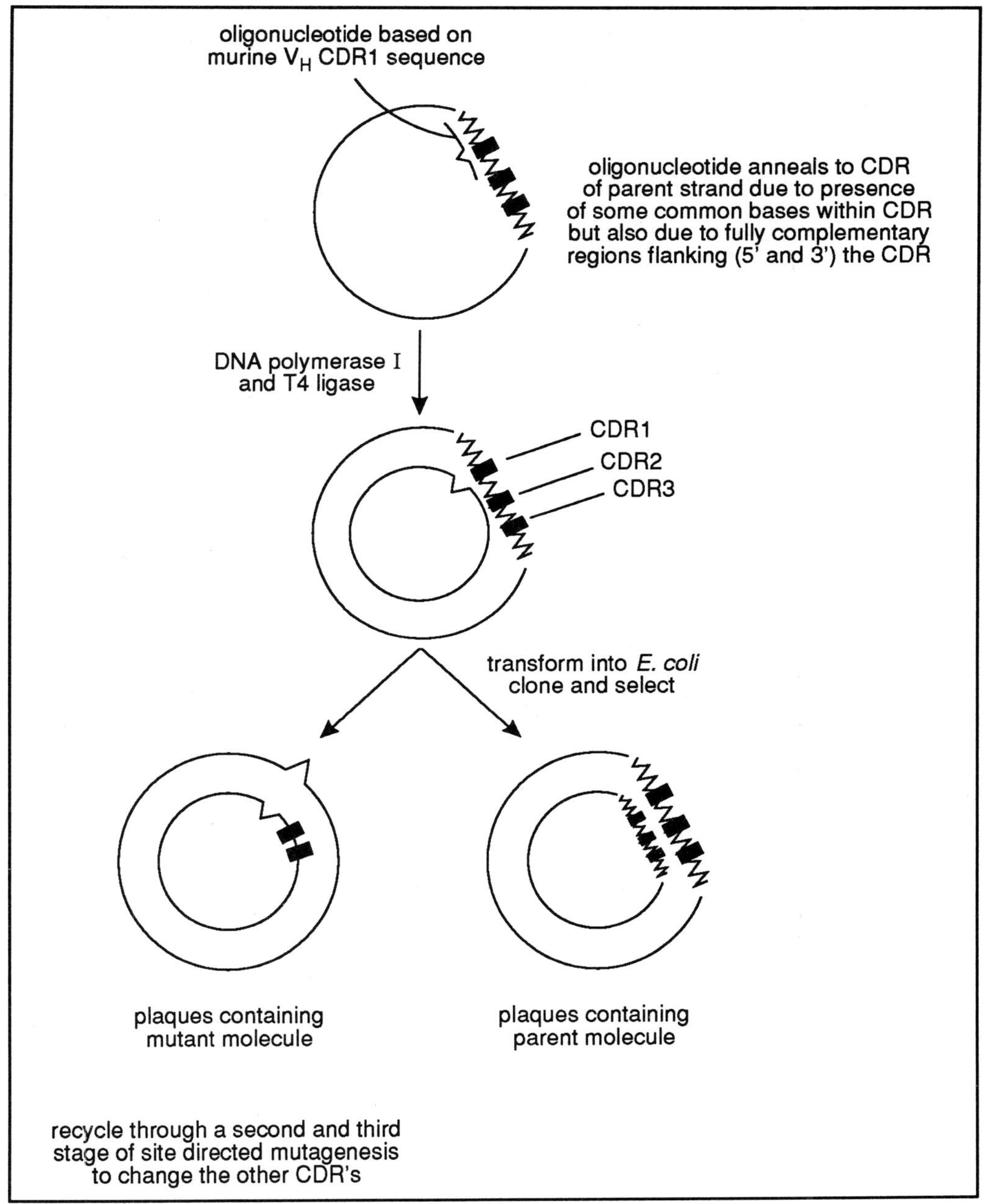

Figure 5.5 Grafting of murine CDRs into human framework regions of V_H and V_K by site directed mutagenesis. An alternative method is to change all 3 CDRs simultaneously.

The reshaped V_H and V_L genes were then ligated into pSV2 based vectors together with inserts for the human constant regions of heavy and light chains. In a protocol similar to that for chimaeric antibodies, the vectors were used to transform *E. coli*, the transformants being selected by ampicillin resistance. The clones were then expanded and used to transfect myeloma cells based on the stepwise transfection first with heavy chain genes then with light chain genes. To remind you of this procedure, take another look at Figure 5.3.

5.3.3 Framework region effects on binding antigen in reshaped antibodies

It has been observed in some reshaped antibodies that grafting of rodent CDRs onto human framework regions resulted in some loss of affinity for the antigen. This could be corrected in some cases by mutations in the flanking framework regions and suggests that interactions between key amino acids in the variable loops and framework regions have to be preserved to maintain the affinity.

Π What general conclusion can you make about the design of antibodies based on site directed mutagenesis?

changing the affinity of antibodies

From what we have seen, it should obviously be possible to make affinity improvements to existing antibodies and even to design new specificities based on this technique. An example of this was reported by Sharon recently when a low affinity p-azophenylarsonate antibody was converted to a high affinity antibody by changing three amino acids in the CDRs of the heavy chain variable domain using site directed mutagenesis.

5.3.4 Clinical applications of reshaped antibodies

Campath-1H

There are a few reports of the use of these reshaped antibodies in therapy. Some antibodies have shown potential as immunotherapeutic agents. For instance, one reshaped rodent antibody Campath-1H with specificity for human leucocytes was found to be effective in the destruction of tumours in two patients and the treatment of an auto-immune disorder in another. Another reshaped antibody, Campath-9H, directed against the CD4 molecule, principally expressed on T helper cells, is undergoing trials for prevention of graft rejection.

respiratory syncytial virus

Respiratory syncytial virus (RSV) is a major cause of respiratory illnesses in young children. A rodent antibody was produced which neutralised infection by RSV in mice and the CDR regions were transferred from this antibody to a human IgG1 antibody using the techniques just described. However, the reshaped antibody lost affinity for RSV but this was restored by alterations to the framework regions of the human IgG1. This antibody was specific for all the clinical isolates of RSV and prevented disease in mice when administered one day before RSV. Perhaps more importantly, it could also successfully treat previously infected animals.

SAQ 5.3

Arrange the steps below in the correct sequence.

1) Site directed mutagenesis.
2) Ligation into pSV2 vector.
3) Immunise mouse.
4) Extract murine mRNAs.
5) Insert human V_H and V_L into M13 vectors.
6) Transfect myeloma cells.
7) Analyse reshaped antibodies.
8) Sequence murine V domain cDNAs.
9) Production of hybridomas.
10) Synthesise oligonucleotides.

5.4 Antibody fragments

5.4.1 Advantages and uses of antibody fragments

As we have said, whole antibody molecules have not yet been expressed in bacteria, but there have been significant developments in bacterial expression of antibody fragments. Since the antigen binding properties of antibodies reside on the amino end of the molecule only, there are many advantages in producing antigen binding fragments devoid of constant regions expressing biological activities, indeed, in some instance it is a disadvantage to possess such biological properties. Have another look at Figure 5.1 to remind you of the different fragments.

Π What advantages would you say such fragments have compared to whole antibodies as immunotherapeutic agents?

less immunogenic

loss of Fc mediated activities

The one characteristic you should have been able to think of is that they are less immunogenic than whole antibodies although this is not so important now that we can switch constant regions. A second characteristic is the loss of Fc mediated activities and this will include the long biological half life of whole IgG molecules. This is important where you want the antibody to exert a function over a short period. An example of this is in the use of radiolabelled antibodies to detect metastatic growths in cancer patients, the so called imaging techniques. Once the patient has been injected with the radiolabelled antibodies and the imaging process completed, you would like rapid destruction (clearance) of the reagent. The loss of Fc function may also be decisive in treatment of cancer patients with immunotoxins which can be constructed from antibody fragments such as scFv with a potent toxin attached. There is a considerable risk that whole antibody immunotoxins may be taken up by cells other than cancer cells which express Fc receptors. This risk is not present using antibody fragment immunotoxins.

small size of antibody fragments

The relatively small size of antibody fragments is also an advantage for tissue penetration; they should be much more effective at penetrating solid tumours than whole antibodies. Additionally, the cost of production of the small antibody fragments will be much less than that of either chimaeric or reshaped antibodies since, as we shall see, some of them can be produced over several days in bacteria without using expensive tissue culture facilities. Finally, minimising the size of antibodies but maintaining their antigen binding function is ideal for engineering a wide variety of fusion proteins which we shall discuss in Chapter 7.

fusion proteins

variable gene libraries

As we shall see, recent developments in the production of variable gene libraries by the polymerase chain reaction coupled with the development of rapid screening methods such as phage display suggest that antibody fragments may be the new coliclonal antibodies of the future and they will replace hybridoma production of whole antibodies. It is anticipated that it will be possible to produce these in a few days once complete libraries of variable domains are available and it may be a better approach to only generate whole antibodies when one or other of the biological activities are required. As we have seen, it is relatively straightforward to link any constant regions to the antigen binding domains and we can predict that antibodies may appear with various numbers of constant regions and even with newly engineered constant regions possessing unique functions.

5.4.2 Repertoire cloning - a variable gene library

The ideal alternative to reshaping of human antibodies is to be able to clone human antibody genes and to express the products in such a way as to be able to select suitable human antibodies with high affinity antigen binding properties. However, since there are considerable technical and ethical barriers to producing human hybridomas, there was, until recently, little suitable source material to clone the human variable domain genes from.

The polymerase chain reaction

polymerase chain reaction (PCR)

Fortunately, with development of the polymerase chain reaction (PCR), the way was clear to attempt repertoire cloning of the human antibody genes. Using this technique, you can selectively amplify any chosen region of a DNA molecule. However, some information on the sequence to be amplified is required since two short oligonucleotides have to be synthesised which, during the reaction, will be annealed to the 5′ and 3′ ends of the two strands of DNA, thus designating the sequence to be amplified.

Π How would you determine the correct sequence for the oligonucleotides which would anneal to all V regions?

The investigators have to choose conserved regions at each end of the heavy and light chain sequences to act as universal primers. These primers also included restriction sites for subsequent cloning.

primer oligonucleotides

Taq polymerase

Figure 5.6 illustrates the major steps and explains the basis of PCR. The basis of the reaction is that the DNA is heated to about 95°C to separate the strands. It is then cooled and the primer oligonucleotides attach to either end of the region to be amplified. The enzyme, *Taq* polymerase, which is unaffected by the heat treatment, then synthesises new strands. The temperature of the reaction mixture is again raised to 95°C to separate the newly synthesised strands from the templates followed by cooling, primer annealing and strand synthesis. This procedure may be repeated many times to produce large amounts of DNA.

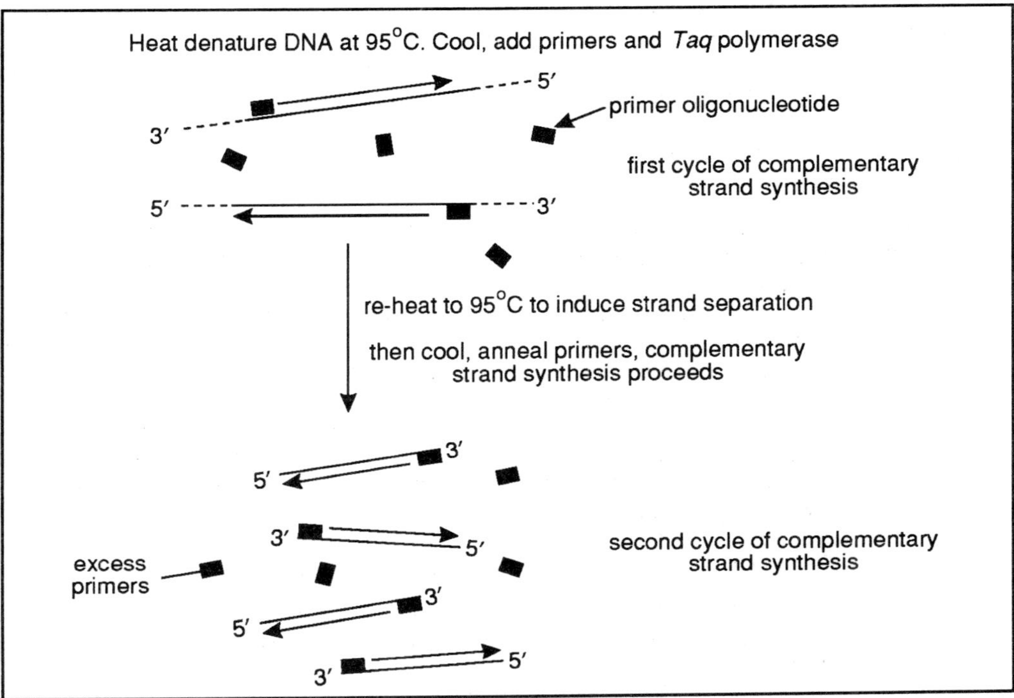

Figure 5.6 The polymerase chain reaction. The procedure is repeated through many cycles to produce large amounts of DNA.

Π How do you think PCR will facilitate the cloning of human variable genes?

Since PCR can amplify sections of a single DNA molecule you could obtain DNA encoding variable regions either from cDNA made from mRNA isolated from B cells or from genomic DNA. Suitable cells include B lymphocytes from the blood stream or from the spleen surgically removed from a patient. These pools of cells will contain either mRNAs (if they are actively synthesising soluble or membrane antibodies) or genomic DNA which has rearranged during development of each B cell (resulting in intron-less VDJ genes).

genomic DNA

Π Could you use the mRNAs directly for PCR amplification?

The PCR method involves the use of *Taq* polymerase which only acts on DNA strands. you would need to generate these from the mRNAs by reverse transcriptase.

Π What do you think is the advantage of using immunised donors as source material for PCR?

One would anticipate that the use of IgG mRNAs from lymphocytes of immunised donors has two advantages. Firstly, it is likely there should be a higher proportion of the total clones which encode specificity for the antigen since the cells with specificity for that antigen would have recently undergone clonal expansion. Secondly, the V genes rescued from the lymphocytes encoding specificity for the antigen are likely to have undergone somatic hypermutation resulting in higher affinity antibodies.

The polymerase chain reaction and expression of single V_H domains

In 1989, Winter's group in Cambridge, UK, first reported cloning of immunoglobulin variable domains using PCR. Later the same year, this group reported the expression of a repertoire of murine V_H domains in *E. coli*.

In these experiments, mice were immunised with either lysozyme or keyhole limpet haemocyanin (KLH). Using PCR, libraries of V_H genes were cloned into plasmid vectors and V_H domains selected from these libraries were expressed in *E. coli*. Using ELISA, the authors detected 21 lysozyme specific and 2 KLH specific clones out of 2000 clones derived from the lysozyme immunised mouse and 14 KLH specific and 2 lysozyme specific clones from a similar number of clones from the KLH immunised mouse.

These results should suggest to you that there are clones specific for a very wide variety of antigens derived from these mice. You can therefore conclude that this sort of approach could be used to produce a V_H library from human cells as well. However, the group found that the single V_H domains, although expressing high affinity for the antigens, exhibited stickiness due to exposed hydrophobic areas which would normally be 'covered' by the light chain domain in an Fv fragment.

Π What other possible deficiency is there in generating single domain (V_H) antibodies?

Without a choice of light chain partners, there are obviously some limits in the potential diversity of these domains.

Repertoire cloning in Man

Later in the same year, Winter's group reported cloning of human V genes by using PCR to harvest rearranged light and heavy chain V genes from B cells of two un-immunised donors. They produced scFv fragments which exhibited specificities to a wide range of antigens including bovine serum albumin, bovine globulin and even self antigens such as the CD4 molecule and the cytokine, tumour necrosis factor.

The combinatorial approach involving light and heavy chain genes

Lerner's group in the USA took a different approach to that of Winter. They immunised mice with antigen, reverse transcribed the mRNAs to DNA and then amplified the Fab portions of both light and heavy chains using PCR. At the same time, restriction sites were attached to each end of the light and heavy chain cDNAs. The light and heavy chain restriction fragments were then ligated into distinct lambda phages and cloned in *E. coli*. Subsequently, by a very complex process, they were excised, mixed and religated into phage as heavy/light gene pairs. Each viable phage now contained both a heavy and light chain fragment resulting from the randomisation due to mixing. The phages were then used to infect *E. coli* and the positive clones (those expressing Fab specific for the antigen) identified (1 in 10 000 clones). Figure 5.7 illustrates the major features of this method.

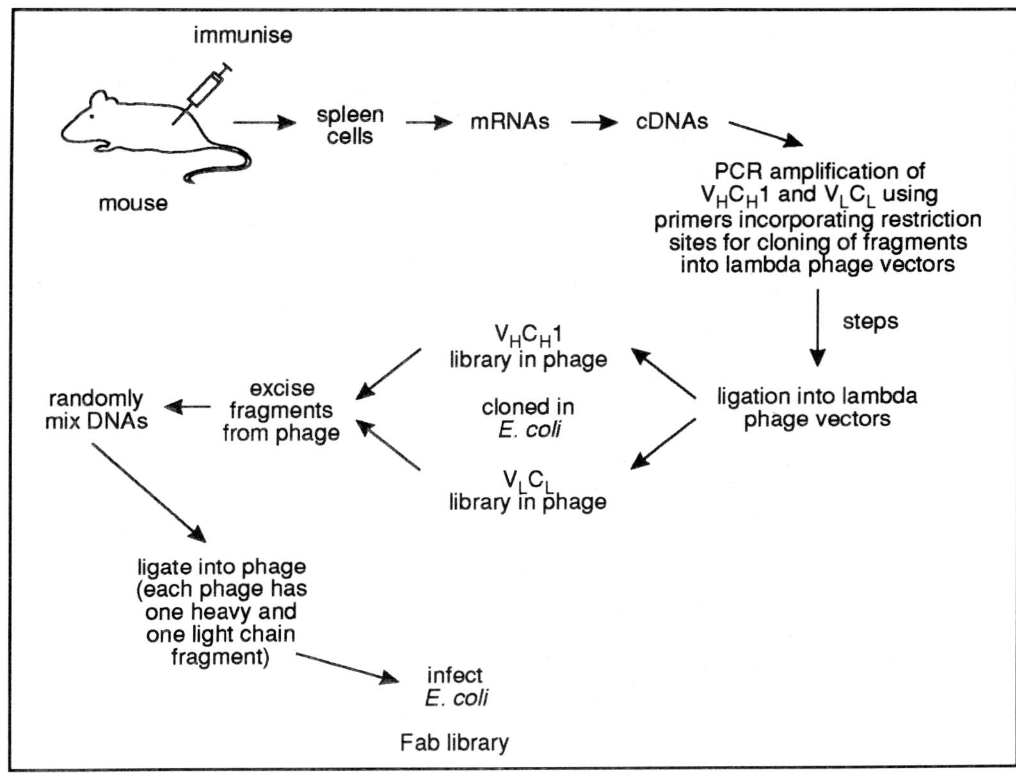

Figure 5.7 Combinatorial approach to repertoire cloning described by Lerner *et al.*

SAQ 5.4

Which one of the following statements is correct?

1) Loss of Fc does not confer any biological disadvantage on Fab fragments.

2) The combinatorial approach to repertoire cloning espoused by Lerner's group is the effective method for generating non-random combinations of heavy and light chains in antigen specific Fab fragments.

3) You must immunise with antigen in order to produce antigen specific antibodies.

4) PCR enables amplification of rearranged V genes from both chromosomes.

5) Synthesis of DNA strands takes place from the 3′ end.

5.4.3 Selection of antibodies by phage display

phage antibodies

The next important development was the development of so called phage antibodies. One of the major difficulties of all the methods so far discussed has been the lack of a rapid method for picking out the right clone from a library of many thousands. This process usually involved a rather long winded examination of all the clones. The rationale behind this development was that if the antibody could be displayed on the surface of phage it would be relatively simple to select those antibodies of choice from a whole library using antigen affinity methods.

First demonstration of phage antibodies

phage library

Winter's group was the first to demonstrate the feasibility of this technique when they successfully recovered phage expressing anti-lysozyme scFv's from a phage library prepared from hybridoma cells. Once again ,the technology is rather complex but the major steps of the procedure are outlined in Figure 5.8. This technique clearly showed that once a whole phage library of human heavy and light variable genes is available, antibodies could be produced in bacteria to any antigen represented in the library within a few days. This contrasts strongly with hybridoma technology and indeed any other method we have discussed thus far.

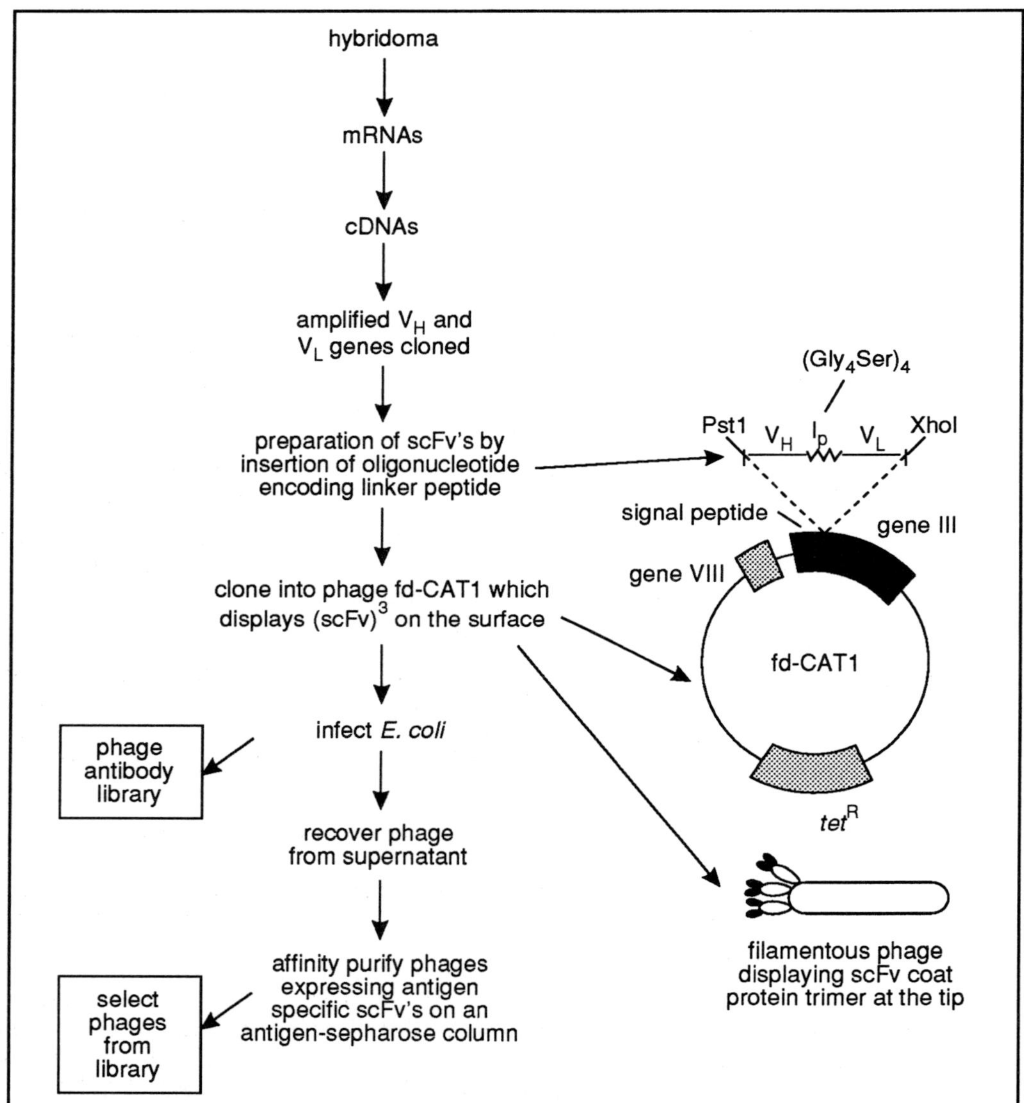

Figure 5.8 The first demonstration of phage antibody display by Winter *et al.* lp = linker peptide, PstI/XhoI = restriction endonuclease sites.

The group has progressed further in developing a phage display technology for Fab fragments rather than scFv's. One of the chains (say V_H - C_H1) is assembled on the surface of the phage linked to coat protein whilst the other chain is secreted into the bacterial periplasm. The group, introduced a signal called an amber mutation between the antibody chain cDNA and the coat protein gene.

amber mutation

Amber mutants are suppressible mutants that result in the creation of a UAG codon in mRNA. This codon normally signifies translation termination so that polypeptide synthesis stops at the 'amber site'. Such mutations can be suppressed in certain strains of *E. coli*.

Thus the gene construct is of this type.

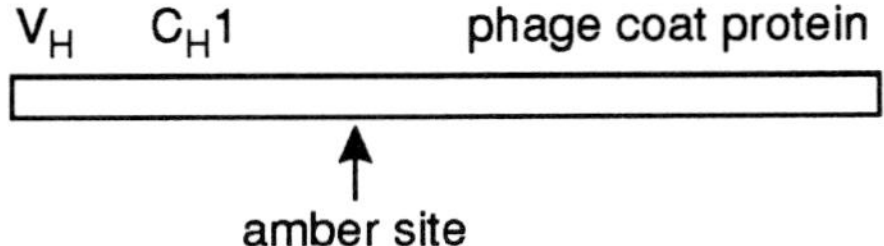

When placed in a normal strain of *E. coli*, the polypeptide that is produced is:

V_H C_H

If, however, the amber mutation is suppressed, the polypeptide that is produced is:

V_H C_H phage coat protein

Thus, by introducing this amber mutation, the group could decide whether to express the Fab on the phage or to produce soluble Fab in the bacterial periplasm.

Based on the rapid progress of the past few years, we can expect more developments in the near future which may well lead to coliclonal antibodies being the major source of antibodies for the majority of applications.

Human phage antibody libraries

blood lymphocytes

Using improvements to the above experiment, Winter's group obtained a large combinatorial library from blood lymphocytes of 2 un-immunised donors and was able to identify antibodies to may different antigens suggesting that this technique could result in a huge display library which could supply human antibody fragments to any antigen on demand.

5.4.4 The equivalent of somatic hypermutation *in vitro*

Π Many of these antibody fragments may be of low affinity especially if the genes are derived from genomic DNA. How could you change these into high affinity antibodies?

in vitro affinity maturation

There is a variety of suggestions of ways to improve the affinities of the antibody fragments derived from phage libraries. Firstly, you could introduce changes in the V regions by site directed mutagenesis (an alternative method is error prone PCR). Secondly, you could switch around the heavy-light chain combinations in the scFv's which could promote better matching. Lastly, phage display and selection using affinity techniques has a tendency to select the higher affinity variants.

Can we return to the question we posed at the start of the chapter, do you think that we are about to dispense with the mouse?

SAQ 5.5

Provide short answers to the following questions.

1) How could you generate antibodies against self components using the technologies discussed?
2) How could you generate antibodies using germ line V segments?
3) Can you replace polyclonal antibodies using these technologies?

5.5 Future developments

5.5.1 Production of antibodies in other than mammalian eukaryotic cells

transgenic sheep

Antibodies have been prepared in a variety of eukaryotic hosts. For instance, transgenic sheep have been used to produce a human antibody to haemophilic factor and this appears in the milk of the ewe. However, the amount of antibody produced is very low. A cytokine, interleukin-2, has been produced in the milk of transgenic rabbits and it is possible that antibodies could be produced in a similar fashion.

baculovirus systems

Baculovirus systems are being investigated for the introduction of antibody genes into insect cells and their subsequent expression. Insect cells, however, are not as easy to grow in culture as bacteria or other eukaryotic cells.

yeast cells

Antibody fragments have been produced in yeast cells. These are easy to grow and have the advantage over bacteria in that the antibodies are glycosylated and fold into the natural form with correct disulphide bridging. However, the antibody products have not been of high quality.

transgenic plants

Antibodies have also been produced in transgenic plants, with considerable success in tobacco. However, the rate of production is obviously a problem and purification of the antibody would be a major drawback.

So you can see that there are attempts to examine alternative production methods for antibodies. However, none of these have the obvious advantages discussed for coliclonal antibodies and we cannot see myeloma cells being replaced by any of these systems in the near future for the production of whole antibodies when they are needed.

5.5.2 Coliclonal, humanised and human monoclonal antibodies

We already have many thousands of different antibodies which are being used as reagents to identify human tissue and cellular antigens. If any of these specificities form the basis of targets for immunotherapy then the decision will have to be made whether to develop coliclonal antibodies or whether it is easier to simply humanise the murine antibodies.

transgenic mice

SCID mice

Some workers have introduced immunoglobulin genes into transgenic mice which could result in production of human antibodies following an immunisation protocol. Apart from the ethics of using animals, this approach will not be as cost effective as using coliclonal antibodies because of the small amount of antibodies derived from a single mouse. Additionally, to immortalise the antibody, immunisation would have to be followed by hybridoma production which is also costly or the genes could be rescued for repertoire cloning. Another approach which is being developed is the growth and stimulation of human cells in SCID mice. These mice possess no mature T or B cells and human cells are not rejected. It is possible that useful human antibodies could be generated using this system. Once again, the antibody producing cells would have to be immortalised or the genes rescued for cloning in bacteria.

This chapter has been mainly concerned with attempts to make human antibodies for therapeutic purposes and there is sufficient evidence that coliclonal human antibody fragments could fill that role. We would anticipate that major advances in technology in the next few years will decide whether this goal is achieved.

Summary and objectives

In this chapter we have reviewed developments in antibody engineering particularly those aimed at producing human antibodies for use as immunotherapeutic agents. We explained the rationale for chimaeric and reshaped antibodies and described how they are expressed in mammalian cells using mammalian expression vectors. We then discussed the advantages of producing antigen specific antibody fragments and showed how we could produce variable gene libraries using the polymerase chain reaction. The development of phage display of antibody fragments leading to rapid selection of specific antibodies from a library and subsequent expression in bacteria may eventually lead to the replacement of hybridoma antibodies by coliclonal antibodies.

Now that you have completed this chapter you should be able to:

- list the major developments in antibody engineering and discuss their advantages and disadvantages;
- list the major characteristics of a mammalian expression vector;
- describe the major steps in production of chimaeric antibodies;
- devise a protocol for reshaping antibodies;
- distinguish between the various antibody fragments used in antibody engineering;
- explain the basis of the polymerase chain reaction and discuss its essential contribution to repertoire cloning;
- describe the various experimental approaches to developing V gene libraries for expression in eukaryotic cells;
- describe what is meant by phage antibodies and discuss their impact on the potential of coliclonal antibodies;
- formulate experimental approaches for the generation of autoreactive antibodies and germ line encoded antibodies based on presently available technology.

Purification of antibodies

Purification of antibodies

6.1 Introduction

6.1.1 Sources of antibodies

In the past three chapters you have learned how to prepare polyclonal antibodies, hybridoma antibodies, bioengineered antibodies and antibody fragments. Having completed these procedures you are left with either antisera, hybridoma supernatants (or ascites fluid) or bacterial pellets containing either insoluble antibody fragments or periplasmic soluble antibody fragments. You will then have to make the decision as to how you are going to purify or extract the antigen specific antibodies or fragments and these purification strategies are the subject of this chapter. This is quite a long chapter, so do not attempt to study it all in one sitting.

antiserum

ascites fluid

bioengineered antibody fragments

hybridoma supernatants

The degree of purity varies considerably with the source of the antibodies. In Table 6.1 you will see that antiserum is the least pure source of antigen specific antibodies, indeed they represent only a small percentage of the total serum protein. The antibodies in ascites fluid are mainly monoclonal since they are secreted by the hybridoma cells in the mouse but there are still some serum proteins present. However, the proportion of antigen specific antibodies is much higher than in serum. After extraction from bacteria, bioengineered antibody fragments represent the only antibodies and must be considered to be at a higher level of purity than ascites fluid. However, the extract does contain bacterial proteins which may exert biological effects in some assays and these should be removed. The purest sources of antibodies are the hybridoma supernatants. The monoclonal antibodies are the only antibodies present even if foetal calf serum has been added since this does not contain antibodies. There will, however, be quite high levels of serum proteins including albumin. However, there are techniques to produce serum-free supernatants so that the only major proteins present are the monoclonal antibodies. One possible problem with these supernatants is that the levels of antibodies are much lower in these than in the other three sources and you may need to incorporate a concentration step in your protocol depending on what you are going to use the antibody for but also to store the antibodies over long periods.

Antibody source	Nature of antibody	Is antibody the major component?	Major contaminants
antiserum	polyclonal	no	other antibodies in excess and all serum proteins
ascites fluid	monoclonal	yes	other antibodies (minor) and serum proteins
bacterially expressed antibodies	monoclonal	yes	bacterial proteins
tissue culture supernatants	monoclonal	yes	non-antibody; foetal calf serum proteins if added

Table 6.1 The sources of antibodies and their relative purity.

Π Are antibodies the only proteins in serum-free hybridoma supernatants?

These supernatants often contain some bovine serum albumin but will also have proteins derived from the cells either due to turnover of membrane proteins or due to cell death.

6.1.2 When to purify antibodies

The decision to purify or, perhaps, partly purify antibodies from either antisera or monoclonal supernatants is dependent on what use you have planned for the antibody. There are many instances where you can use the impure product as it is without any further purification.

For instance, we learned in Chapter 3 that crude antiserum, suitably absorbed to remove unwanted cross-reactive specificities is often used for many immunoprecipitation reactions such as for measuring levels of IgG in serum using gel diffusion, or to detect serum proteins in immunoelectrophoresis as well as identifying cell surface antigens. Of course, it is advisable to prepare polyclonal antisera using a pure immunogen to reduce the risk that the serum will contain antibodies to impurities.

As for antisera, ascites fluid is often used without further purification for many assays; an additional bonus, of course, with ascites fluid is that the antibodies are so concentrated and represent about 90% of all antibodies present. Thus ascites fluid can be used at extremely high dilutions, resulting in too low a concentration for the impurities including the background antibodies to exert any effect in the assay.

Hybridoma supernatants can be used as capture antibodies in ELISA and as the second antibody to identify captured antigen but it would be advisable to purify it before using it to prepare enzyme antibody conjugates.

Π Before you read on, can you think of any reason why the antibody would need to be purified for this purpose?

When attaching the enzyme label, some of this will become conjugated to the impurities present and this will result in a high background (high signal to noise ratio) in the ELISA. Most experiments using immunoprecipitation of antigen can also be done without purification of the hybridoma antibody.

However, as indicated above, it is advisable to purify any antibodies, irrespective of the source, if you wish to attach radio isotopes, enzymes or prepare other conjugates. As has been suggested, bacterially expressed antibodies are routinely purified to remove bacterial proteins. Additionally, any antibodies, irrespective of the source, would need to be purified before being used as immunotoxins in Man.

6.1.3 The purity of antibody samples and purification protocols

Examine Figure 6.1 where we have shown the relative purity of each source of antibodies and suggested minimum protocols for their purification if you decide that your antibodies were not suitable in their crude form.

antiserum

antibody fraction

Antiserum is relatively impure in that the actual antigen specific antibodies are a small percentage of the total antibodies present and there are also high levels of serum proteins such as albumin which is present at about three times the concentration of the total serum antibodies. To ensure high recovery of antibodies, therefore, the purification protocol would normally include at least one step to separate out the bulk of the unwanted proteins from the antibody fraction. If you wish to obtain antigen specific antibodies, this would require an additional affinity separation using the antigen.

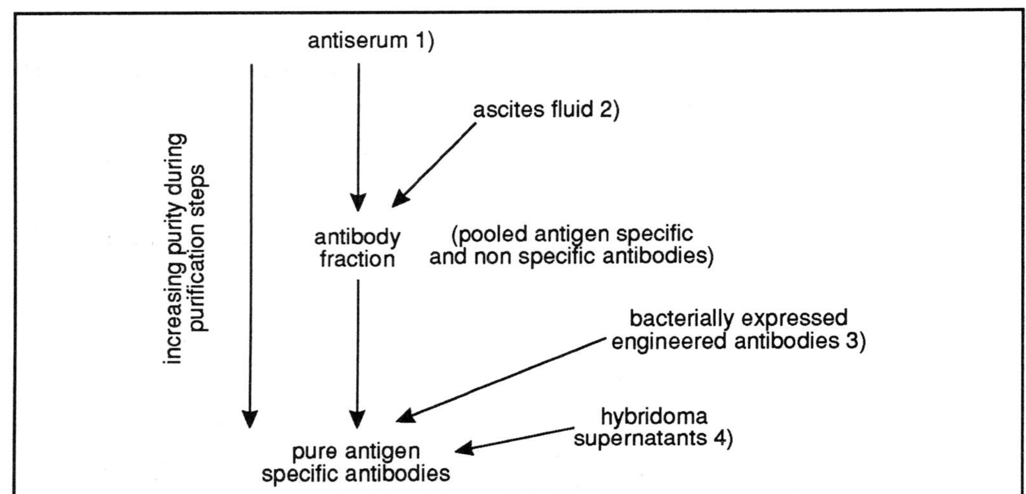

Figure 6.1 The major sources of antibody reagents and their degree of purity. For example, you will conclude that of the four primary sources of antibodies (numbered), antiserum is the least pure and hybridoma supernatants the most pure.

ascites fluid

Ascites fluid would also normally be subjected to at least a 2 stage purification procedure, an initial isolation of the antibody fraction followed by isolation of antigen specific antibodies.

bacterially expressed engineered antibodies

hybridoma supernatants

Bacterially expressed engineered antibodies (antibody fragments), once extracted from the bacteria and renatured, can be purified by a single step, resulting in monoclonal antigen specific antibodies. Similarly, antigen specific antibodies from hybridoma supernatants (and transfectoma supernatants) can be purified in one step. Having established these general factors, we have to emphasise that it is very much up to the individual laboratory what approach they take to purify the antigen specific antibodies. There will be times when the investigator thinks it worthwhile to do a one step purification of antigen specific antibodies from crude antiserum - there are ways to do that to produce extremely small amounts of antibodies.

preliminary steps

It must be emphasised that cost is a major consideration before any protocol is worked out and any form of affinity separation, whether it be protein A/G for purification of IgG or antigen affinity chromatography, for antigen specific antibodies, is expensive. It is, therefore, often prudent to proceed through preliminary steps before using these techniques or to use other less costly techniques. Having said that, there is no doubt that purification on protein A or G beads, which bind the Fc regions of selected IgG classes, is the method of choice by most investigators. Starting from antiserum or ascites fluid it will often be preceded by some other method such as ammonium sulphate precipitation and column chromatography. For hybridoma supernatants, one step purification with protein A/G beads or anti-immunoglobulin affinity chromatography is the method of choice.

SAQ 6.1

Which one of the following statements is correct?

1) Hybridoma supernatants require an antigen specific purification step.

2) Protein A affinity purification results in the isolation of antigen specific antibodies.

3) Antigen specific bioengineered fragments expressed in *E. coli* can be purified using protein A.

4) Hybridoma supernatants are the least likely of the four sources of antigen specific antibodies to require purification.

5) The sources of antibodies which contain serum need to be purified from antigen non specific antibodies.

6.1.4 Purification methods for IgM and IgG antibodies

Antibodies and immunoglobulins

You should realise at this stage that there are two major types of antibodies being purified. One purification procedure is where the antibody specificity is irrelevant, ie we simply need purified IgM or IgG antibodies as, for example, immunogens in the preparation of anti-isotype antisera or for subsequent preparation of subunits (heavy or light chains) or fragments of antibodies for structural studies. We often refer to these antibodies as immunoglobulins to avoid confusion; you will, of course, realise that there is no need for the final antigen affinity purification step for preparation of immunoglobulins.

When we refer to antibodies we mean antigen specific antibodies which are intended for use, for example, in antigen specific assays, for isolation of antigens or for immunotherapy. In the majority of cases, these antibodies are IgG and we shall confine our discussions mainly to the purification of this isotype. Except for the final purification step, the protocols for both immunoglobulins and antibodies are very similar. Therefore, in describing each purification method we shall refer to both of them simply as antibodies.

Summary of purification steps

We are going to describe various procedures to you which are used in the majority of laboratories for the purification of IgG antibodies. These steps are summarised in Figure 6.2. Examine this figure carefully as it contains a lot of information. You might find it helpful to draw out this scheme for yourself as this will ensure that you have seen all the information in this figure.

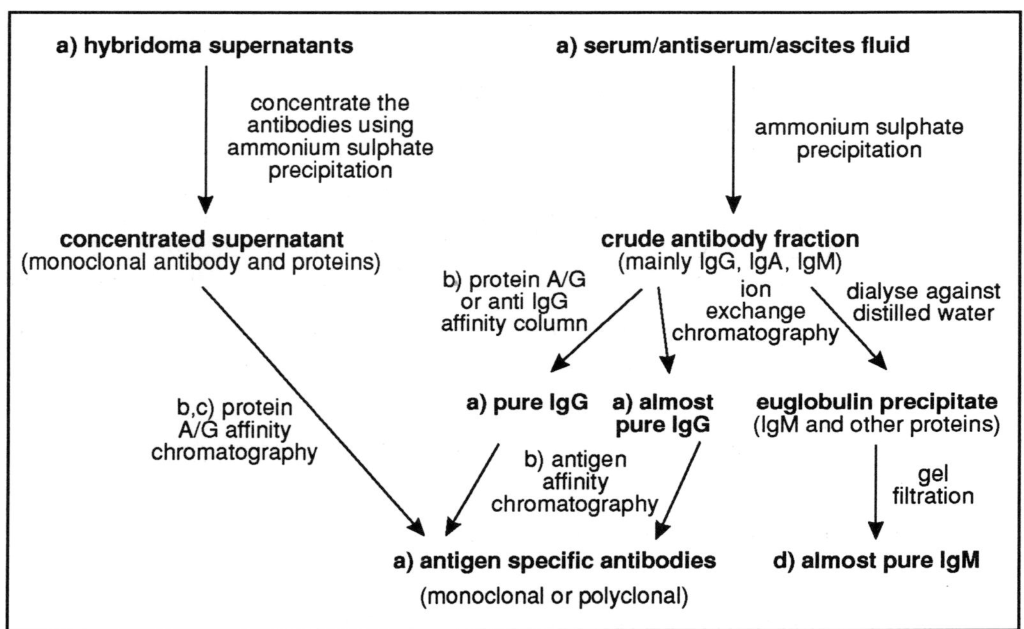

Figure 6.2 Purification steps for hybridoma antibodies and IgG from other sources. Almost pure = more than 95% purity. a) - all these reagents are used in assays, the level of purity being determined by the nature of the assay, b) - costly procedures, c) - not all hybridoma antibodies bind to protein A or G; you should also note that this method is specific for IgG, d) - see also Section 6.2.7.

Normally, fractionation of serum and ascites fluid begins with salt fractionation using ammonium sulphate which results in the precipitation of all the antibody classes leaving most other proteins including albumin in solution. IgG purification then proceeds using either affinity chromatography or ion exchange chromatography to produce a pure or almost pure IgG fraction respectively. The final step, if needed, is to use affinity chromatography to isolate antigen specific antibodies from the IgG fraction. Highly purified IgM can be isolated from the initial precipitate by dialysis of the reconstituted precipitate against distilled water followed by gel filtration. To purify hybridoma antibodies, it may be advantageous to concentrate the supernatant before purifying the antibodies in a single step using protein A or G affinity chromatography.

We shall now look at the individual methods involved in these purification steps.

SAQ 6.2

Identify the processes in the right column which result in the products shown in the left column. Use each item only once.

1) crude serum antibody fraction	a) AS + PAG
2) pure serum IgG	b) AS + IEC
3) IgM	c) AS
4) IgG with minor contaminants	d) DI + GF

AS = ammonium sulphate precipitation, IEC = ion exchange chromatography, PAG = protein A or G affinity chromatography, DI = dialysis, GF = gel filtration.

6.2 Purification methods

6.2.1 Salt fractionation using ammonium sulphate

Selectively precipitating proteins using salts (usually ammonium sulphate) is a well established biochemical procedure and we will not elaborate on the principles here (they are dealt with in the BIOTOL text 'Analysis of Amino Acids, Proteins and Nucleic Acids'). We will however give a brief outline of the practical steps involved.

A saturated solution of ammonium sulphate is prepared by attempting to dissolve about 1 kg of ammonium sulphate crystals in one litre of water.

Π How would you know the solution was saturated?

When these solutions are prepared, there should always be some crystals of ammonium sulphate left undissolved in the bottle indicating saturation.

The serum or other antibody solution is slowly stirred and the ammonium sulphate is added dropwise until 40% saturation is reached. The suspension is stirred for one hour in the cold and then centrifuged at 2-10 000 g. The precipitate is then dissolved in physiological saline and the antibody fractions reprecipitated.

Π What is the purpose of this step?

During the first precipitation, other proteins may become trapped in the precipitate. The effect of redissolving the antibody fraction and reprecipitating it reduces the amount of these impurities. This second precipitate is then redissolved in a minimal amount of physiological saline and dialysed extensively against saline or the buffer that will be used in the next purification step.

concentration step

You will observe that this process is also a concentration step and this procedure is often used to concentrate antibodies from hybridoma supernatants or chromatography column eluates before storage (see storage requirements in Section 6.5.1).

SAQ 6.3

You need to isolate a mouse antibody fraction using ammonium sulphate precipitation at 45% saturation. What volume of saturated ammonium sulphate solution will you need to add to 4 ml of mouse serum? Choose the correct answer from the list below.

1) 1.45 ml; 2) 3.27 ml; 3) 2.78 ml; 4) 1.56 ml; 5) 3.92 ml.

6.2.2 The euglobulin fraction

euglobulin fraction

The euglobulin fraction can be obtained from either whole serum or the antibody fraction produced by ammonium sulphate precipitation. The protein solution is placed in a dialysis sac and placed in a beaker of distilled water in the cold overnight. A precipitate forms (the euglobulin fraction) which consists of the majority of IgM molecules together with small amounts of plasminogen, complement proteins and some IgG. The precipitate is recovered by centrifugation and redissolved in buffers containing high salt. The IgM is then further purified using gel filtration (Section 6.2.4).

6.2.3 Ion exchange chromatography

Principal of the method

Here we have assumed that you are familiar with the principles of ion exchange chromatography as applied to the separation of protein. If not, we recommend the BIOTOL texts 'Analysis of Amino Acids, Proteins and Nucleic Acids' or 'The Molecular Fabric of Cells'. Ion exchange chromatography is an excellent method for the production of almost pure IgG.

Anion exchangers are used to purify IgG

diethyl-aminoethyl (DEAE) groups

The most popular method for the separation of antibodies is anion exchange chromatography in which the matrix carries either aminoethyl (AE-), diethylaminoethyl (DEAE-) or quaternary aminoethyl (QAE-) positively charged groups. The most common anion exchanger used for separation of antibodies in the past and quite often now is DEAE-cellulose, indeed anion exchange is often inappropriately named DEAE cellulose chromatography. This is because this anion exchanger has been replaced, because of its low capacity and poor flow properties by DEAE Sephacel which is a beaded form of cellulose possessing high capacity and high flow properties. Other commonly used anion exchangers are DEAE Sephadex, DEAE Sepharose and QAE Sephadex. The structural features of these exchangers are shown in Figure 6.3.

isoelectric point (PI)

Proteins at the pH of their isoelectric point, pI, have an overall charge of zero. Below their pI they possess a net positive charge due to protonated amino and side (R) groups whereas above their pI they possess a net negative charge due to unprotonated carboxyl and side (R) groups. Antibodies have more basic pI's than other serum proteins, their pIs are in the range pH 6-8 and therefore antibodies can be separated from other serum proteins very successfully using ion exchange chromatography.

The batch and chromatographic methods for anion exchange

There are two major protocols for the separation of IgG by anion exchange, the batch method and the chromatographic method. In the batch method the pH of the buffer is just below the isoelectric point of IgG antibodies.

Π Before you read on, would you expect the IgG antibodies to bind to the exchanger at this pH?

batch method

Under these conditions, the IgG does not bind to the exchanger whereas all the other serum proteins do. Hence, if a slurry of DEAE Sephacel is added to buffered serum, it will bind all proteins except IgG.

IgG containing supernatant

To separate IgG by this method, serum, ascites fluid or the reconstituted ammonium sulphate precipitated antibody fraction is dialysed against 5 mmol l^{-1} phosphate buffer pH 6.5. The previously prepared DEAE Sephacel or other DEAE matrix is equilibrated in the same buffer and is subsequently mixed with the protein sample after which the resulting slurry is stirred for 1-2 hours. The matrix is then separated from the IgG containing supernatant by either centrifugation or filtration and the IgG fraction processed for storage and for affinity chromatography.

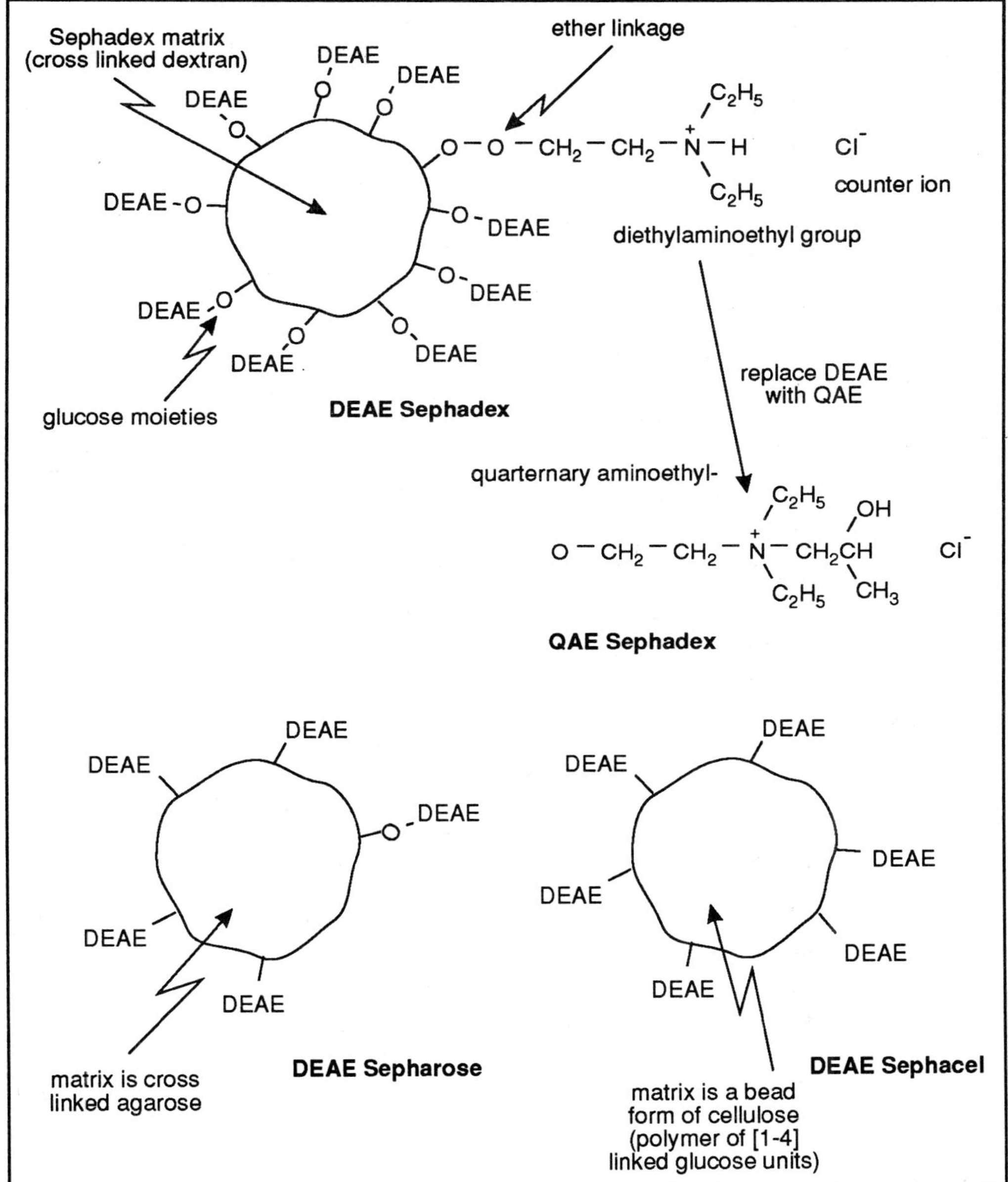

Figure 6.3 Major structural differences between common anion exchangers.

chromatographic method

In the chromatographic method, the pH of the buffer is above the pI of IgG. If serum proteins in such a buffer are applied to a column of DEAE Sephacel, all proteins will initially bind to the column. However, by gradually increasing the ionic strength of the eluting buffer, IgG will elute first and so purification will have been achieved.

The essential practical details of this method are as follows. A column is made of the anion exchanger equilibrated with a buffer of low ionic strength (say 5 mmol l^{-1} phosphate) and a high pH (about 8.5). The antibody containing sample is dialysed against the equilibrating buffer and then applied to the column. The column is then washed extensively with equilibrating buffer to remove any unbound substances and then the ionic strength of the eluting buffer is either gradually increased using a gradient mixer or is increased stepwise. The ionic strength is most easily increased by increasing the concentration of salt (eg sodium chloride providing Cl^- counter ions) in the buffer.

Π What other approach apart from increasing the ionic strength of the buffer could you take to elute the IgG?

You could gradually decrease the pH of the buffer which would have the same effect or you could use a combination of both (as is described in Figure 6.4).

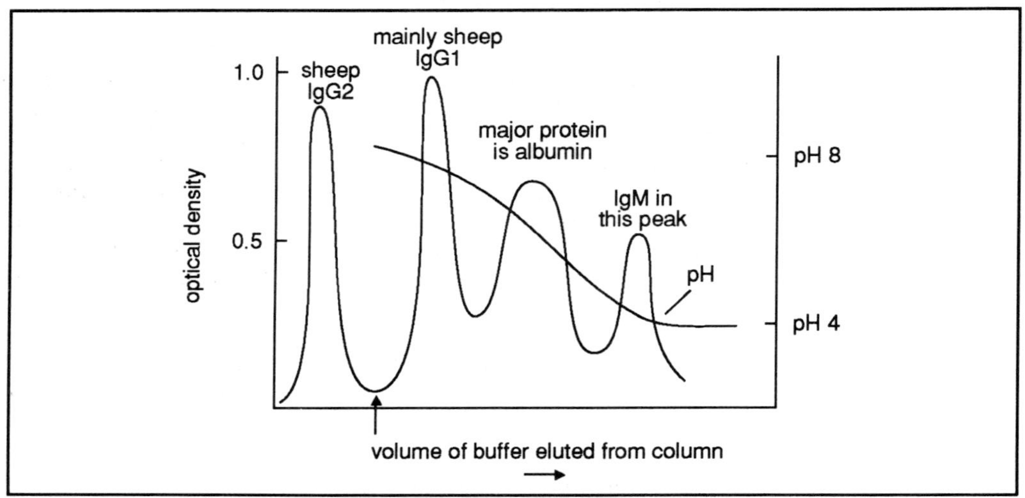

Figure 6.4 A typical elution profile of DEAE cellulose chromatography of sheep serum using a starting buffer of 0.01 mol l^{-1} phosphate pH 8.0 and a limit buffer of 0.3 mol l^{-1} phosphate pH 4 using a gradient mixer. This procedure involved both changes in ionic strength and pH. The sheep IgG2 did not bind to the column, the gradient elution was started as indicated by the arrow.

pure IgG2

A typical chromatographic separation is shown in Figure 6.4. You can see that this is a very good method for producing almost pure IgG. In fact, in the separation of sheep IgG's under the conditions shown in Figure 6.4, the first peak is pure IgG2, the second peak almost pure IgG1.

Π Is it possible to use the batch method instead of the chromatographic method to separate several different proteins?

stepwise desorption

One of the major uses of the batch method is to separate antibodies which do not bind to the anion exchanger since there is no real need to go through the procedure of setting up a column for this. You can, of course, also use the batch method for separating proteins which do adsorb to the exchanger if you can develop a suitable stepwise method for elution of the individual proteins - it would obviously not be possible to use a gradient procedure. Such a method would involve mixing of the DEAE matrix with the sample in initial buffer resulting in adsorption of the individual proteins followed by stepwise desorption of individual proteins using buffers of increasing ionic strength or decreasing pH. Each step would involve centrifugation to recover your antibody containing fraction.

Regeneration of the anion exchanger

regenerate the column

If you have prepared a large column for separation of antibodies or other proteins of interest, you will not wish to dismantle the column after each separation. This is not necessary as it is normal practice to regenerate the column usually by washing the column extensively with high salt buffer which should contain the counter ion to facilitate equilibration. This treatment removes any remaining substances bound to the exchanger. Chromatography of serum samples often results in deposits of lipids in the matrix and these can be removed by washing the exchanger with alkali or alcohol. Following regeneration, the exchanger is finally equilibrated with the initial buffer for the next separation or can be dried for storage.

SAQ 6.4

Complete the following paragraph using words from the selection given below (use each word or phrase once).

[ ] exchange is a popular method for the purification of [ ] from serum. In this method, [ ] groups covalently linked to an insoluble matrix have negatively charged [ ] which can be exchanged for [ ] proteins which are being separated. If the protein, however, is [ ], it will not bind to the matrix; this is the basis of the [ ] method for separation of IgG. In the [ ] method, the IgG, along with all the other proteins in the sample, are [ ] and adsorb to the column. By [ ] the [ ] or [ ] the pH of the eluting buffer, you can selectively elute IgG from the column before the other proteins since antibodies possess more basic [ ] than the other serum proteins.

Word list: ionic strength, anion, chromatographic, IgG, positively charged, isoelectric points, negatively charged, counter ions, increasing positively charged, batch, negatively charged, lowering.

6.2.4 Size exclusion chromatography

Size exclusion chromatography, more commonly called gel filtration separates molecules on the basis of molecular size. You should be aware of the existence of a variety of gels used for this type of chromatography (for example Sephadex, Sepharose, Polyacrylamide) and the principles involved in the separation of molecules using these types of materials. (If you are unfamiliar with these we would recommend the BIOTOL text 'Analysis of Amino Acids, Proteins and Nucleic Acids'). As a reminder we have illustrated the basis of gel filtration in Figure 6.5.

Π Sephadex G25 has a fractionation range from 1000 to 5000 Daltons whereas Sephadex 200 has a fractionation range of 5000 to 60 000 Daltons. Will either of these gels separate IgM, IgG, albumin (66 000 Daltons) and decapeptides?

desalting procedures

The G25 column would only separate the decapeptides from the remaining components which would all be excluded from the column. One of the major uses of such a column is in desalting procedures. For instance you could remove the ammonium sulphate from the antibody fraction produced by precipitation of serum with 40% salt. The G200 column would separate all components from each other since the molecular weight of the IgM is above the exclusion limit and the decapeptides are below the lower limit fractionation.

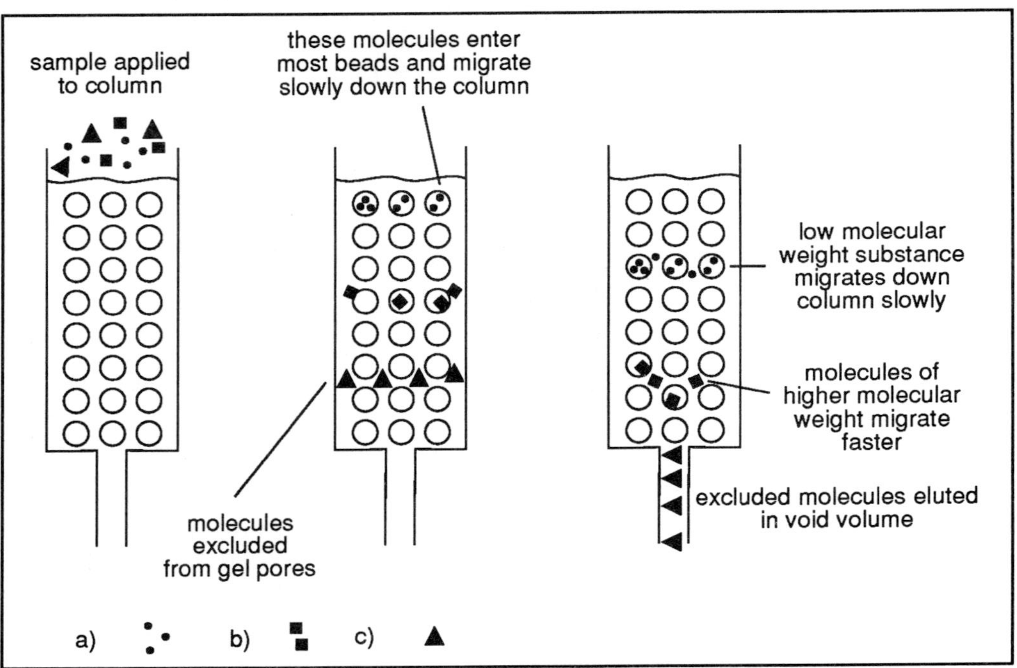

Figure 6.5 Basis of gel filtration. a) and b) represent two proteins whose molecular weights fall within the fractionation range of the column. The molecular weight of b) is greater than that of a). c) represents proteins of a high molecular weight which are excluded from the gel and are eluted in the void volume.

However, you should note that this gel is not suitable for purification of IgM from, say, serum proteins, since they would be eluted along with many other proteins of high molecular weights. We would obtain a more highly purified IgM by subjecting the euglobulin fraction to Sepharose 6B chromatography which has a fractionation range of 10 000 to 4 million Daltons.

SAQ 6.5

Even this purification step will not result in pure IgM. Write down the possible ways to produce pure IgM.

Molecular weight determination of globular proteins

Size exclusion chromatography can be used to derive molecular weights of the fractionated proteins by calibrating the column with a number of molecular weight standards.

The fractionation of serum

Sephadex gels need to be pre-swollen for about three days in buffer prior to packing the column, each gram of dry gel will usually swell to 30-40 cm^3 after this period. Most of the other gel types are provided ready for use. The equipment needed for gel exclusion chromatography is shown in Figure 6.6 and a similar arrangement can be set up for anion exchange chromatography.

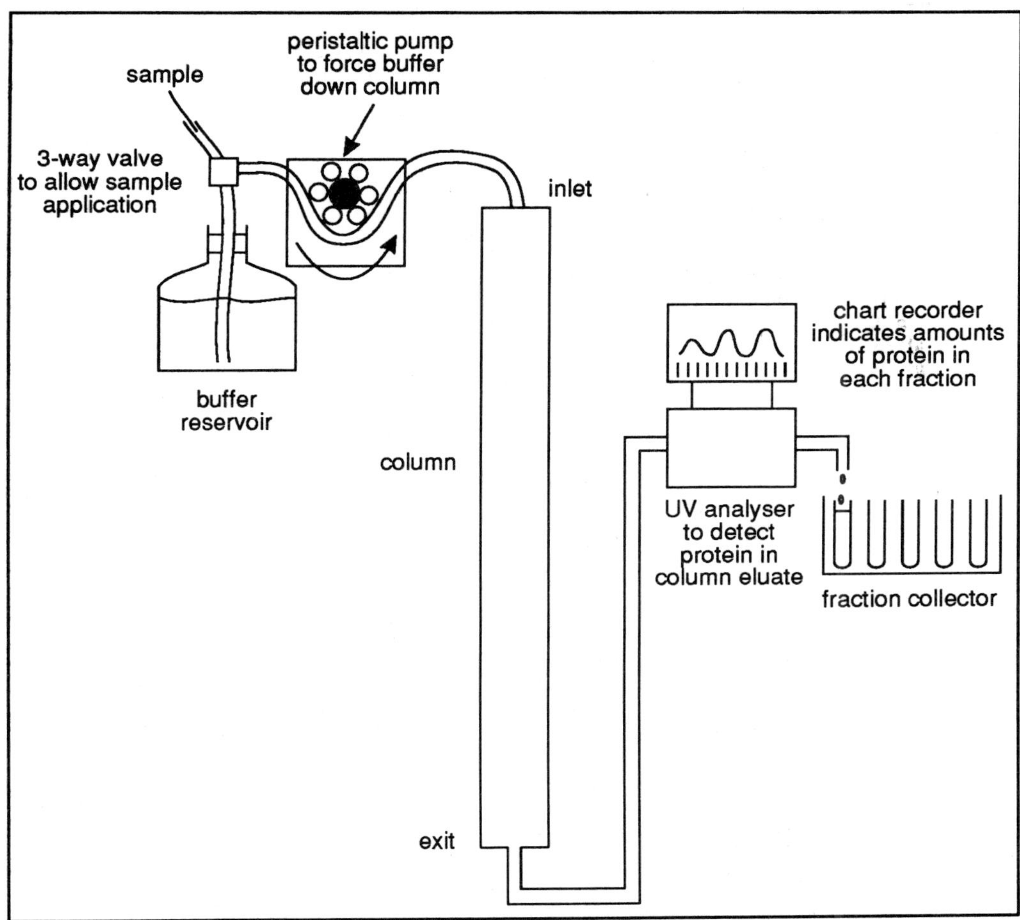

Figure 6.6 Apparatus for gel filtration chromatography.

After the column is packed and equilibrated with buffer, the sample is applied, and buffer is then continually pumped through the column. The eluate exits from the column and passes through a UV spectrophotometer which measures the absorbance usually at 280 nm and the absorbances are then transmitted to a chart recorder where they are recorded in graph form. The eluate is collected in small volume fractions by a fraction collector. Once the separation procedure is finished, fractions representing each peak are pooled and further processed.

The main disadvantage with size exclusion chromatography is that it results in dilution of the sample so the separated proteins may need concentrating. This can be done using ammonium sulphate precipitation.

high performance liquid chromatography

The speed and resolution of gel exclusion chromatography has until recently been limited by the time needed by molecules of various sizes to diffuse into and out of the gel pores. This problem of solute transfer was partly due to the large size of the individual beads which were more than 50 micrometers in diameter. However, due to advances in instrument technology and synthesis of the column packing materials, it became possible to prepare columns with 5 micrometer diameter beads. This increased the total surface area of the stationary phase (gel pores) giving the columns high capacity but also resulted in very dense packing of the column material. This necessitated the use of high pressure using special pumps to force the buffer through the column and the use of metal columns. This in turn allows high flow rates through narrow bore columns and allows extremely good separation of proteins applied to the column in small amounts. This technique is usually referred to as HPLC (High Performance Liquid Chromatography).

SAQ 6.6

We have said that you can determine molecular weights of proteins by size exclusion chromatography. You end up with what we call the relative molecular mass or apparent molecular weight of the protein. This is estimated by plotting log molecular weights of the protein standards against their elution volume and you can use this information to determine the molecular weight of the unknown based on its observed elution volume. You can substitute fraction number for elution volume since the volumes of the fractions are constant. Use the data below to determine the apparent molecular weight of the unknown antibody.

Protein	Molecular weight (D)	Fraction number
cytochrome C	12 400	41
chymotrypsinogen	25 000	38
ovalbumin	43 000	34
bovine serum albumin	66 000	30
lactic dehydrogenase	135 000	26
catalase	240 000	21
unknown antibody		25

6.2.5 Affinity chromatography using protein A or protein G

Background

Protein A and protein G are derived from bacterial cell walls. These proteins are extremely useful in the purification of IgG classes from many species by possessing the ability to bind to the Fc regions of the antibodies.

protein A

protein G

Protein A is found in the cell walls of *Staphylococcus aureus* strains. It is a 42 000 Dalton polypeptide possessing 4 Fc binding sites of which it can use two at any one time. The binding sites on Fc are found in the second and third constant regions of the IgG heavy chains. Protein G, a more recent discovery, is a polypeptide of about 30 000 Daltons found in the cell walls of some strains of beta haemolytic streptococci. The genes for both proteins have now been cloned and the proteins are available as bacterially expressed products. Protein G also possesses a binding site for albumin which could be a disadvantage for the purification of IgG. However, this site has now been engineered out.

We have listed the IgG antibodies from different species which can be purified using protein A or G affinity chromatography in Table 6.2. You will notice that protein A does not bind well to sheep or goat IgG which are sources of many polyclonal antisera nor to IgG from rats which are an important source of monoclonal antibodies. One small disadvantage with the use of protein A to separate human IgG is that it also binds to a lesser extent IgM and IgA.

Species	Affinity for protein A	Affinity for protein G
Man	high	high
mouse	medium	medium
horse	medium	high
sheep	poor	medium
rat	poor	medium
goat	no binding	medium

Table 6.2 Binding affinities of IgG from various species for protein A and protein G. You should note that there may be subclass differences (see Table 6.3) and the affinities shown represent average affinities of all IgG antibodies.

murine IgG1

rat monoclonals

We have supplied more details of the IgG subclasses which bind to proteins A and G in Table 6.3. As we said earlier, affinity chromatography using protein A is the most popular method for recovery of IgG monoclonal antibodies from hybridoma supernatants. It is interesting to note that murine IgG1 has little affinity for protein A but binds well to protein G. In contrast, these affinity methods have little use for purifying rat monoclonals. You will now appreciate why we emphasised early the importance of determining (see Section 4.4) the isotype of monoclonals prior to their purification.

Species and subclass		Affinity for protein A	Affinity for protein G
mouse	IgG1	poor	high
	IgG2a	high	high
	IgG2b	medium high	medium high
	IgG3	medium	medium high
Man	IgG1	high	high
	IgG2	high	high
	IgG3	no binding	high
	IgG4	high	high
Rat	IgG1	no binding	poor
	IgG2a	no binding	high
	IgG2b	no binding	medium
	IgG2c	poor	medium

Table 6.3 The binding affinities of mouse, human and rat IgG antibodies for protein A and G.

Purification of IgG antibodies using protein A/G affinity chromatography

protein A Sepharose

This is relatively simple procedure. The antibody preparation is adjusted to pH 8 and applied to the column of protein A Sepharose which is available commercially (the method for protein G is similar). You can normally load about 10-20 mg antibody (about the total IgG in 1-2 ml of mouse serum) on to a 1 ml column. The column is now extensively washed with a buffer of pH 8 (10 mmol l^{-1} Tris would be suitable) to eliminate any non-specific binding.

Π How could you check whether all the non-specific binding proteins have been washed off the column?

The efficiency of this process can be checked by examining the absorbance at 280 nm of the eluate during washing. When it returns to zero it can be assumed that all the non-specific binding contaminants have been washed off the column. The antibodies are then eluted with 0.1 mol l^{-1} glycine buffer with a pH of about 3 although some antibodies will elute with much more gentle elution conditions than this. The antibodies need to be collected into tubes containing high molarity buffer of about pH 8 to correct the pH back to neutral. We have illustrated the major steps in protein A affinity chromatography in Figure 6.7.

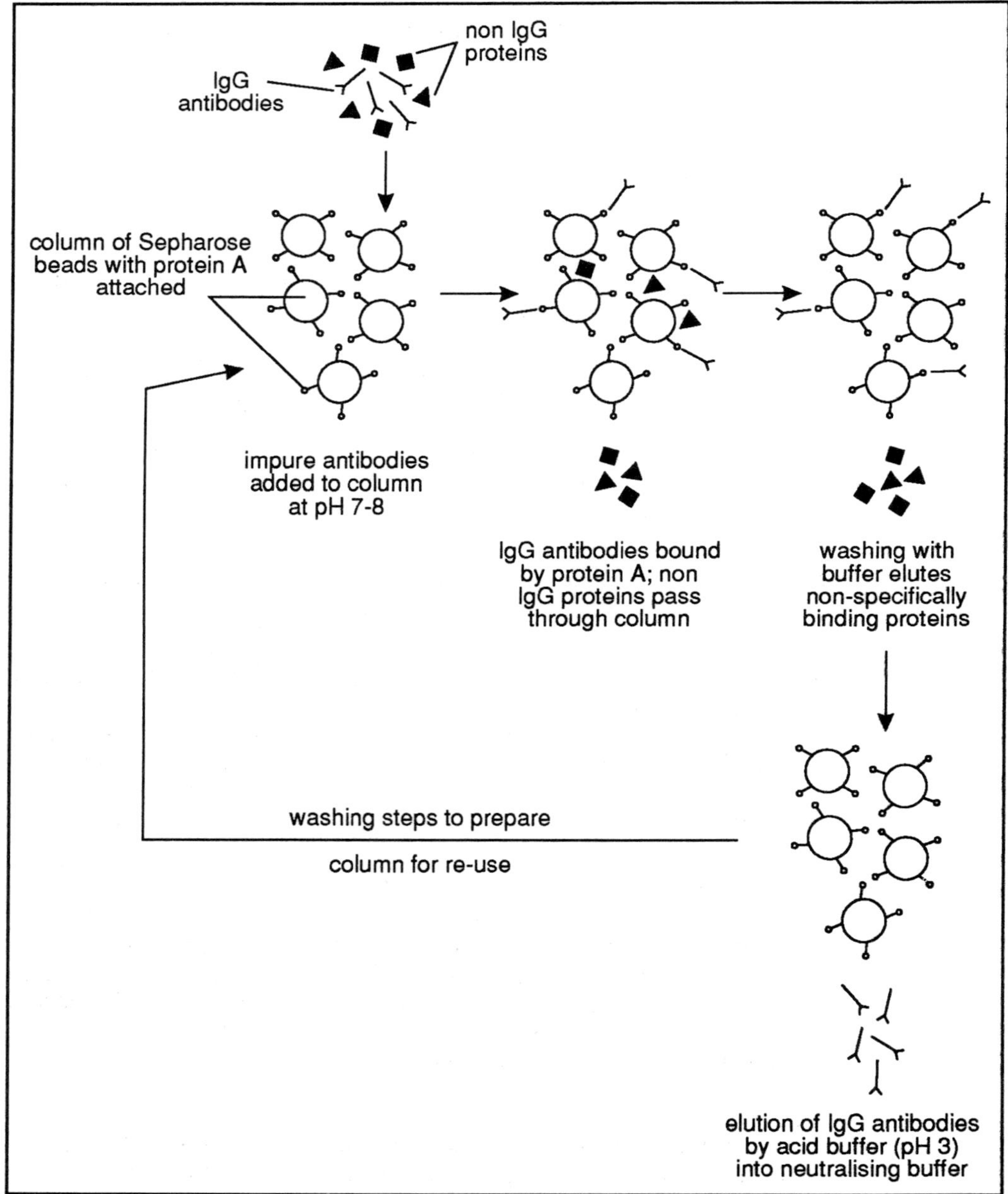

Figure 6.7 Purification of IgG antibodies using protein A affinity chromatography. The procedure using protein G is similar.

Π To make sure you know this procedure, close your text and draw your own diagram of the stages involved in protein A affinity chromatography. When you have completed the drawing, check it against Figure 6.7.

It is also possible to selectively elute human and murine IgG subclasses from a protein A column using different buffers.

It is normal procedure to dialyse the purified IgG antibodies against saline to remove buffer salts and often the sample is concentrated prior to storage.

SAQ 6.7

Which of the following statements are correct?

1) You cannot purify human IgG3 from a purified IgG fraction using protein A affinity chromatography.

2) Sheep and goat IgG antibodies do not bind to protein A or G.

3) Protein A and G affinity chromatography is best suited for purification of human IgG subclasses.

4) Elution conditions for purification of antibodies from protein A affinity columns vary from antibody to antibody.

5) One ml of mouse serum from a hyperimmune mouse may contain 1 mg of antigen specific antibodies. All of these could be purified from the antiserum in a single step on a protein A column of volume 0.2 ml.

6.2.6 Purification of antibodies using anti-isotype antisera

anti-isotype antibodies

This procedure using anti-isotype antibodies coupled to beads is an alternative to protein A or G affinity chromatography especially where you have a plentiful supply of antiserum since this method will be less costly. For isolating those antibodies which do not bind very well to protein A or G, this is basically your only choice apart from antigen affinity chromatography. You should realise, however, that antigen specific purification methods, which include purification using anti-isotype antisera, employ rather extreme conditions to elute the antibodies and the more gentle, less potentially damaging purification methods using protein A or G, if available, may be preferred.

extreme conditions

The purity of the anti-isotype antiserum

The anti-isotype polyclonal antibodies will need to be purified from the antiserum by one of the methods we have just discussed and must then be conjugated to Sepharose beads. Before you proceed further with this purification you should think about the possible quality of the antiserum and about the further purification steps which are needed before using these antibodies as affinity reagents. A lot of difficulties can be avoided by purchasing these reagents.

Π Before you read on write down the sort of problems you would expect to encounter with using polyclonal anti-isotype antiserum.

preselect the anti-immunoglobulin antibodies

You may expect to have cross reacting antibodies to other isotypes present in the antiserum due perhaps to the impurity of the antigen used to immunise the rabbits or due to using whole antibody as immunogen. If this is so, it will be worthwhile to preselect the anti-immunoglobulin antibodies on an affinity column before proceeding with the preparation. Otherwise, the affinity purification will not result in single isotype antibodies.

Π If you find cross reactivity in your antiserum with other isotypes and you suspect the presence of light chain antibodies, how would you proceed to purify the antiserum?

pooling of anti-isotype Mabs

If you suspect that there are anti-light chain antibodies in the antiserum then you could use an affinity column of either light chains or antibodies of a different isotype to remove these antibodies. If you still anticipate problems you may wish to contemplate the production of monoclonal antibodies (Mabs) to the antibody class since many of these will be monospecific for the heavy chains of the antibodies. By pooling these antibodies you can produce a very good 'polyclonal reagent'.

Π During coupling of the anti-isotype antibodies to the activated beads to make an affinity column, considerable loss of antibody binding activity may occur. Why is this and how could you avoid it (think of protein A and its properties). We will provide an answer in the next section.

Covalent coupling of antibodies to protein A Sepharose

antibody protein A Sepharose complexes

In coupling antibodies to matrices, the coupling reagents are not specific for the Fc end of the antibodies and many antibodies may become attached by their antigen binding end to the beads preventing subsequent binding of the IgG antibodies that must be purified. This can be avoided by preparing antibody protein A Sepharose complexes. If the anti-isotype antibodies are first reacted with protein A Sepharose, they become attached by their Fc ends leaving the antigen binding ends free for subsequent binding of IgG molecules. Addition of a suitable linking agent such as disuccinimydyl suberate covalently links protein A to the anti-isotype antibodies thus forming a new matrix for affinity chromatography (Figure 6.8).

Affinity purification of antibodies of a single isotype

The antiserum is prepared as described in Chapter 3. It may be worthwhile to look over that section to remind you of the steps in the preparation of the antiserum. The antiserum should then be analysed for cross reactivity with other isotypes using the

Ouchterlony

Ouchterlony method. If you have cross reactivity then you will need to eliminate it as we have suggested. The IgG antibody fraction is then isolated from the antiserum using the methods just described and the resulting antibodies are conjugated to activated beads.

pre-cycled with serum

A column is then constructed with the antibody conjugated beads and extensively washed with low ionic strength buffer of physiological pH (about 7.5). Ideally, the column should then be pre-cycled with serum from the same animal species which was used to prepare the antiserum.

Π What is the purpose of this procedure?

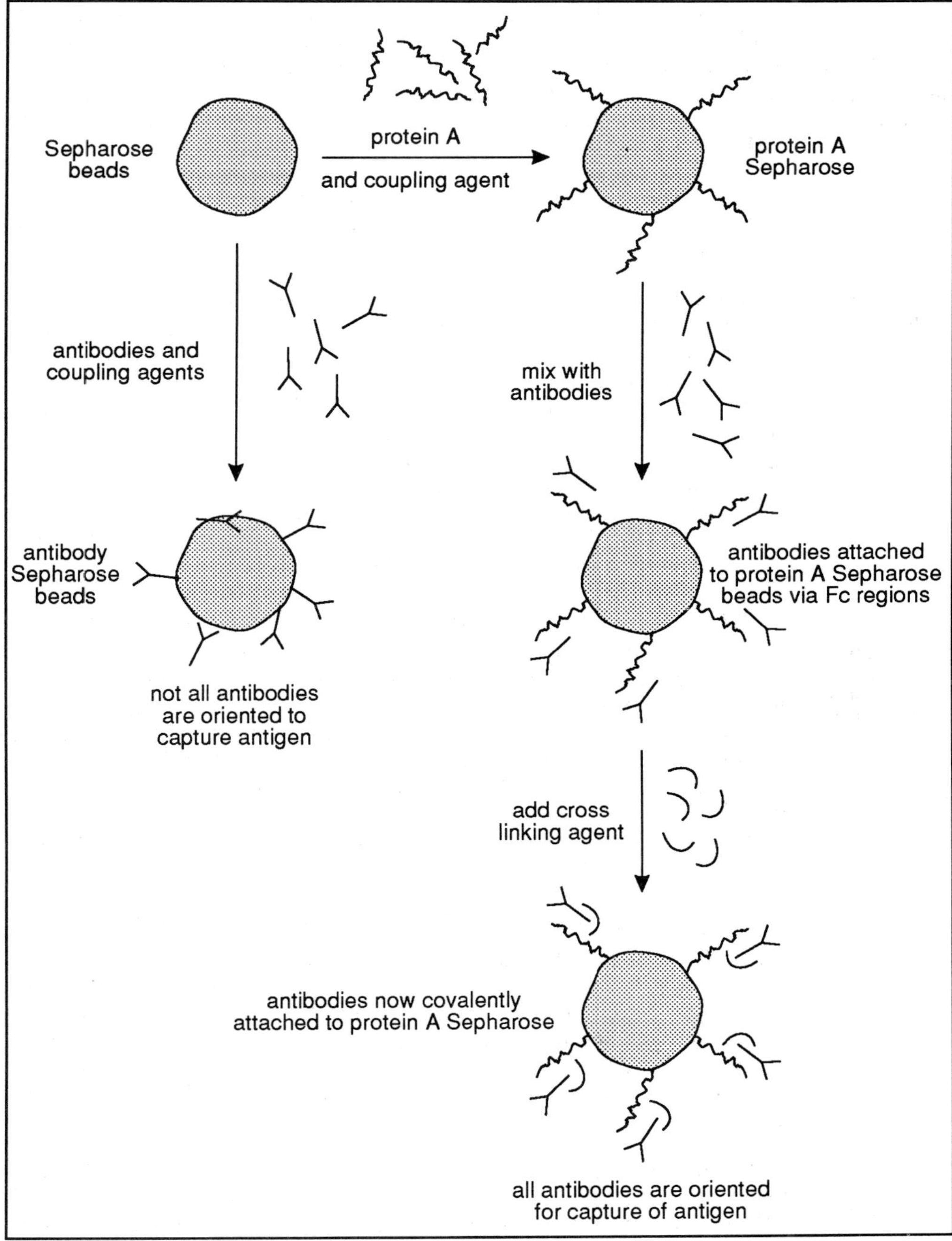

Figure 6.8 Covalent binding of antibodies to protein A Sepharose to improve antigen binding activity.

irreversible binding

pre-cycle the column

Some of the antibodies in the antiserum will be of such a high affinity for the antibodies to be purified that they will irreversibly bind the antibodies being purified. To prevent loss of purified antibodies to the column, it is advisable to block these high affinity antibodies. To do this you pre-cycle the column with normal serum which, of course, also contains antibodies. To do this, normal serum is applied to the column and after washing, the antibodies in the serum are eluted from the column using eluting buffers as described below. The column is then washed extensively with the pH 7.5 buffer.

elution of antibodies with acid and alkaline buffers

The sample containing the antibodies to be purified, such as an ammonium sulphate fraction of serum or a hybridoma supernatant, is now applied to the column followed by extensive washing to remove any non-specific binding. The antibodies are then eluted in a stepwise fashion, first with a strongly acidic buffer and then with an alkaline buffer.

Π What is the purpose of such a two stage elution?

The antibodies will be bound by non-covalent interactions which may involve positively or negatively charge moieties which can be sensitive to acidic or basic buffers and hence treatment with both results in recovery of all antibodies bound to the column.

A typical example of an elution profile would be to elute the antibodies with 100 mmol l^{-1} glycine pH 2.5 followed by washing of the column in alkaline buffer (say 10 mmol l^{-1} Tris pH 8). Antibodies remaining attached to the column are now eluted with 100 mmol l^{-1} triethylamine buffer pH 11.

Π You are subjecting the purified antibodies to extreme pH conditions, how do you prevent their denaturation?

During the elution of antibodies at extreme acidic and alkaline pH you collect them in tubes containing neutralising buffers. Thus, the acid and alkaline fractions may be collected into tubes containing strong Tris buffer at about neutral pH which neutralises the eluting buffers.

Finally, the antibody fractions collected are pooled and dialysed against physiological saline. Since the antibodies have undergone considerable dilution during the elution procedure you may wish to concentrate them using ammonium sulphate precipitation or by pressure dialysis (ultrafiltration).

We have illustrated all the major steps comprising this procedure in Figure 6.9. Having read the text, now make sure you are familiar with each step and the reason for performing it. Then attempt the next SAQ without further reference to the text or the figure.

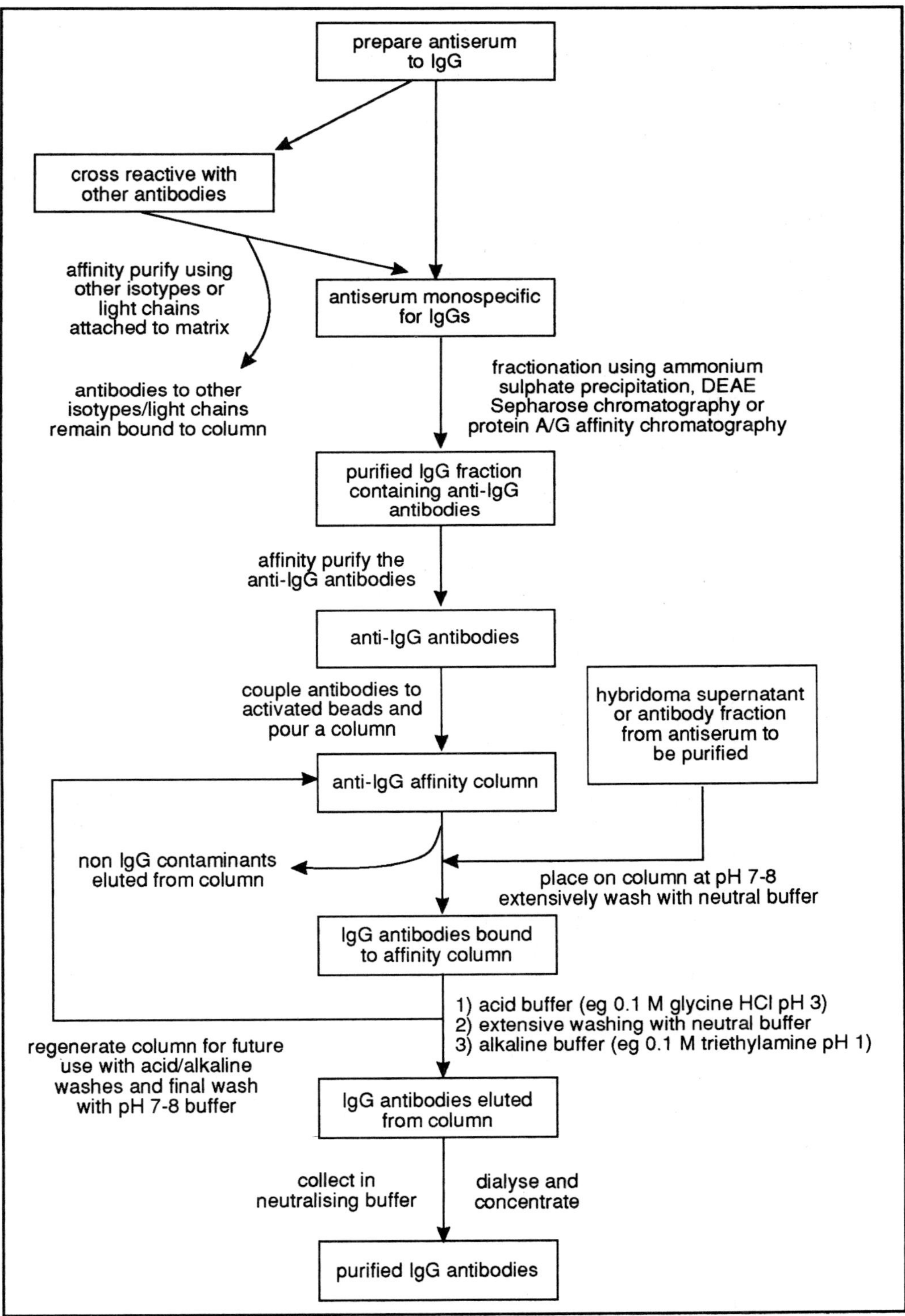

Figure 6.9 Purification of IgG antibodies using anti-isotype affinity chromatography. A typical protocol is shown of which there are many variations especially for the elution of the antibodies.

SAQ 6.8

The collection of items in the left column below comprise the major steps (but no washing steps) in affinity purification of antibodies of a single isotype. Place them in the correct order and then match them with the reasons for performing each step from the right column (Abs = antibodies).

1) Precycle with normal serum	a) Adsorption step
2) Elute with alkaline buffer	b) Are the Abs monospecific?
3) Prepare column	c) Recover first lot of Abs
4) Prepare affinity matrix	d) Make ready for storage
5) Affinity purification of anti-isotype Abs	e) Remove non specific binding
6) Prepare antiserum	f) Block high affinity Abs
7) Dialyse and concentrate	g) Produce anti-isotype Abs
8) Wash with neutral buffer	h) Support the beads
9) Elute with acid buffer	i) Eliminate cross reactivity
10) Check cross reactions	j) Recover second lot of antibodies
11) Apply sample to column	k) Conjugate Abs to beads

6.2.7 Purification of IgA and IgM using affinity chromatography

IgM and IgA

protamine

Jacalin

The substance protamine can be coupled to Sepharose using the cyanogen bromide reaction and protamine-Sepharose can be used to isolate IgM. IgM can subsequently be further purified using gel filtration. The lectin Jacalin from the seeds of the jackfruit, is available commercially, attached to agarose for purification of human IgA1. Once bound, the IgA can be eluted from the column with galactose or melibiose.

6.2.8 Affinity purification of antibodies on antigen columns

This procedure is essentially identical to that described for anti-isotype affinity chromatography except that in this case the antigen is coupled to the column matrix. The binding conditions are the same but the conditions for elution will depend on the nature of the antibody. You will realise, of course, that the same rules and general conditions apply whether you are purifying antibodies or antigens using affinity chromatography. We shall concentrate on the use of antigen affinity columns since we are dealing with the purification of antibodies in this chapter.

Π What sort of antibodies would you wish to purify by antigen affinity methods?

Antigen affinity methods are not commonly used these days since a majority of antibodies produced are monoclonal antibodies and you do not need to use such methods to purify them since they only express a single specificity. This means that you can isolate the antigen specific antibodies using a non-antigen specific method such as protein A affinity purification. The major use for the antigen affinity method is in extracting antigen specific antibodies from a mix of polyclonal antibodies such as is found in antisera.

Preparation of immobilised antibodies or antigens

agarose or polyacrylamide beads

The same methods are used for preparing both antibody and antigen affinity columns. We are not going to give experimental details for the preparation of activated beads since many of these are available from manufacturers. Most of the products are either agarose or polyacrylamide beads which have been activated by various reagents producing derivatives of the matrix which then react with amino groups on the protein or antibody to form a covalent linkage. These include carbonyldiimidazole, cyanogen bromide, glutaraldehyde, hydroxysuccinimide and tosyl chloride. Note that many more examples of reagents and techniques for attaching proteins to supports are given in the BIOTOL text 'Technological Applications of Biocatalysts'.

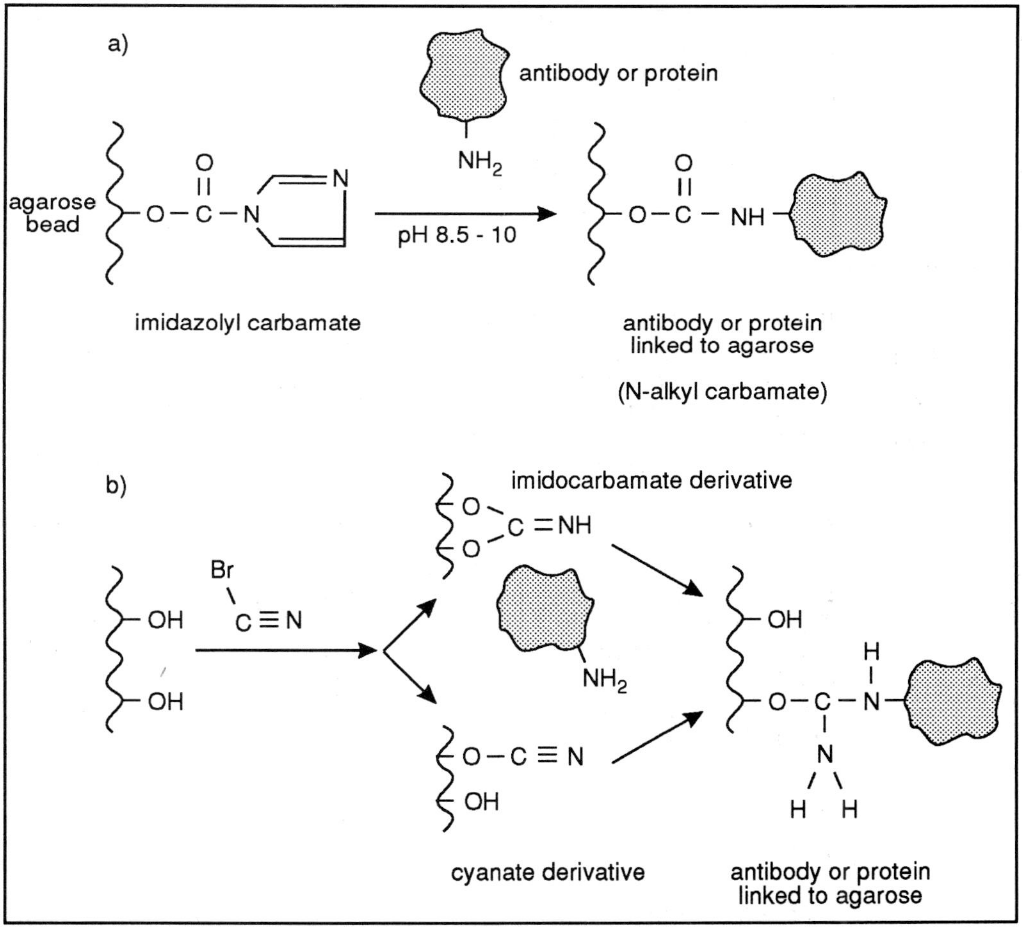

Figure 6.10 Coupling of antibodies or antigens to activated beads.

activated beads

Reacti-Gel

CNBr activated Sepharose

Two of the more commonly used examples of activated beads for coupling to antibodies or proteins available commercially are Reacti-Gel and CNBr activated Sepharose. Reacti-Gel (Pierce Chemicals) is a cross linked beaded agarose which has been reacted with 1,1'-carbonyldiimodazole to produce the agarose derivative imidazolyl carbamate which forms a stable N-alkyl carbamate with available amino groups (such as the epsilon group of lysine) at pH 8.5 - 10. CNBr activated Sepharose (Pharmacia), an extremely popular reagent, is formed by the action of CNBr on hydroxyl groups (vicinol diols) on agarose resulting in imidocarbonate and cyanate intermediates which then react with primary amino groups on proteins to give isourea derivatives (Figure 6.10). The actual coupling procedure is summarised in Figure 6.11. Note that sites on the activated beads not filled with antigen are 'blocked' by reacting them with ethanolamine.

In using any of these activated beads it is advisable to pre-cycle the columns before use with the acid eluting buffer in case there has been any leaching of the antibody or antigen.

Once affinity columns have been prepared, they can be repeatedly used. However they have to be protected from drying and microbial contamination. To achieve this, they are normally sealed and kept in the refrigerator in a neutral buffer supplemented with merthiolate (antimicrobial agent).

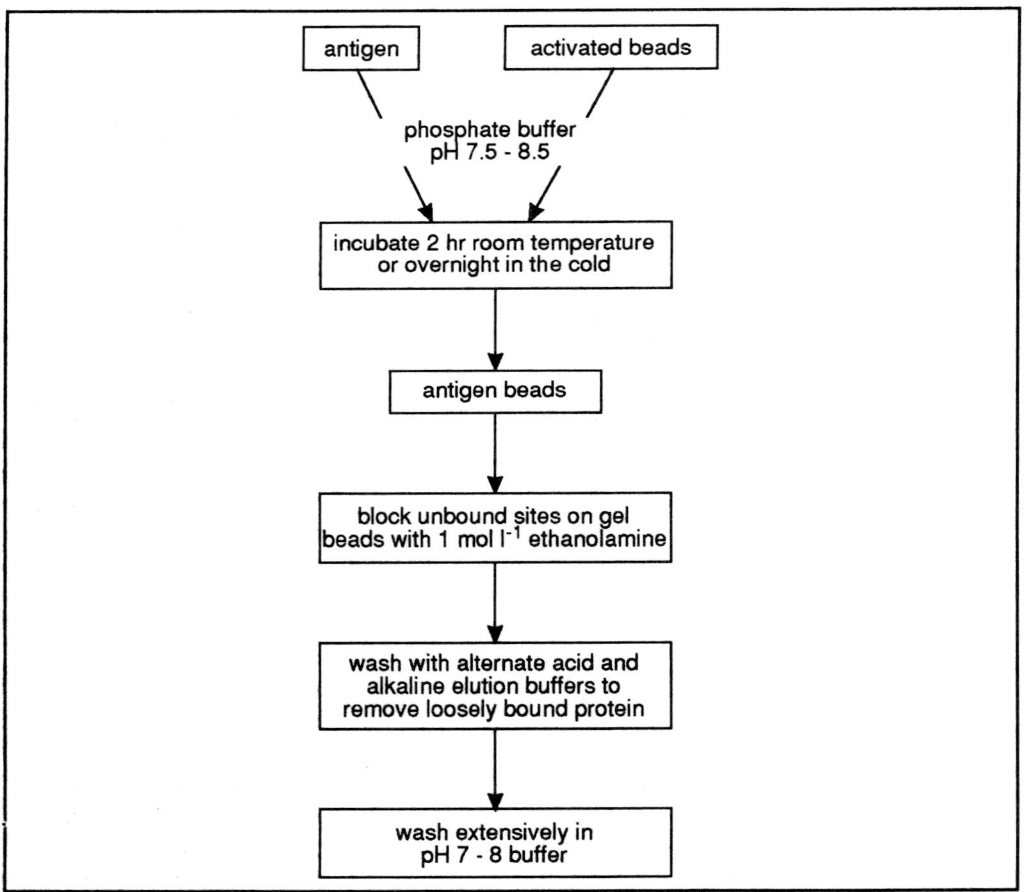

Figure 6.11 Principal steps in coupling of antigen (or antibodies) to activated beads to prepare an immunoaffinity reagent.

Elution of antibodies from antigen affinity columns

The usual method for eluting antibodies from antigen columns (or antigens from antibody affinity columns) is to elute with buffers of extreme pH. The most used method is elution with glycine HCl of pH 2 - 3. We have seen that some antibodies may elute with alkaline buffers. What happens if neither treatment causes elution or if these treatments damage the antibodies or antigens?

chaotropic ions

The second approach is often the use of chaotropic ions. Chaotropes are water structure breakers and are effective in disrupting antigen-antibody interactions. The most commonly used chaotrope is thiocyanate which is used at high molarity (up to 3.5 mol l^{-1}) to elute antibodies or antigens. If this approach does not work, the final approach is to try to elute with agents such as high molar guanidine HCl or urea but you run the risk of denaturing the antibodies.

antibodies of medium affinity

It should be remembered that some antibodies or antigens will elute in less extreme conditions than even glycine HCl buffer and it is a good idea to develop a protocol starting with more gentle conditions before proceeding with buffers of extreme pH. Affinity chromatography is best carried out using antibodies of medium affinity to avoid problems of elution.

SAQ 6.9

Which one of the following statements is correct?

1) Fractionation of human serum using the sequential steps of ammonium sulphate precipitation, gel exclusion and protein A affinity chromatography would result in pure IgG.

2) Antigen affinity chromatography of hybridoma supernatants is never an option to produce pure monoclonal antibodies.

3) Recycling of antigen affinity column means pretreatment with normal serum.

4) Purification of monoclonal IgG from ascites fluid by ammonium sulphate precipitation and anion exchange chromatography results in almost pure Mabs.

5) Treatment of antigen affinity columns with buffers of extreme pH results in elution of all antibodies.

6.3 Subunits of antibodies

During the early investigations into the structure of antibodies, classical experiments showed that antibodies could be reduced to heavy and light chains and cleavage with enzymes resulted in antibody fragments Fab, $F(ab')_2$ and Fc. With the advances in molecular biology and antibody engineering, it would appear likely that these fragments, when needed, will be produced in bacteria thus consigning these earlier methods to the history books. However, we shall briefly describe their production by conventional methods.

6.3.1 Heavy and light chains

Reduction of IgG

This procedure is relatively straightforward. The purified or monoclonal IgG is reconstituted in 0.5 M Tris buffer pH 8 and the reducing agent 2-mercaptoethanol is added until the molarity of the reaction mixture is 0.75 M. This reduction separates the heavy and light chains. To prevent reformation of disulphide bridges the alkylating agent iodoacetamide is added, the pH being maintained with triethylamine at pH 8. On completion of the reaction, the mixture is first dialysed against phosphate buffered saline pH 7.5 and then against 1 mol l^{-1} acetic acid before being fractionated on a Sephadex G100 column (Figure 6.12).

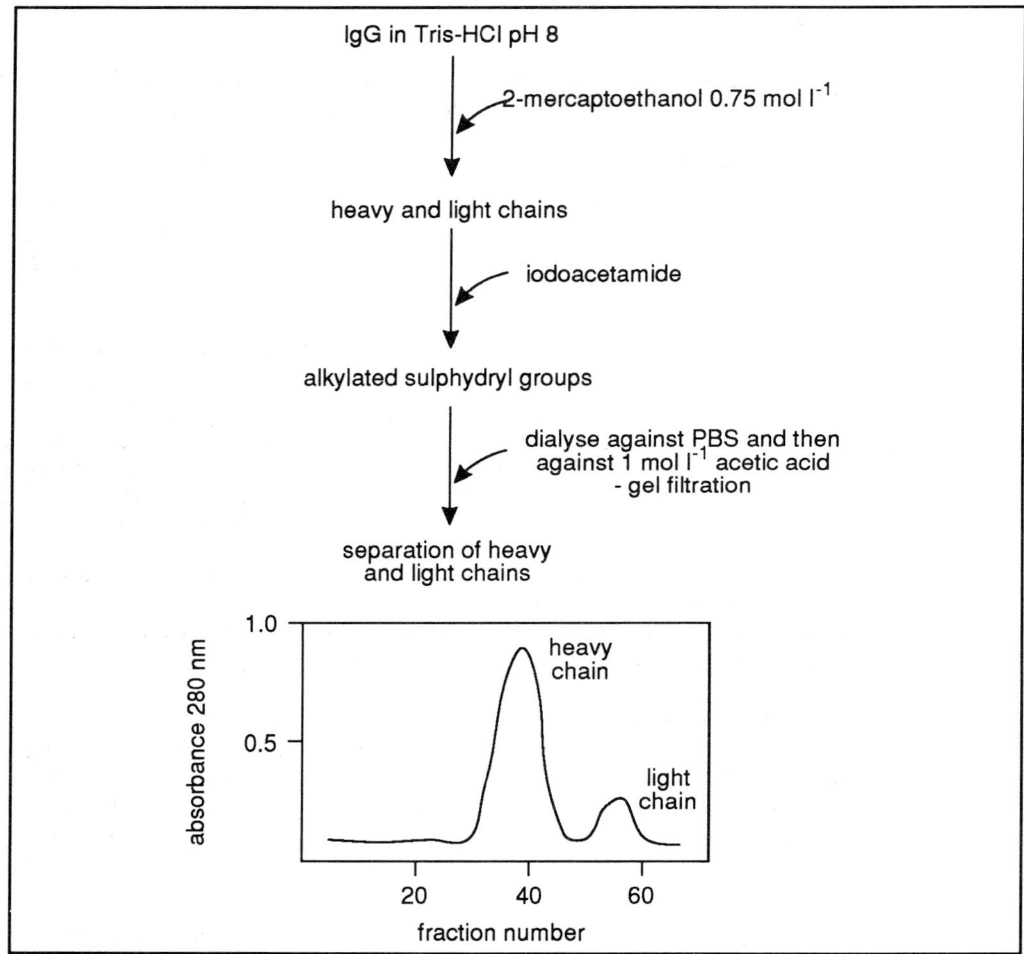

Figure 6.12 Reduction of IgG to heavy and light chains showing the principal steps and a typical elution profile following gel filtration of the reduction products.

6.3.2 Proteolytic cleavage of IgG

pepsin digestion

Pepsin digestion of IgG is normally carried out overnight at 37°C in 0.1 mol l^{-1} sodium acetate corrected to pH 4.5 with acetic acid. To terminate the reaction, the pH is adjusted to pH 7.5. The reaction mixture is then fractionated on a Sephadex G150 column using phosphate buffered saline resulting in absorbance peaks corresponding to undigested IgG (150 000 Daltons), ($F(ab')_2$ (100 000 Daltons), pFc′ and smaller peptides.

papain digestion

Papain digestion of human IgG is carried out at 37°C for 16 hours in phosphate buffered saline pH 7.4 supplemented with cysteine and ethylenediaminetetraacetic acid (EDTA) to activate the enzyme. The reaction is stopped by dialysis against water, the digest is then dialysed against phosphate buffered saline pH 7.4 and fractionated on a Sephadex G150 column to remove IgG which appears in the void volume. The second peak contains both Fab and Fc fragments (50 000 Daltons) which are then passed through a protein A Sepharose column.

Π Can you describe the behaviour of the fragments on this protein A Sepharose column?

The Fab fragments pass through the column and are collected. The Fc fragments bind to the column and are eluted from the column in acid buffer (glycine HCl pH 3).

6.3.3 Bacterially expressed antibody fragments

We discussed at length in Chapter 5 the bioengineering of antibody fragments such as Fab's and scFv's which may eventually replace whole antibodies. We looked at ways in which these proteins could be made in *E. coli* and we will now briefly look at how the gene products are purified.

Site of accumulation of the heterologous proteins in bacteria

inclusion bodies

periplasmic space

Proteins expressed in *E. coli* may be found in a soluble form in the cytoplasm or in insoluble form in inclusion bodies. Alternatively they may be transported across the cytoplasmic membrane and be deposited in the periplasmic space beneath the cell wall. In the latter case, the antibody fragments are often found to fold into their natural state and accomplish the correct disulphide bonding although the levels of production are generally lower than when they are produced in the cytoplasm as inclusion bodies.

In direct contrast, antibody fragments which accumulate in the cytoplasm of *E. coli*, especially in inclusion bodies due to high level expression of cloned genes, are denatured and reduced and once extracted from the bacteria, these proteins need a refolding procedure which promotes correct disulphide bonding.

The ideal solution would be to induce the bacteria to secrete the properly folded antibody fragment into the culture medium from where it could be purified. We shall look briefly at some of these approaches to give an idea how bacterial expression systems for the production of antibody fragments are progressing, a detailed discussion of the experimental approaches is, however, beyond the scope of this book.

Inclusion bodies

There are three steps in the renaturation of antibody fragments from inclusion bodies. Firstly you have to disrupt the bacterial cells, then to solubilise the proteins and finally to promote protein folding.

To disrupt the cells, they can be sonicated on ice in deionised water and the inclusion bodies recovered by high speed centrifugation. After washing, the inclusion body paste is dissolved in either 6 mol l^{-1} urea or guanidine HCl which may include a reducing agent to ensure that all the proteins are in the reduced form. The solubilised proteins may then be subjected to purification steps such as ion exchange chromatography or gel filtration before being dialysed against phosphate buffered saline to remove the denaturant and to effect folding of the protein.

Periplasmic antibody fragments

signal sequence

Pluckthun's laboratory has designed an expression system which eliminates the need for refolding procedures. They constructed a vector which includes special signal sequences which are spliced to the heavy and light variable genes. When expressed in *E. coli*, precursor proteins for both V_H and V_L, each fused to a signal sequence, are synthesised in reduced form in the cytoplasm. The signal sequences mediate the translocation of the precursor proteins into the periplasm where the signal sequences are cleaved off. The variable heavy and light domains then assemble and the correct disulphide bonds are formed in the oxidising environment of the periplasm. Fab fragments have been assembled using a similar system. Purification of the fragments then simply involve lysis of the bacteria, concentration and antigen affinity chromatography. This represents the single step purification of bacterially expressed antibody fragments that we referred to in Section 6.1.3.

protein A antibody fragment fusion protein

Another approach has been to construct a fusion protein of an antibody fragment and protein A. The rationale behind this approach is that protein A is a periplasmic protein and it will direct the fusion product into the periplasm. However, you then need to be able to cleave the protein A from the antibody fragment.

Secretion of antibody fragments by *E. coli*

Π Based on what we discussed so far, can you think of a method by which antibody fragments could be actually secreted by the bacteria?

Since production of fusion proteins made of antibody chains and periplasmic proteins resulted in the antibody fragments being deposited in the periplasm, a fusion protein including an actual secretory protein should result in export of the antibody fragments.

This sort of approach is presently being investigated by various laboratories and uses the *E. coli* haemolysin (Hly) secretory process. The secretion signal sequence has been localised to the last 40 amino acids or so at the carboxyl terminus of haemolysin so fusion proteins including such a signal peptide should promote secretion of antibody fragments. At the moment, it is the actual fusion protein that is secreted from the bacteria so a cleavage step is required to produce the antibody fragments. Also the amount of secreted product is much lower than the amount found in inclusion bodies or periplasmic extracts. Once the fusion protein is cleaved, antigen affinity chromatography should suffice for the purification step.

We have summarised these approaches in Figure 6.13.

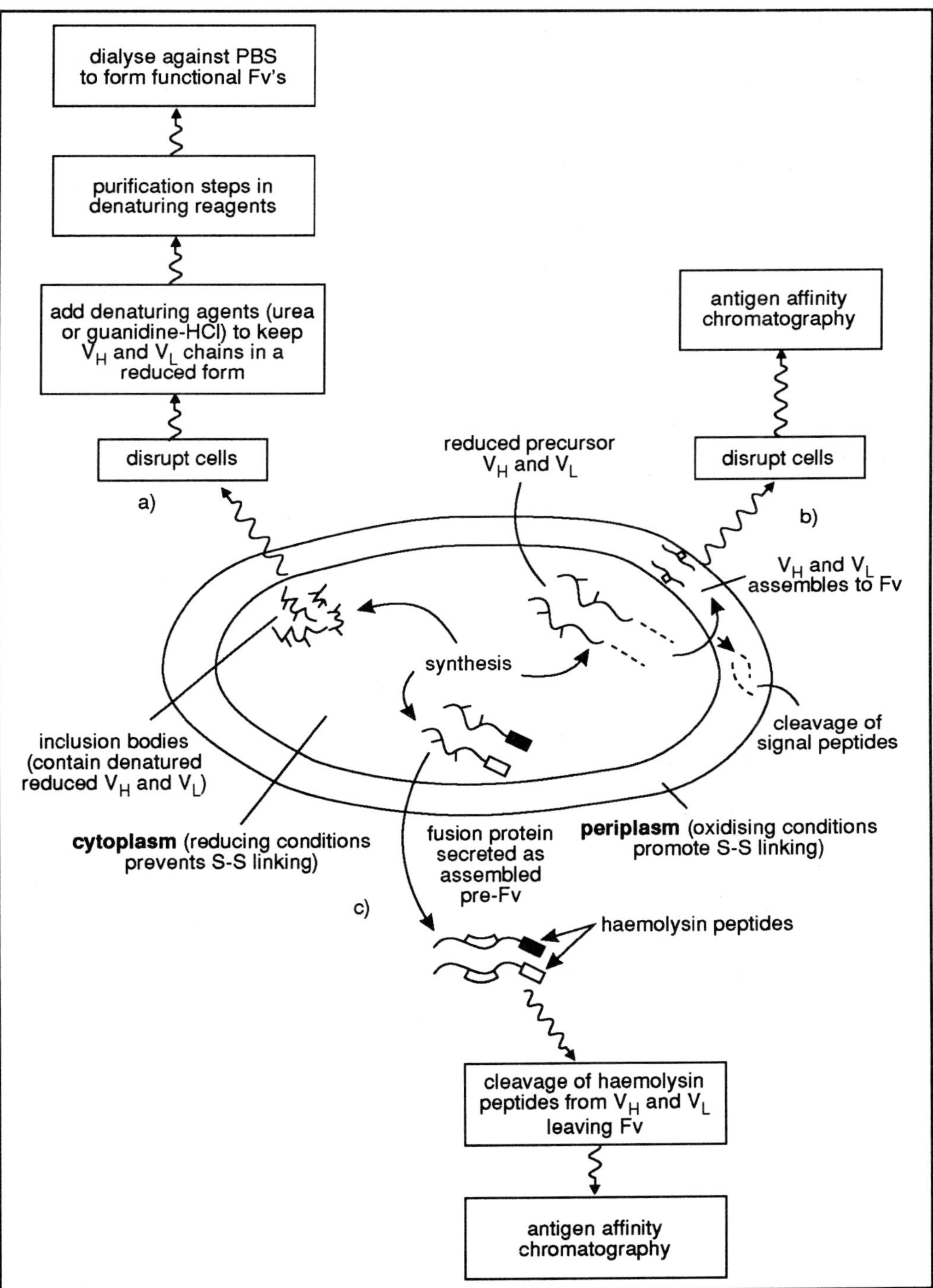

Figure 6.13 Production of functional antibody fragments from *E. coli*. Schematic diagram showing different expression strategies for production of functional Fv fragments, similar approaches could result in $F(ab')_2$ and scFv fragments. a) Represents a current approach to produce antibody fragments from inclusion bodies. b) Shows a more recent development to produce functional Fv from the periplasm. c) Shows the latest attempt to produce secreted functional Fv from bacteria. Boxes and wavy arrows indicate the laboratory steps to purify the antibody fragments. Note that in c) chimaeric (fusion) proteins are produced containing V_H and V_L joined to haemolysin peptides.

SAQ 6.10

Examine Figure 6.13 and make sure you understand the three different approaches depicted; then attempt the question. Which of the following conclusions are supported by the experimental evidence depicted in Figure 6.13?

1) Functional Fv's are not created in the cytoplasm.

2) Conditions in the periplasm promote correct assembling of antibody fragments.

3) Protocols b) and c) avoid use of denaturing reagents.

4) The yield of functional Fv's per unit weight of expressed protein is lowest in protocol a).

5) The haemolysin signal peptides are not cleaved from Fv during secretion.

6.4 General methods for handling purified antibodies

We shall conclude this chapter with some essential information on how to quantitate antibodies, determine their degree of purity and how to store the antibodies that have been purified. Most of these methods can be found in detail in good biochemistry practical books.

6.4.1 Quantitation of antibodies

Absorbance at 280 nm

This method is based on the absorbance of ultraviolet irradiation by proteins which is maximal at 280 nm due mainly to the two amino acids tryptophan and tyrosine.

Π Would you expect the absorbance to vary from protein to protein?

Since the absorbance of UV is mainly due to the presence of tryptophan and tyrosine, it will depend on how many of these amino acid residues each protein possesses. For instance, a 1 mg ml^{-1} solution of bovine serum albumin gives an absorbance reading of 0.7 on the spectrophotometer whereas IgG at the same concentration gives a reading of 1.35. Since these readings have been taken using a 1cm path length cell, these figures represent the absorbance coefficients for these two proteins.

absorbance coefficients

You should only use this assay for highly purified antibodies. The assay has a range of sensitivity from 20 μg to 3 mg ml^{-1}.

Colorimetric assays

The Bradford assay is one of the most reliable and accurate proteins assays. Others include the Lowry assay and the Bicinchoninic acid assay.

Bradford assay

The Bradford assay is a colorimetric assay based on the staining of the proteins by Coumassie brilliant blue G-250 dye. 100 mg of the dye are dissolved into 100 ml 85%

phosphoric acid and 50 ml of 95% ethanol, the volume is then made to 1 litre with cold water. Before use, 0.05 ml of 1 mol l^{-1} NaOH is added per ml of reagent. 1 ml is then added to 0.02 ml of antibody sample in phosphate buffered saline and after 5 minutes the colour is read at 590 nm on the spectrophotometer in glass or polystyrene cuvettes. The antibody concentration is determined by reference to a standard curve which is linear between concentrations of 0.2 mg and 1.5 mg ml^{-1}.

The ELISA

You will notice that the methods described are not very sensitive techniques and will measure the total protein in the sample, including impurities. Once you have a pure sample of antibody or antibody fragments development of an ELISA assay is advantageous since it can detect much lower levels of the antibodies and it is possible to measure the actual antibody concentration in the presence of impurities. If, therefore, the total protein concentration is known an ELISA enables us to determine the percentage purity of the sample.

6.4.2 Determining the purity of antibodies

definition of purity

When you have purified the antibodies or antibody fragments you will need to determine whether there are any impurities present. What we mean by purity is that there are no detectable impurities present in the sample and this definition obviously depends on the method by which you assess purity. This assessment is also dependent on the amount of sample used in the test. If a test has a lower limit of sensitivity of 20 μg and you assay 40 μg of sample for impurities, these will not be detected if the impurities amount to only 1 μg. Often, we do not need pure antibodies, we simply need to ensure that the minor impurities do not adversely affect the particular assay we are using the antibody for.

SDS polyacrylamide gel electrophoresis

sodium dodecyl sulphate (SDS) polyacrylamide gel electrophoresis (SDS-PAGE)

One of the most reliable assays for determining the purity of antibody samples and eminently suitable for assaying the purified antibodies we have been discussing, is sodium dodecyl sulphate (SDS) polyacrylamide gel electrophoresis (SDS-PAGE). Since nanogram quantities of protein can be detected using this method, it is the method of choice for determining the purity of antibodies. As is known, the distribution of negative and positive charges on different proteins varies enormously and when they are subjected to electrophoresis under alkaline conditions they move towards the anode at different rates fairly independent of their molecular weights. If, however, you impose an overwhelming negative charge on all the proteins, the charge-mass ratio becomes constant and the mobility in the gels will be inversely proportional to their molecular weight. This is the basis of SDS-PAGE.

charge-mass ratio

SDS-PAGE is based on the observation that when proteins are heated to 100°C in the presence of SDS and a reducing agent such as 2-mercaptoethanol or dithiothreitol, the proteins unfold and bind SDS in the ratio of about 1 SDS molecule for every two amino acids (about 1.4g SDS per gram of protein). Because of this high ratio of SDS to protein, the charge-mass ratio is constant for most proteins and there is a linear relationship between distance travelled in the gel and the logarithm of the molecular weight.

1.4g SDS per gram of protein

Π How would you expect IgG to behave in SDS-PAGE?

Under the reducing conditions, IgG will be reduced to heavy and light chains and we observe two bands in the gel of respectively about 25 kD and 50 kD. If we have a pure sample of IgG, these will be the only two bands in the gel.

TEMED

ammonium persulphate

Polyacrylamide is a polymer of acrylamide monomer stabilised by cross links of N, N'-methylene bis-acrylamide. Formation of the gel matrix is catalysed by TEMED (tetramethylethylene diamine) which causes the generation of free radicals from ammonium persulphate which then catalyse gel polymerisation. The gel acts as a molecular sieve, the pore size decreasing as you increase the concentration of acrylamide. Thus, by changing the acrylamide concentration you can make gels of various molecular weight fractionation ranges.

stacking and resolving gels, gel slabs, molecular weight standards

SDS-PAGE is done in gel slabs which allow analysis of multiple samples and the inclusion of sets of molecular weight standards to calibrate the gels. Thus the molecular weights of the antibodies or the antibody fragments can be estimated from their migration distances in the gel. Separation of the proteins is done through two gels, a lower resolving gel and an upper stacking gel, a system developed by Laemmli.

stacking phenomenon

The samples are applied to the top of the stacking gel in wells formed during polymerisation.

Π Examine Figure 6.14. You will notice that the acrylamide content of the stacking gel is low. What is the significance of this?

chloride ions

glycinate ions

Since this is a stacking gel, we want to avoid any sieving action, since this would retard the higher molecular weight proteins so we make the acrylamide content low. At the start of the electrophoretic run, the proteins enter the stacking gel preceded by fast moving chloride ions from the gel buffer and followed by glycinate ions from the electrode buffer. Development of a voltage gradient between these two layers of ions causes the SDS proteins to be concentrated into a thin zone or stack. Thus, irrespective of the sample volume, all the proteins enter the resolving gel together and begin separation at the same time. You will notice that the resolving gel has high acrylamide content which varies depending on the fractionation range of the gel.

Once in the resolving gel, the progress of the proteins is retarded due to the smaller pore size, the larger size proteins being slowed more than the smaller ones, resulting in separation based on their molecular weights.

The practical aspects of SDA-PAGE are dealt with in the BIOTOL text 'Analysis of Amino Acids, Proteins and Nucleic Acids'.

On completion of electrophoresis, the proteins are detected by staining the gel with Coumassie Blue, silver stain or copper stain. The distances migrated by each protein band from the top of the resolving gel are measured and molecular weights determined by reference to a calibration curve - Figure 6.14d).

You will realise that the bands represent migration of single protein chains due to the reducing conditions of the sample buffer. You can omit the reducing agent from the sample buffer to obtain molecular weights of whole molecules, for instance whole IgG.

Π Would you expect the molecular weights of proteins obtained under non-reducing conditions to be as accurate as those obtained under reducing conditions?

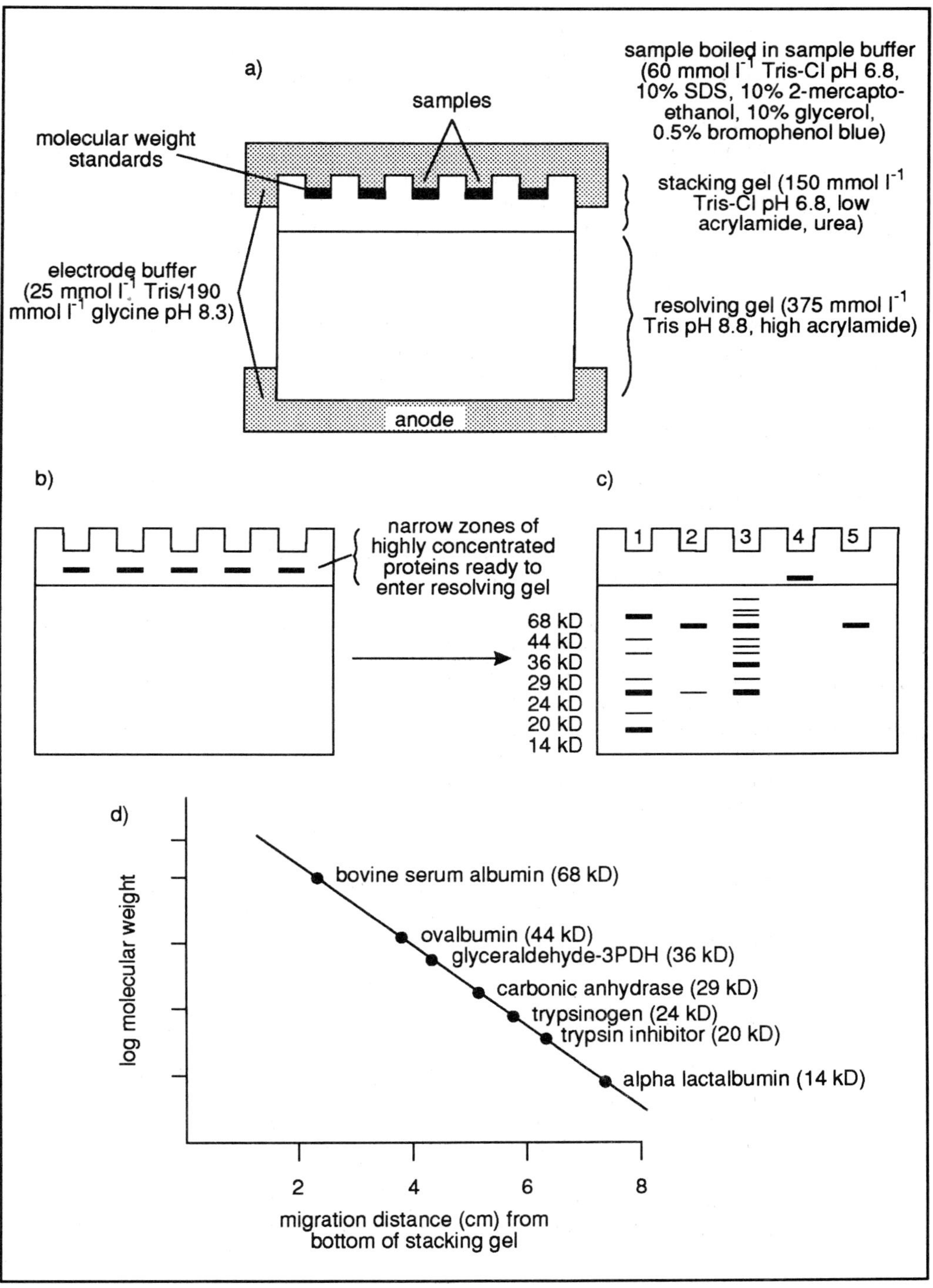

Figure 6.14 SDS-PAGE showing typical experimental conditions using a 10% acrylamide gel (fractionation range 20 - 100 000 Daltons). a) A gel set-up at the start of electrophoresis with samples and standards in wells, major components of individual buffers shown. b) Formation of narrow zones of high protein density in stacking gel in the early stages of electrophoresis. c) Band patterns of sample and standard marker proteins on completion of electrophoresis and gel staining. Well 1 - molecular weight standards. Well 2 - pure IgG. Well 3 - 40% ammonium sulphate precipitate of serum. Well 4 - non reduced IgG. Well 5 - scFv. d) Calibration curve using log molecular weights of standards versus distance travelled in the resolving gel. The stacking and resolving gels and the electrode buffer all contain low amounts of SDS.

Since the proteins will not fully unfold unless the S-S bridges are cleaved, the amount of SDS binding to individual molecules will vary. This will affect the distance migrated through the gel and therefore the accuracy of the molecular weights derived under non-reducing conditions is less than that of those determined under reducing conditions.

SAQ 6.11

Examine Figure 6.14 and then explain the reasons for the inclusion of the following reagents.

1) Urea in the stacking gel.

2) Glycerol in the sample buffer.

3) Bromophenol blue in the sample buffer.

4) Glycine in the electrode buffer.

6.5 Storage of antibodies

6.5.1 Denaturation, degradation and contamination

After we have spent many hours purifying antibodies or antibody fragments, we need to be able to store them for long periods without significant deterioration since a single preparation is likely to last a long time as so little is used for each assay.

The likely problems we will face in storing antibodies is their denaturation, their degradation with proteases and contamination with bacteria and fungi.

Denaturation

If we expose antibodies to denaturing reagents such as urea or guanidine HCl or chaotropic agents we should ensure that these are removed by dialysis as soon as possible. Similarly, extremes of pH which denature some antibodies quite rapidly, should be corrected quickly, preferably using neutralising buffers. Most antibodies can survive heating up to 60°C but we should avoid higher temperatures which induce irreversible damage.

One of the commonest causes of denaturation is something that can be avoided - that of repeated freezing and thawing which results in irreversible aggregation with many antibodies. Antibodies should always be stored in small aliquots. Once a small sample of antibody is thawed, it should be stored in the refrigerator and used until it is exhausted. Most antibody solutions survive in refrigerator temperatures for up to 6 months. This avoids denaturation due to freezing and thawing.

The general rules for storage are to store at neutral pH, preferably in concentrated solutions (1-10 mg ml^{-1}). If this cannot be done, add 1% bovine serum albumin. The pH of hybridoma supernatants should be corrected by addition of 1 mol l^{-1} Tris pH 8.

Degradation and contamination by microbes

proteases

Serum samples may have residual protease activity although this is unlikely to be a problem with purified polyclonal or hybridoma antibodies. However, proteases can also arise due to contamination from microbes. We can combat these by adding sodium azide (0.02%) or methiolate (0.005%) to the samples or by filter sterilising them followed by freezing to -20°C or preferably -70°C. Lastly, antibody conjugates are best stored at 4°C since they are very susceptible to aggregation on freezing.

Summary and objectives

You have now learned how to purify antibodies from a variety of sources including serum, hybridoma supernatants and bacterial extracts. We described ammonium sulphate precipitation which results in the partial purification of antibodies from other serum proteins and which is also used to concentrate antibodies such as are found in hybridoma supernatants. We discussed the advantages of anion exchange methods and affinity chromatography for the purification of IgG, size exclusion chromatography particularly for IgM and the use of antigen affinity methods to produce monospecific antibodies. We also examined methods for the production of subunits of IgG and bacterially expressed antibody fragments. Finally, we described the best approaches for the storage of antibodies.

Now that you have completed this chapter you should be able to:

- develop protocols for purification of different antibody classes based on their distinctive properties;
- calculate the amount of saturated ammonium sulphate which will precipitate antibodies from a given volume of serum or supernatant;
- describe the use of anion exchange chromatography for the purification of IgG;
- develop a protocol for purification of IgM;
- manipulate data from size exclusion chromatography to calculate apparent molecular weights of separated proteins;
- list the major steps in protein A/G affinity chromatography and discuss the limitations of this method for separation of IgG subclasses from Man, rat and mouse;
- develop protocols for the use of antigen affinity columns for purification of monospecific antibodies and antibody isotypes;
- describe the various approaches for production and purification of bacterially expressed functional antibody fragments;
- discuss the use of SDS-PAGE to determine the purity of antibodies;
- describe ways of storing antibodies to prevent denaturation, degradation and contamination.

Cytokines

Cytokines

7.1 Introduction - the basic aspects of cytokines

Up to this point we have learned a lot about antibodies, their structure and biological function and about the way they are produced. In the last chapter we shall discuss their uses and potential mainly as immunotherapeutic agents in combatting disease. However, the text would not be complete without reference to a second major class of immunochemicals, the cytokines, a much more recent discovery, which appear to have as much potential in immunotherapy. In this chapter, we shall make you familiar with the basic characteristics of cytokines, their functions and the way they are produced and assayed. In Chapter 8 we will discuss their potential along with antibodies in immunotherapy.

This is, again, a long chapter so do not attempt to complete it all in one sitting.

7.1.1 Supernatants from activated cells contained cytokines

soluble factors

cytokines

In the 1960s, the first reports appeared in the literature suggesting that following activation by antigens or mitogens, T cells secreted soluble factors into the culture medium *in vitro* which mediated a variety of effects on immune cells. We now call these factors cytokines.

macrophage migration inhibition factor

One cytokine inhibited the migration of macrophages in culture dishes and was called macrophage migration inhibition factor or MIF (see Figure 7.1). The same supernatant contained another cytokine which caused the non-antigen specific stimulation of lymphocytes causing them to divide (mitogenesis) and this factor was called mitogenic factor (MF).

The demonstration of MIF is based on using peritoneal exudate cells (PEC). These are placed in a capillary tube in a petri dish containing antigen embedded in agar. Macrophages present in the PEC normally migrate out of the tube forming a fan of cells which can be seen at the end of each tube. MIF prevents this migration of macrophages and they remain inside of the capillary tube.

The presence of mitogenic factor (MF) is based on examining the proliferation of lymphocytes. In Figure 7.1 we show that both MIF and MF can be demonstrated in the supernatants of lymphocytes which have been inactivated with antigen *in vitro*.

Within the space of a few years, experiments reported in the literature suggested that there could be more than 100 cytokines based on distinct biological activities in supernatants derived from not only T cells but other lymphoid and non-lymphoid cells as well.

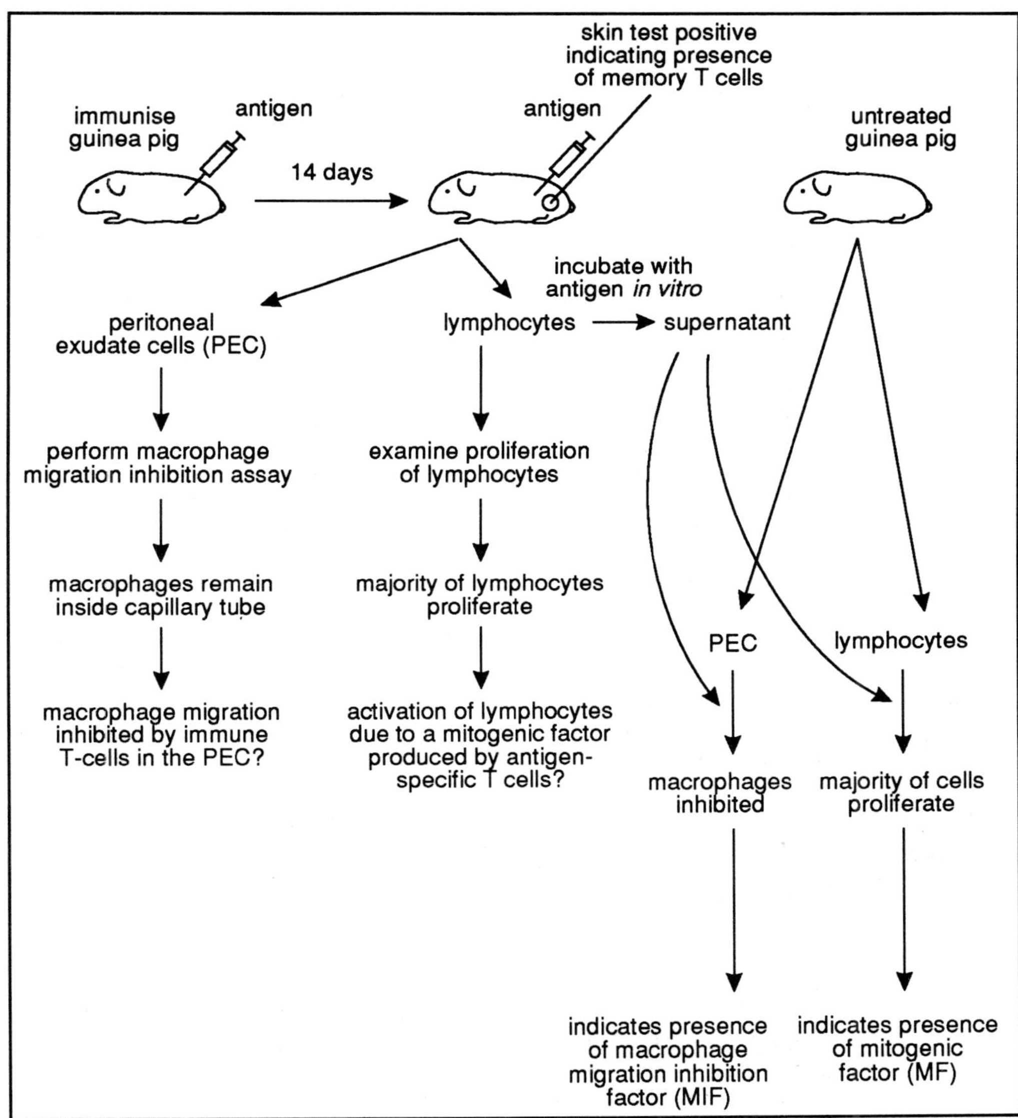

Figure 7.1 The first demonstration of cytokines. The detection of MIF and MF produced by memory T cells in supernatants.

SAQ 7.1

Which of the following conclusions can be made from the experimental data in Figure 7.1?

1) Lymphocytes from un-immunised guinea pigs produce mitogenic factor.

2) Lymphocytes removed from immunised guinea pigs produce active supernatants without further stimulation.

3) Antigen stimulation results in some expression of antigen non-specific effects.

4) The experiment showing macrophage migration inhibition of PEC from immunised guinea pigs demonstrates the production of MIF.

5) Experiments using supernatants from immune lymphocytes demonstrate the presence of MIF and MF released by T cells.

cytokines are produced in small quantities, an non-antigen specific, biological activity

You should remember that the majority of these factors acted in an antigen non-specific manner although they were produced by antigen specific T cells. At this stage there was no information on the biochemical nature of the cytokines as they were present in the supernatants in extremely small quantities and the only way to detect them was by some biological activity they exerted on a particular cell type.

7.1.2 Cell lines, monoclonal antibodies and gene cloning

Π Before you read on, what do you think were the major problems associated with further characterisation of individual cytokines found in supernatants?

As you will have gathered so far, there were major obstacles to further characterisation of the cytokines. Firstly, as we implied earlier, the presence of many different biological activities in the supernatants does not necessarily reflect the presence of the same number of cytokines and some cytokines could possibly enhance or suppress the biological activity of other cytokines. So the major objective was purification of individual cytokines.

This leads us to the second obstacle. The concentration of individual cytokines in supernatants was extremely low and could not be detected biochemically. This means that there were potentially only very small amounts of each cytokine available for analysis even after extensive purification and concentration steps, and remember, each purification step has to be accompanied by analysis of each fraction for biological activity. You will appreciate that this was a mammoth task!

monoclonal antibodies

gene cloning

However, the discovery of cell lines which produced much higher amounts of some cytokines facilitated the production of monoclonal antibodies to cytokines using hybridomas, a technology also introduced at about this time. This meant that cytokines and their receptors could by affinity purified and information on their amino acid sequences could then be obtained. Another important development was gene cloning techniques which led to the cloning, sequencing and expression of cytokine genes and a veritable explosion of knowledge on cytokines.

7.1.3 Cytokines are involved in all aspects of immunity

Irrespective of the number of individual cytokines, we now accept that these are intercellular messengers which control communication between immune cells in all aspects of immunity. Hence they control the development of cells in the bone marrow (haematopoiesis), the trafficking of immune cells, the activation and clonal expansion of T and B cells, the production of antibodies by B cells, cytotoxic activities of T cells and the activities of macrophages and other lymphoid cells.

Their involvement in all aspects of immunity is further emphasised in Table 7.1 where we have designated the principal immune activities with which each cytokine is associated. The usefulness of this information is two fold.

Firstly, it introduces you to the wide variety of cytokines and their names and secondly, gives you an idea of what sorts of roles the cytokines play in immunity. You will notice that many cytokines perform more than one function.

Immune activity	**Cytokine involvement**
haematopoiesis (development of blood cells including T/B cells, polymorphs, red cells and platelets in the bone marrow)	interleukin 3 (IL-3) interleukin 6 (IL-6) interleukin 7 (IL-7) interleukin 11 (IL-11) interleukin 9 (IL-9)? macrophage colony stimulating factor (M-CSF) granulocyte CSF (G-CSF) granulocyte macrophage CSF (GM-CSF)
thymic development of T cells	interleukin 2 (IL-2)
mediate natural immunity	interferon-alpha (IFN-α) interferon-beta (IFN-β) interleukin 1 (IL-1) interleukin 8 (IL-8) tumour necrosis factor - alpha (TNF-α) tumour necrosis factor - beta (lymphotoxin - TNF-β)
activates inflammatory cells such as macrophages, neutrophils, eosinophils, mast cells	interferon-gamma (IFN-γ) TNF-α TNF-β interleukin 5 (IL-5) macrophage migration inhibition factor (MIF)
activates/regulates T cells (T_H, T_C) and B cells	IFN-γ interleukins 1,2,4,5,6,9,10,12 TNF-α
regulate class switching in B cells	IFN-γ IL-2, IL-4, IL-5, IL-12, IL-13 transforming growth factor - beta (TGF-β)
regulates MHC expression	
increased expression of Class I	IFN-α, IFN-β, IFN-γ, TNF-β
increased expression of Class II	IFN-γ, IL-4

Table 7.1 Some examples of cytokine involvement in immune mechanisms. Many of these purported involvements are mainly based on *in vitro* experiments and may not have any relevance to the physiological situation. Additionally, many cytokines synergise with each other but not directly promote these activities. The list only includes those cytokines we are fairly sure about. Current literature may suggest other cytokines as well.

T helper cell

immuno-therapeutic agents

The major cytokine producer is the T helper cell but many other cells produce them as well. Because of their central role in all aspects of immunity, cytokines are being presently targeted as immunotherapeutic agents and once you have learned about the basic characteristics of cytokines and the ways they are produced and assayed, we shall introduce you to some of their potential clinical uses in the last chapter.

7.1.4 Nomenclature of cytokines

lymphokine

cytokine

monokine

Lymphokine was the first name which was coined to describe these biological activities because the majority were found in supernatants of lymphocytes. Two other names you should be aware of are monokine and cytokine, the former describing soluble mediators produced by macrophages (monocytes), the latter describing all mediators produced by lymphoid and non-lymphoid cells. More recently, it has been shown that some of the so called lymphokines can be produced by cells other than lymphocytes and the name cytokine is probably the most acceptable description. As you can see by the heading of this chapter, it is the name we shall adopt.

acronyms

The individual cytokines were originally given a name which reflected the biological activity they exerted on a target cell and these names were abbreviated to acronyms. Examples of these are MIF (macrophage migration inhibition factor) and MF (mitogenic factor) we described in Section 7.1.1. Other examples include LAF (lymphocyte activation factor), produced by macrophages which activate lymphocytes, quite a few TRFs (T cell replacing factors) which could substitute for T cells in *in vitro* assays and TCGF (T cell growth factor) which supported the clonal expansion of T cells.

interleukin

In 1979 it was decided to introduce a new term, interleukin, to describe some cytokine activities as it was realised that each biological activity reported in the literature was not due to different cytokines but as different effects mediated by the same cytokine. For instance, TCGF, killer helper factor (KHF), co-stimulator (CS) and thymocyte mitogenic factor (TMF) were four cytokines reported by different laboratories which all exerted effects on T cells resulting in clonal expansion. It was decided that these activities were all due to the same cytokine and this was called interleukin 2 (IL-2).

Other changes in the nomenclature which gave rise to the first 6 interleukins are shown in Table 7.2 and the rules governing naming of any future interleukin are shown in Table 7.3. Examine these tables carefully before reading on.

Cytokine	Molecular weight kD	Major activities	Other previous names
interleukin 1 (IL-1) (two types IL-1α and IL-1β) source: macrophages, fibroblasts, B cells and other cells	17	1) promotes T/B cell growth 2) induces IL-2 production 3) many other effects on a wide spectrum of cells/tissues 4) major role in tissue trauma	lymphocyte activating factor (LAF) mitogenic protein (MP) T cell replacing factor III (TRF-III) endogenous pyrogen (EP) B cell differentiation factor (BCDF) etc
interleukin 2 (IL-2) source: T cells	15	1) supports long term growth of T cells 2) promotes thymic maturation of T cells 3) supports growth / differentiation of B cells 4) promotes growth and killer activity of natural killer cells	T cell growth factor (TCGF) killer helper factor (KHF) co-stimulatory factor (CS) thymocyte mitogenic factor (TMF)
interleukin 3 (IL-3) source: T cells, mast cells, stromal cells	14-28	1) major role in haematopoiesis	multi-potential colony stimulating factor haemopoietic cell growth factor (HPGF) mast cell growth factor (MCGF) haemopoietin 2 (HP-2)
interleukin 4 (IL-4) source: T cells, mast cells, stromal cells	20	1) promotes T/B cell growth 2) growth factor for mast cells 3) up-regulates MHC II	B cell stimulation factor 1 (BSF-1) T cell growth factor II (TCGF-II) mast cell growth factor II (MCGF-II)
interleukin 5 (IL-5) source: T cells, mast cells	45	1) promotes murine B cell proliferation and differentiation to plasma cells 2) supports eosinophil development in bone marrow	T cell replacing factor (TRF) B cell growth factor (BCGF-II) eosinophil differentiation factor (EDF)
Interleukin 6 (IL-6) source: T cells, mast cells, macrophages, fibroblasts	21-28	1) stimulates antibody synthesis 2) promotes T cell activation 3) major role in tissue trauma	B cell stimulation factor 2 (BSF-2) B cell differentiation factor (BCDF) hybridoma/plasmacytoma growth factor (HPGF)

Table 7.2 Nomenclature of cytokines showing some of the previous names of interleukins 1-6 and their major activities. The major sources of each cytokine are also shown in the left column.

SAQ 7.2

Study Table 7.1 and 7.2 and then match the groups of cytokines in the left column with an immune activity from the right column with which each cytokine in the group is associated. Use each item only once.

1) IL-3, IL-7, IL-11	a) antibody synthesis
2) IFN-γ, IL-4	b) inflammation
3) IL-2, IL-4, IL-6	c) MHC II expression
4) IL-4, IL-5, IL-6, IL-13	d) haematopoiesis
5) IL-1, IL-6, TNF-α	e) T cell growth

1)	The molecule shall have been purified, cloned and expressed.
2)	The nucleotide and inferred amino acid sequence must be different from that of any other interleukin or molecule.
3)	The molecule must be a product of immune cells and must mediate an important function in immunity (if the molecule only mediates a single function it may be more appropriate to assign a non interleukin descriptive name).
4)	If the molecule is a member of a family of molecules which exert their main influence outside the immune system, its designation should be consistent with the other family names rather than being described as an interleukin.

Table 7.3 The recommended criteria for the naming of interleukins as established by the World Health Organisation - International Union of Immunology Societies.

The name interleukin was intended to describe those cytokines which were produced by leukocytes and acted on leukocytes. For instance, IL-2 is produced by T cells and acts primarily on T cells and B cells. There are now 13 interleukins which have been characterised and we shall briefly introduce you to each one of them. Additionally, there are other cytokines we shall talk about such as interferons, colony stimulating factors, transforming growth factors and tumour necrosis factors all of which play central roles in some aspect(s) of immunity.

7.2 General properties of cytokines

Π Based on what you have read so far, make a list of what you consider would be the principal characteristics of cytokines in their roles as intercellular messengers. You can check your response with the list below.

7.2.1 Common properties of all cytokines

Cytokines exhibit quite a number of common features which are listed as follows:

- they are produced during the effector phases of natural and acquired immunity;
- they regulate development, activation, differentiation and effector functions of immune cells;
- they are produced by a variety of cells and act on more than one type of cell;

- they are produced in extremely small quantities (nanomolar or less) and are effective at extremely low concentrations;
- cytokine action is mediated through ligand receptor interactions;
- T cells (especially T helper cells) are the major producers of cytokines but many other cells also produce them.

7.2.2 Modes of action of cytokines

cytokines are pleotropic

synergism

There are some definitions relating to the modes of action of cytokines which you should know. Most cytokines are pleotropic which means they act on more than one cell type. The majority of cytokines also exhibit synergism which means that cytokines act in concert with each other. For instance, a cell produces one cytokine which binds to a receptor on another cell and this induces production of another cytokine or expression of a receptor for a second cytokine.

autocrine and paracrine activities

Cytokines generally act only over short distances and exhibit autocrine or paracrine activities. Autocrine activity is when the cytokine acts on the cell which produces it whereas paracrine activity involves acting on a neighbouring cell. Cytokines rarely exhibit endocrine activity which means acting on cells distant from the producer cell. A possible exception is IL-1 which is the most hormone-like cytokine we know.

most cytokines are non-antigen specific

Another important feature of cytokine action is that they are not antigen specific. For instance, once activated, a T helper cell produces a battery of cytokines which act on B cells promoting their activation, clonal expansion and differentiation into antibody producing cells. The same cytokines are released by many T helper cells irrespective of their antigen specificity. Some of these same cytokines can also be found affecting different cell mediated responses or even development of blood cells in the bone marrow which, by the way, is a process totally independent of antigen.

There are, however, soluble mediators produced by antigen specific T helper and suppressor cells which only act on cells responding to that particular antigen - these are the so called antigen specific factors. Despite their discovery some time ago, the role(s) of such factors have still not been elucidated and they are not generally regarded as cytokines.

SAQ 7.3

Which one of the following statements is correct?

1) Cytokines are only produced following antigen stimulation of T cells.

2) Cytokines exhibit pleotropism, that is, many cell types produce the same cytokine.

3) Most cytokines act in an endocrine fashion.

4) Many T helper cells secrete the same battery of cytokines irrespective of their antigen specificity.

5) A cytokine which has been purified from a supernatant can now have an interleukin designation.

7.2.3 Cytokines and immunity

Before we go on to discuss the production and assay of cytokines we must introduce you to most of the players - we have only introduced you to 6 of them so far (IL-1 to IL-6). However, it is obviously impossible to describe the biochemistry and biology of each cytokine in depth within the confines of this one chapter so we have simply indicated the sizes and major sources of each cytokine and the major biological properties of each in Tables 7.2 (interleukins 1-6), 7.4 and 7.5 (the remaining cytokines).

Cytokine	Molecular weight kD	Major source	Major activities
interleukin 7 (IL-7)	25 kD	stromal cells	B cell development in bone marrow
interleukin 11 (IL-11)	23 kD	stromal cells	B cell development mega-karyocyte development with IL-3? (this cell gives rise to platelets)
granulocyte-CSF (G-CSF)	20 kD	macrophages endothelial cells fibroblasts	granulocyte development enhances granulocyte activity eg phagocytic activity in neutrophils
macrophage-CSF (M-CSF)	40-50 kD	T cells endothelial cells fibroblasts macrophages	monocyte development and activates mature macrophages
granulocyte-macrophage-CSF (GM-CSF)	14-35 kD	T cells macrophages fibroblasts endothelial cells mast cells	development of granulocytes and monocytes at an earlier stage than M-CSF or G-CSF also activates mature macrophages
interleukin 9 (IL-9)	30-40 kD	T cells (TH2 in mouse)	involved in T cell growth? development of mast cells?
interleukin 8 (IL-8)	9 kD	T cells macrophages fibroblasts endothelial cells etc	chemotactic for T cells and polymorphs particularly neutrophils and monocytes (?)
interleukin 10 (IL-10)	18 kD	T helper cells (TH2) in mouse some B cells and monocytes	inhibits cytokine synthesis particularly IFN-γ by TH1 cells promotes mast cell activation promotes proliferation of some T cells
interleukin 12 (IL-12)	35 kD/40 kD heterodimer	B cells and macrophages	induces IFN-γ and GM-CSF by T cells and natural killer cells enhances cytotoxic activity of NK
interleukin 13 (IL-13)	10 kD	T helper cells TH2 in the mouse	promotes monocyte differentiation promotes B cell proliferation and induces IgM and IgG synthesis

Table 7.4 Colony stimulating factors and interleukins 7 - 13. The first 6 cytokines listed are mainly involved in the development of blood cells from stem cells in the bone marrow. Note CSF = colony stimulating factor.

Cytokine	Molecular weight kD	Major source	Major activities
transforming growth factor-beta (TGF-β)	25 kD	Tcells, B cells, macrophages	generally anti-proliferative promotes Ig class switching to IgA whilst inhibiting production of other isotypes
interferon-alpha (IFN-α) (about 18 functional species)	20 kD	wide variety of cells	antiviral activity, enhance MHC I expression, stimulate natural killer cells inhibits IgE production
interferon-beta (IFN-β)	23 kD	mainly fibroblasts	similar to IFN-α
interferon-gamma (IFN-γ) (immune interferon)	17 -25 kD (homodimer)	T cells and NK cells	modulates nearly all phases of immunity, possesses little antiviral activity, induces/increases MHC I/II on many cells but inhibits MHC II expression on B cells, blocks TH2 responses in mice, activates macrophages resulting in microbicidal activity and cytokine production, inhibits IgE production, induces class switching to IgG_{2a} in mice, enhances TNF-α production
tumour necrosis factor alpha (TNF-α) (cachectin)	17 kD (trimer 51 kD)	T cells, B cells, macrophages, mast cells	activates macrophages along with IFN-γ, increases response of activated T cells, promotes chemotaxis of neutrophils and macrophages
tumour necrosis factor beta (TNF-β) (lymphotoxin)	25 kD	T cells, macrophages	non specific killer of a wide variety of cells some similar effects as TNF-α

Table 7.5 The interferon family of cytokines, tumour necrosis factors and transforming growth factor-beta. TH = T helper cells.

For a detailed discussion of the role(s) of cytokines in such activities as T and B cell activation, cell mediated immunity and the inflammatory response we refer you to an earlier book in this series ('Cellular Interaction and Immunobiology').

7.3 TH1 and TH2 T helper subsets

7.3.1 Cytokine production by TH1 and TH2

two major subsets of T helper cells

To give you a sample of the complex cytokine interactions in immunity we refer you to Figure 7.2 which depicts the major activities of the two major subsets of T helper cells which have been discovered in mice. There is some evidence for a similar division of labour in Man but it is not yet proven. These two subsets TH1 and TH2 are of immense interest since they each possess a distinct battery of cytokines. TH1 produces IFN-γ, IL-2 and TNF whereas TH2 produces IL-4, IL-5, IL-10 and, by recent reports, a new cytokine IL-13. Both subsets produce IL-3 but the significance of that for TH is not known.

Π Look at Figure 7.2 and read its legend carefully. What can you conclude about the types of response TH1 and TH2 appear to be responsible for?

Although it is not totally clear cut, you can conclude that TH1 cells are responsible for cell mediated immune responses and TH2 for a majority of humoral responses.

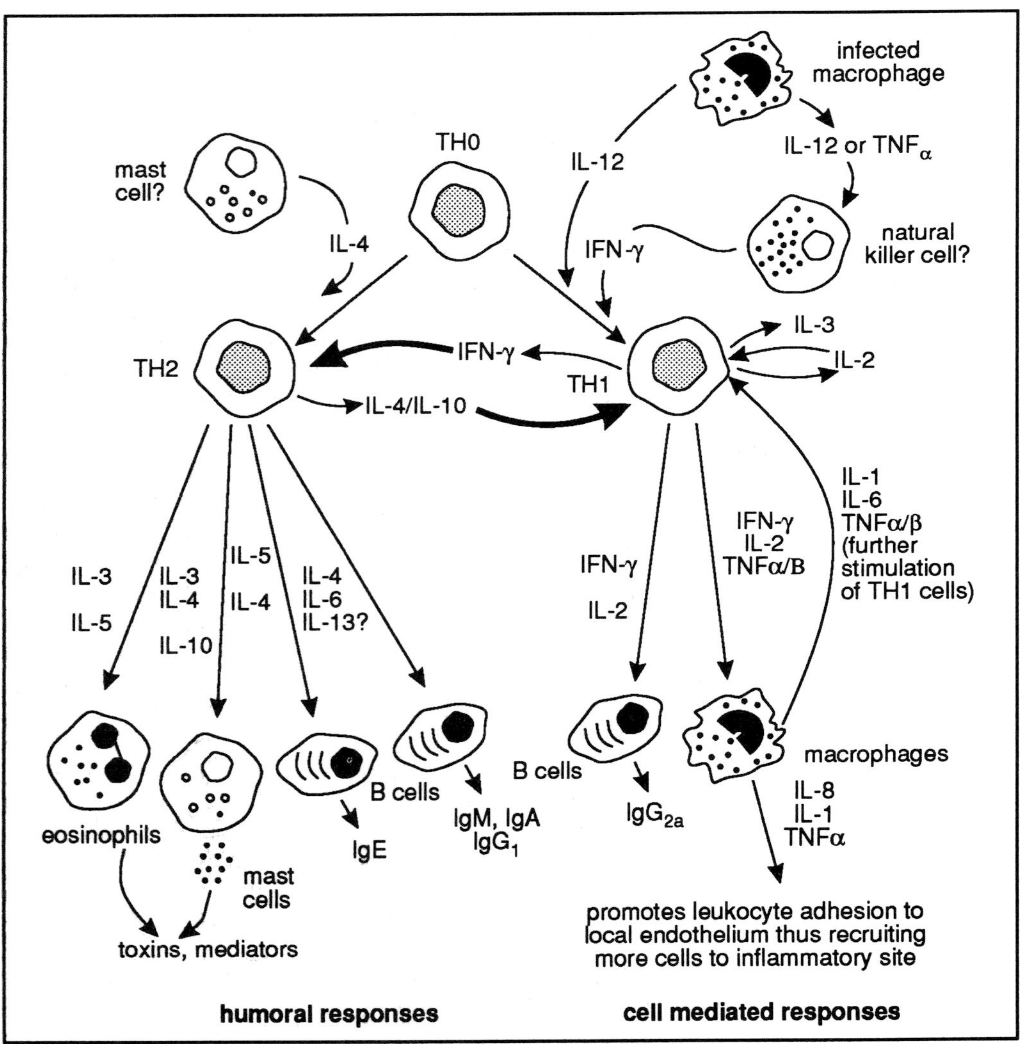

Figure 7.2 Cytokine production and immune activities of the two major subsets of murine T helper cells. This scheme demonstrates synergistic activities of cytokines ie how cytokine release induces production of other cytokines. The scheme shows that TH1 produces IFN-γ, IL-2 and TNFs whereas TH2 produces IL-4, IL-5, IL-10 and IL-13. As can be seen this leads to mainly cell mediated immunity when TH1 cells are activated and a humoral response when TH2 cells are activated. This scheme suggests how TH1 and TH2 cells could be activated although this is not proven. TH1 and TH2 subsets appear to be mutually inhibitory since IFN-γ inhibits activation of TH2 and IL-4 and IL-10 (previously called cytokine synthesis inhibitor factor) inhibit production of IFN-γ (thus blocking all TH1 responses). The inhibition is shown as thick arrows.

7.3.2 TH1 and cell mediated immunity

IFN-γ, IL-2, TNF

You will need to keep referring to Figure 7.2 as you read the text. You will notice that on stimulation by IFN-γ and IL-12, TH1 cells produce IFN-γ, IL-2 and TNFs. These can then act on local macrophages which may be infected with bacteria and they become activated. IFN-γ plays a major part here (it used to be called macrophage activating factor, MAF) inducing bactericidal activity inside the cells possibly by production of nitric oxide and other metabolites and by increasing the numbers of MHC II molecules expressed on the cell.

Π Before you read on, can you think what is the significance of this increased expression of MHC II?

Increasing the numbers of MHC II molecules (and therefore foreign peptides) on the macrophages promotes the further stimulation of TH1 cells. This then causes further production of TH1 derived cytokines which further stimulate the macrophages and so on.

macrophage derived IL-1, IL-6, TNF, IL-8

The macrophage is also stimulated to produce its own set of cytokines which includes IL-1, IL-6 and TNF. These, together with IL-2 produced by the T cells themselves, cause further activation of TH1 cells. Increased numbers of TH1 cells are also recruited to the inflammatory site from the blood stream by macrophage derived cytokines such as IL-8, the chemotactic factor for leukocytes including T cells, and IL-1 and TNF-α both of which are involved in promoting adhesion of leukocytes to the blood vessel endothelium, an essential step prior to migration from the blood stream under the influence of IL-8 towards the inflammatory site.

You will notice that Figure 7.2 suggests that TH1 and TH2 cells are derived from a precursor TH0. A suggested mechanism for conversion of TH0 to TH1 cells is shown in the figure and involves the production of IL-12 by infected macrophages or indirectly by IFN-γ produced by natural killer cells which are themselves stimulated by IL-12 (originally called natural killer cells stimulatory factor, NKCSF).

SAQ 7.4

Identify the cytokines best described by each statement. Use Tables 7.2, 7.4 and 7.5 and Figure 7.2 to help you.

1) Does not support proliferative activities in immune cells but promotes protection of the mucosal surface.

2) Five interleukins which regulate the response of B cells.

3) The 4 interleukins which are principally haematopoietic factors.

4) The two cytokines which play a major role in the activation and long term growth of T cells.

5) A cytokine which increases the expression of MHC II on B cells.

7.3.3 TH2 cells and humoral response

allergen stimulated mast cells IL-4, IL-5, IL-10 and IL-13

If IL-4 is preferentially induced locally, say by allergen stimulated mast cells, then the TH2 subset will develop and produce IL-4, IL-5, IL-10 and IL-13. Some of these activate eosinophils and mast cells, some of them induce production of certain antibody classes or subclasses and a predominantly humoral response results.

7.3.4 TH1 and TH2 subsets are mutually inhibitory

Due to the production of distinct cytokines, the TH1 or TH2 subsets will be preferentially activated resulting in a predominantly cell mediated response or a predominantly humoral response respectively.

IFN-γ produced by TH1 cells is known to block the activation of TH2 cells and IL-4 and IL-10 from activated TH2 cells have a similar effect on TH1 cells possibly by blocking antigen presentation by macrophages resulting in no production of IFN-γ.

You will probably conclude that cytokine biology is rather complex and you would be correct. However, its complexity is partly due to our ignorance of cytokine biology at the moment as we have little idea of what goes on in the body. Sometime in the future, we shall have a clearer picture which hopefully will explain why we need so many cytokines. We think the scheme on TH1 and TH2 cells is an indication that the picture is slowly becoming clearer.

SAQ 7.5

Complete the table below indicating with a + where you think a cytokine has a major role in the activity shown in the left column.

Activity	IL-1	IL-2	IL-3	IL-4	IL-5	IL-6	IL-8	IL-10	IFN-γ
1) MHC expression									
2) T cell activation									
3) Chemotactic factor									
4) T/NK product									
5) Ab production									
6) Macrophage product									

7.4 Production of cytokines using cell culture methods

We are now going to review the major methods for production of cytokines. To do this we shall take a historical perspective starting with the preparation of supernatants followed by attempts to produce cytokines from cell lines, hybridomas and T cell clones and we shall examine the methods used to purify the cytokines from these sources. With the advent of hybridomas, purification of cytokines became much easier and we shall briefly review the methods used. Finally, in Section 7.5 we shall take a fairly detailed look at cloning and expression of cytokine genes which has almost totally replaced all cell culture methods for bulk production of cytokines.

7.4.1 Supernatant derived cytokines

In the earliest experiments, mouse spleen cell suspensions were prepared or peripheral blood cells were obtained from human volunteers. The human lymphocytes and monocytes were separated from the red cells and polymorphs on Ficoll-hypaque gradients and the red cells in the mouse cell suspensions were usually removed by a lysis method. The enriched mononuclear cells were then placed in cell culture medium and a stimulant added.

enriched mononuclear cells

T cell cytokines

phytohaemagglutinin (PHA)

phorbol esters

If you wished to induce T cell cytokines you would use known T cell stimulants such as the plant lectins phytohaemagglutinin (PHA) or concanavalin A (Con A) or you could use phorbol esters such as phorbol myristic acetate. Any of these reagents would activate a majority of the T cells as they are mitogens. Allogeneic cells (from a genetically non-identical individual) would activate from 5-20% of the T cells of an individual whereas use of antigen would only stimulate antigen specific memory T cells (a minority of the total cells) from a previously immunised animal or individual.

monokines

endotoxin

If you intended to produce macrophage derived cytokines (monokines) then some of the T cell stimulants would suffice or you could use potent macrophage activators such as endotoxin (lipopolysaccharide) from Gram negative bacteria.

conditioned medium

Most cytokines are produced within a few hours of stimulation and a 1-2 day culture is sufficient to obtain what is called conditioned medium. This would be subjected to numerous biological assays to determine the presence of individual cytokines. Following detection of a particular cytokine, it would probably be concentrated by salt precipitation or ultrafiltration and attempts would be made to purify it from other cytokines using the sort of techniques we described in the last chapter. At each purification step, all fractions had to be assayed for the presence of the cytokine using the bioassay for it. Once the fractions has been identified, they are pooled and concentrated and then subjected to a further purification step and so on. A typical protocol describing purification of a cytokine from conditioned medium is shown in Figure 7.3 and very similar approaches were used to isolate individual cytokines from the supernatants of cell lines, hybridomas and T cell clones.

problems with purification of cytokines

In many instances, there were considerable problems which prevented purification of cytokines from these supernatants to homogeneity. Firstly, the minute amounts of cytokine recovered after the purification steps made further biochemical characterisation impossible. Secondly, the presence of multiple cytokines in supernatants meant that even when a particular cytokine had been partly purified, you could still not be certain that the biological activity was not due to another cytokine which had been co-purified with your cytokine nor that another cytokine might be needed for the expression of the biological activity of the purified cytokine. In this case, of course, your bioassay would yield nothing.

The investigators of that time, therefore, looked at ways of getting around these problems mainly by searching for cell lines which made much greater amounts of the required cytokine and for cell lines which synthesised a much narrower spectrum of cytokines.

SAQ 7.6

Investigators often calculate the recovery and purification factor of the cytokine following multi-step purification methods. In one purification of tumour necrosis factor, crude serum containing TNF was processed by salt precipitation and a combination of ion exchange and gel filtration methods. Using the data in the table, calculate the specific activity SA (units activity per μg), the % recovery R and the purification factor PF (number of times the cytokine has been purified) of the initial and final fractions.

Fraction	**Volume ml**	Concentration **μg ml^{-1}**	Activity units/ml^{-1} **(x 10^{-6})**	SA	**R**	**PF**
crude serum	1000	38 000	3.9	?	100	?
final step	5	40	4.4	?	?	?

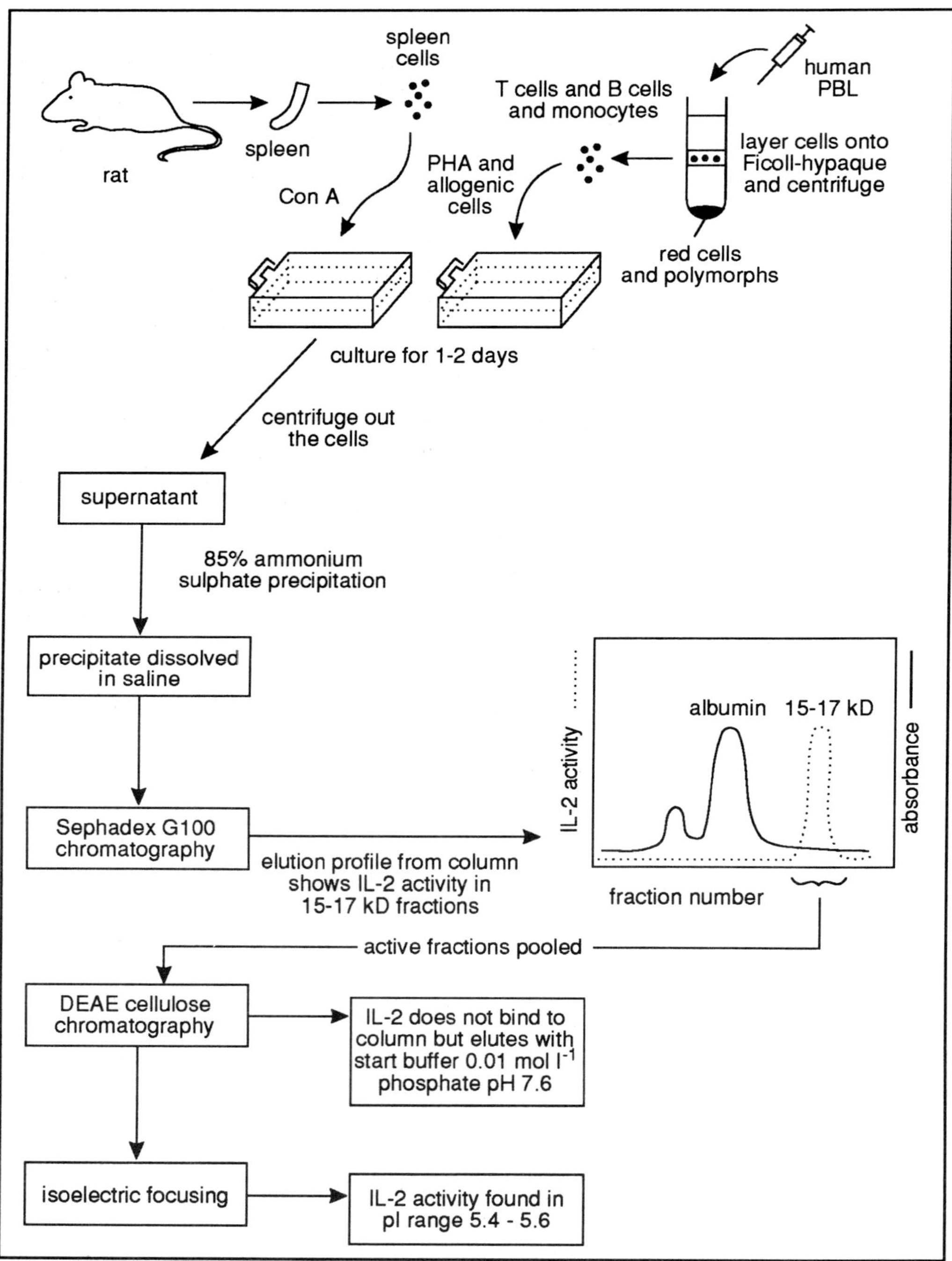

Figure 7.3 Purification of IL-2 from supernatants of Con A stimulated rat spleen cells as described by Gillis, Smith and Watson (1980). The authors also performed a similar purification of human IL-2 prepared by PHA stimulation of PBL in the presence of allogeneic cells (provides extra stimulation). The mononuclear cells (T, B and monocytes) were separated from the red cells and polymorphs on a Ficoll-hypaque gradient. The more dense cells pelleted in the tube on centrifugation while the less dense mononuclear cells remained at the interface and could be removed for culture. Isoelectric focusing (IEF) has not been previously described in the text.

7.4.2 Cell lines

constitutive production

As an example we can follow the progress made in the early 1980s in searching for cell lines which produced IL-2. Gillis and Watson screened a large number of both murine and human T cell leukaemia and lymphoma cell lines for both constitutive and mitogen induced IL-2 production. Constitutive production is when cells continually secrete the cytokine. Such cell lines were rare; most cells required stimulation with reagents such as Con A, PHA or PMA.

LBRM-33

Jurkat-FHCRC

Out of all the cell lines tested, only one mouse and one human line were found to produce IL-2 when stimulated with mitogens; no constitutive producers were detected. The murine cell line LBRM-33 was found to produce 1000-5000 times the amount of IL-2 released by activated murine spleen cells and the human cell line Jurkat-FHCRC produced 100-300 times the IL-2 found in mitogen stimulated human PBL. So both the cell lines provided an excellent source of IL-2 but notice it still had to be purified.

American Type Culture Collection, European Collection of Animal Cultures

Examine Table 7.6 which shows a selection of cell lines which have been used successfully to produce reasonable amounts of different cytokines. Some of them produced large amounts of more than one cytokine. Many of these cell lines are available from the American Type Culture Collection (ATCC) or the European Collection of Animal Cultures (ECACC).

Cell line	Source	Man	Mouse	Cytokines produced	Stimulus required?
JURKAT-FHCRC	leukaemia	+		IL-2	mitogens PHA/PMA
LBRM-33	leukaemia		+	IL-2	mitogens PHA/PMA
MLA-144	lymphoblastoid	(gibbon)		IL-2	mitogens PHA/PMA
EL-4	thymoma		+	IL-2	mitogens PHA/PMA
				IL-3	mitogens PHA/PMA
				IL-4	mitogens PHA/PMA
WEHI-3B	myelomonocytic		+	IL-1	endotoxin (LPS)
				IL-3	constitutive
P388D1	leukaemia		+	IL-1	endotoxin or lymphocytes + PHA
TCL-Nal	leukaemia virus transformed line	+		IL-6	constitutive
HL-60	myelomonocytic	+		TNF-α	endotoxin

Table 7.6 Examples of cell lines which produce, either constitutively or on stimulation, large amounts of the individual cytokines. Note that none of these cell lines only produce one cytokine - they are simply a major source of the cytokines indicated.

Π In terms of purification and assays for cytokines, what is one of the problems with using the majority of cell lines shown in Table 7.6?

The problem with most cell lines is that they do not constitutively produce the cytokine and some mitogen or other stimulus has to be added to activate the cells. This is then a contaminant of the resulting supernatant and it can interfere with the bioassays for some cytokines. For instance, it could lead to a false positive in fractions being assayed for IL-2 since mitogens stimulate the growth of T cells and this is what the bioassay measures.

Another problem with both LBRM-33 and Jurkat-FHCRC lines was that when mitogen was added, the cells produced IL-2 for a short period and then died. This meant a new supply of cells were needed to produce each batch of IL-2.

SAQ 7.7

You wish to characterise and purify chicken IL-2 and you decide to use the protocol described in Figure 7.3 after obtaining your conditioned medium by Con A stimulation of chicken PBL. After completion of the work you have the following data. The bioassay used was to test each fraction after each purification step for its effects on the growth of a mouse T cell line.

Sephadex G100 chromatography: 2 peaks (30 kD and 24 kD) with IL-2 activity.

DEAE cellulose chromatography: IL-2 fractions did not bind to the column.

Isoelectric focusing: 4 peaks of IL-2 activity at pH 4, 5, 5.5 and 6.

Which of the following statements could explain your results?

1) There may be another cytokine as well as IL-2 present which supports the growth of T cells.

2) The IL-2 has undergone post-synthetic modification, eg attachment of sialic acid residues.

3) The supernatant did not contain chicken IL-2.

7.4.3 T cell hybridomas

T hybridomas

Following development of hybridoma technology, a cell line BW5147 derived from a T cell thymoma was developed which could be fused to activated T cells to produce T hybridomas. The technology is very similar to that of generating B cell hybrids although not as successful since the resulting hybrids were not as stable as B cell hybrids and cytokine production of many of them ceased within a few months of fusion. The greater use of this technology was in generating T helper and suppressor clones and to identify antigen specific products. Many of these, however, also produced a variety of non-specific cytokines including IL-2.

Some attempts were made to fuse the IL-2 producing cell lines to tumour cells in order to obtain constitutive production so avoiding the problems we have just discussed above. This was, for the most part, restricted to the production of murine IL-2 as stable human IL-2 producing hybridomas could not be obtained. One such hybridoma,

IA8C3C10, produced in Gillis's laboratory from the LBRM-33 parent line, did produce IL-2 constitutively, but at much lower levels than the parent cells.

BW5147 cells

selected in HAT

One way to produce mouse T hybridomas was to remove the spleen from a mouse, prepare a cell suspension and then culture these cells in the presence of mitogenic amounts of Con A +/- PMA for 1-2 days before fusing with BW5147 cells. The hybrids were then selected in HAT and the supernatants tested for the presence of IL-2. The IL-2 producer cells were then cloned and expanded. It was generally found that the IL-2 titres of hybridoma supernatants were much lower than those of the parent cells. However, some hybridomas were reported to produce only a single cytokine which obviously would make purification much easier.

7.4.4 T cell clones

cloning individual T cells

limiting dilution

The third approach to produce cytokines was by cloning individual T cells. In this method, mice were immunised with antigen and T cells from the spleen were then cultured in small numbers in microculture plates in the presence of antigen pulsed antigen presenting cells and/or IL-2. The cells were then cloned by limiting dilution in the presence of feeder cells (which may be antigen presenting cells) and IL-2. In order to test for the different cytokines the individual T cell clones were producing, aliquots of cells were then cultured in small volumes of culture fluid in the presence of antigen pulsed antigen presenting cells or were stimulated with mitogen or IL-2 and the supernatants tested for cytokines.

The general finding was that the clones produced quite a wide spectrum of cytokines which included IL-2, IL-4, IL-6, colony stimulating factors and interferon-gamma among others and that individual clones secreted different combinations of cytokines. The actual combination depended on what sort of stimulus the cells had received at the start of the culture. However, the amounts of each cytokine produced was not as high as that produced by cell lines. There was also questions as to whether long term clones really represented 'normal' T cells.

Even though T cell hybrids and T cell clones have not provided a significant source of bulk cytokines, they have proved to be very important developments which have greatly increased our knowledge of lymphocyte biology.

For instance, T cell hybrids were instrumental in the characterisation of the T cell receptor and T cell clones have been used to investigate interactions with B cells both *in vivo* and *in vitro* and murine T cell subsets and cytokine production were identified using cloning techniques.

Another important application is in the identification of autoreactive T cell clones from lesions in various diseases in Man and animal models such as rheumatoid arthritis and experimental allergic encephalomyelitis (possibly the murine equivalent of multiple sclerosis in Man) respectively and the patterns of cytokine production by these cells. This sort of information is steadily leading us to an in-depth understanding of the mechanisms in leading to tissue damage involved and, hopefully, to a cure for some of these diseases.

Finally, the availability of cell lines, T hybridomas and T cell clones provided the material for subsequent cloning of cytokine genes which has finally resulted in the production of large amounts of each cytokine for further study.

SAQ 7.8

Which one of the following statements is correct?

1) BW5147 is sensitive to HAT.

2) T cell clones are antigen specific and only produce one cytokine.

3) Biological assays for cytokines give an indication of the purity of the cytokine.

4) Biological assays of supernatants from most cell lines may be unreliable.

5) T cell cloning is not a good approach for the production of cytokines.

7.4.5 Monoclonal antibodies for the purification of cytokines

The early investigators faced some difficulties suitable obtaining monoclonal antibodies due to the extremely low amounts of cytokines available as immunogens. However, with the advent of cytokines from tumour cell lines and hybridomas it became possible to generate monoclonal antibodies which reacted with cytokines.

Π Before you read on, make a list of the possible uses of antibodies to cytokines for the researcher. Check your list with Figures 7.4 and 7.5.

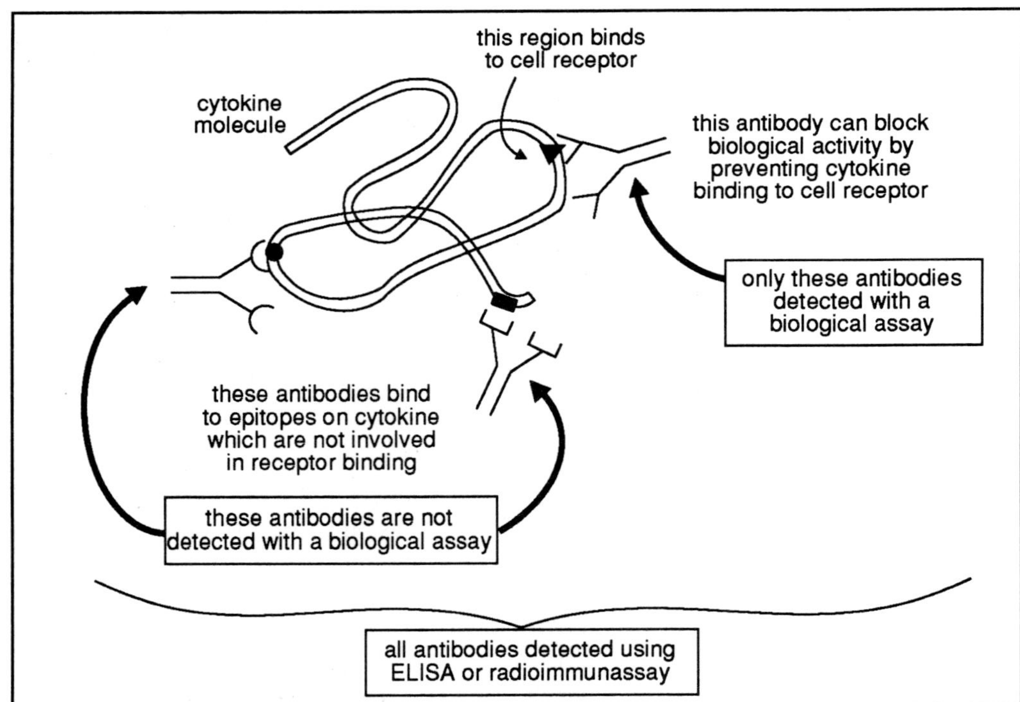

Figure 7.4 Monoclonal antibodies from hybridoma supernatants are directed against various epitopes on the cytokine and can be used for epitope analysis. A few of these antibodies will be specific for epitopes within the region which binds the cell receptor, the majority will be for other epitopes not involved in receptor binding.

The main uses are for measuring/detecting cytokines, for blocking the activities of cytokines and for purifying cytokines. In the use of antibodies to assay cytokines, the advantage with the generation of monoclonal antibodies is that the cytokine does not have to have been purified to homogeneity as long as an assay is available which is specific for that one cytokine. Again, with reference to IL-2, there was a problem with the assay since the only detection system for IL-2 at that time was the biological activity and there was little pure IL-2 for use in any assay.

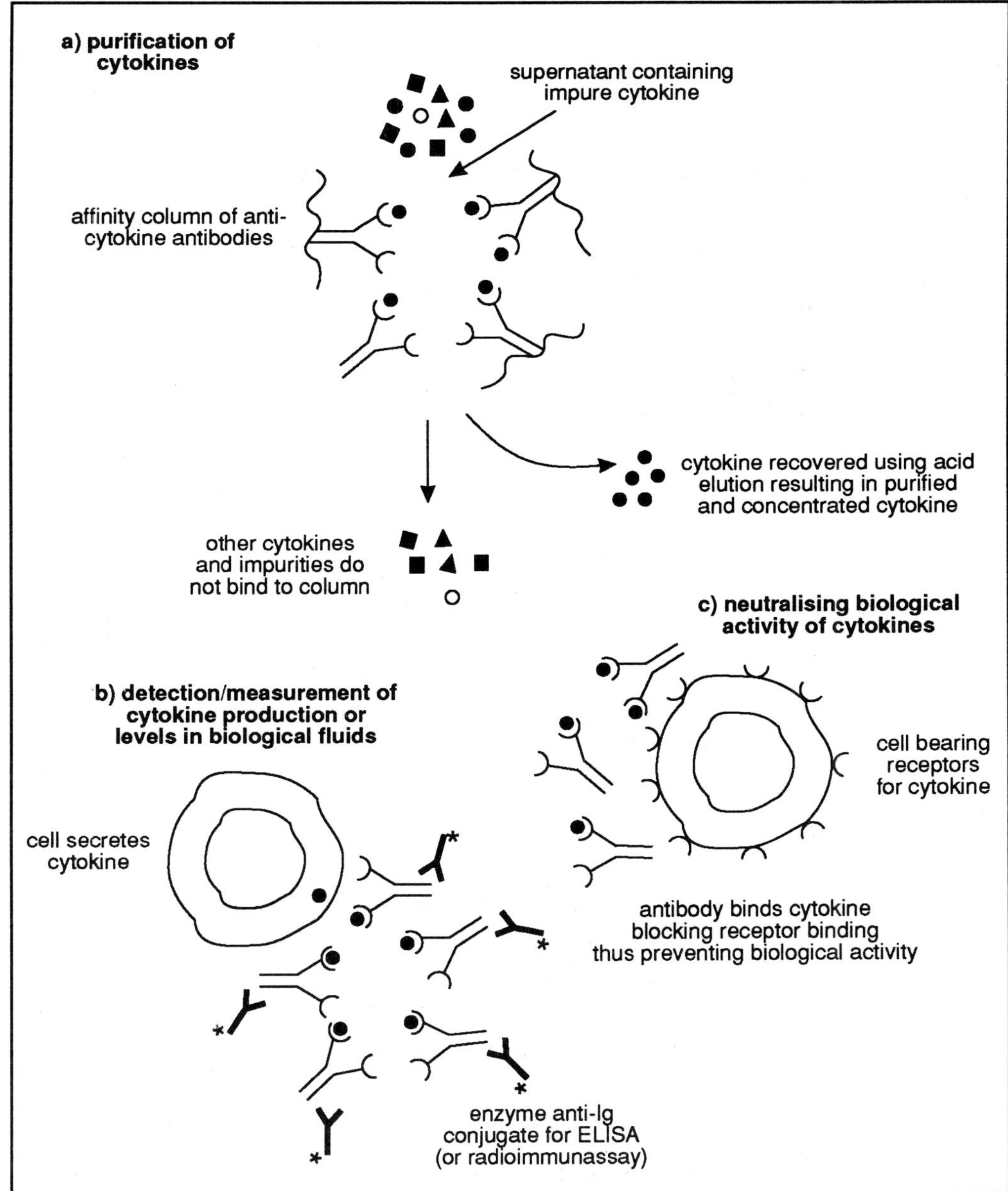

Figure 7.5 Three major uses for anti-cytokine antibodies. a) purification of cytokines, b) detection and measurement of cytokines, c) neutralisation of cytokines.

Π If you are developing an antibody to IL-2, what is the problem with using the biological assay as an assay for the antibody? (See Figure 7.4)

You will remember that all cytokines act through receptors expressed on the responding cells and the biological assay therefore depends on the binding of the cytokines with their receptors. This could be the basis of an assay for antibodies in hybridoma supernatants since if the antibodies bind to the receptor binding portion of IL-2 then this will block the binding of the cytokine to the receptor. In the case of IL-2, this would result in no growth of the T cells and would therefore indicate the presence of anti-IL-2 antibodies in those supernatants. We would call this assay a neutralisation assay. Such an assay could not detect antibodies which are directed to epitopes on IL-2 not involved with the binding site. Additionally, if the cytokines possess higher binding affinities than the antibodies, which in many cases they do, then this assay would not detect the antibodies except when they are in excess. The conclusion we have to draw is that such as assay would not detect the majority of monoclonal antibodies with specificity for IL-2.

neutralisation assay

Π Can you suggest a suitable assay with limited availability of IL-2?

ELISA

The obvious answer, based on our earlier discussions, would be the ELISA method which can be set up with extremely low amounts of IL-2. In this technique, IL-2 would be absorbed to the plastic wells and undiluted hybridoma supernatants would be added. After a suitable incubation time, an anti-mouse immunoglobulin antibody enzyme conjugate would be added followed by the substrate. The wells showing the highest colour intensity would indicate the clones producing the greatest amounts of anti-IL-2 (Figure 7.6 shows the principal steps in this method).

Π It may be a good idea to return to Sections 2.6.3 and 2.6.4 to remind you of the details of this extremely important assay as applied to cytokine methodology.

Once the antibody has been prepared, it can then be used for the immunoaffinity purification of the cytokine - what we call one step purification. This would involve going through the antibody purification procedures discussed in Chapter 6 for hybridoma supernatants and then conjugating the antibodies to beads for preparation of an affinity column.

Π Can you remember the different approaches you could take to purify and concentrate the monoclonal antibodies from the hybridoma supernatants. Write a list of the possible methods.

Your list should have included ammonium sulphate precipitation to concentrate the antibodies followed by either protein A or G affinity chromatography. You would obviously need to know the isotype of the antibodies before you started (Section 4.4). Alternative methods after concentration are either DEAE cellulose chromatography or anti-isotype affinity chromatography. Once the anti-cytokine antibody has been purified, it can be bound to Sepharose (or other suitable support) using the linking agents we described in earlier chapters.

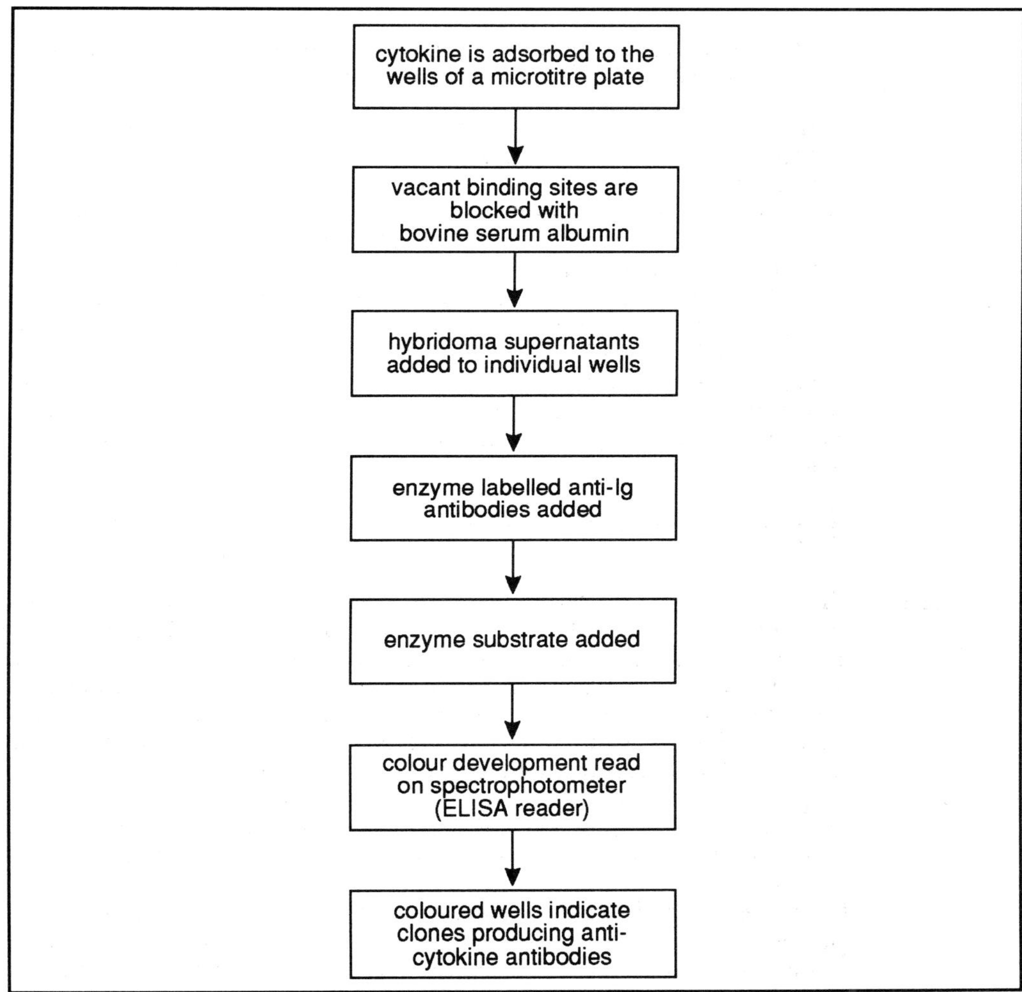

Figure 7.6 The major steps in assaying for anti-cytokine antibodies in hybridoma supernatants using the ELISA assay.

To get back to our cytokine purification, following pre-cycling with the elution buffer, the supernatants containing cytokine can now be passed through the anti-cytokine antibody Sepharose affinity column. After suitable washing steps to remove non-specific binding components, the cytokine is eluted with an acid buffer.

Π Apart from purification what will you also have achieved with this method?

A typical purification procedure is as follows. 4 litres of Jurkat - derived supernatant is passed down a 1 ml immunoaffinity column consisting of 10-20 mg coupled anti-IL-2 antibodies attached to an Affigel support. The IL-2 is recovered from the column using 2 ml of the eluting buffer thus resulting in a 2000 fold concentration of the IL-2. Since the recovery of IL-2 is considerably less than 100%, the molarity of the IL-2 in the eluate is only 1000 fold higher than that in the supernatant. So you also achieve a concentration of your cytokine as well as purification.

Using this technology you could be reasonably assured that you had a pure cytokine due to the specificity of the antibodies but there will still be limited amounts of it.

7.5 Cloning and expression of cytokine genes

The tedious and labour intensive procedures of purification of cytokines from supernatants has been largely replaced by cloning of cytokine genes and their expression in bacteria, yeast or mammalian cells and there is no longer any shortage of pure cytokines since they can be produced in large amounts using this technology. We shall concentrate our discussions on just the cloning of cytokine genes since their expression and subsequent purification of the cytokine products is essentially the same as described for antibody fragments.

7.5.1 Cloning cytokine genes using differential hybridisation screening

So how does one go about cloning a gene for a cytokine? If you simply want the protein product then you will need to produce cDNA which is made up of the nucleotide sequence of the leader peptide and the mature protein. If you wish to study the regulation of cytokine gene expression, you will need to produce genomic or chromosomal DNA which also includes the intronic sequences. We shall concentrate on the production of cytokines only in this section so we will be dealing with cDNA.

poly-A mRNA/*Xenopus laevis* oocytes

The first successful attempts at cloning cytokine genes followed a protocol something like this. Keep referring to Figure 7.7 while you are reading the text. This shows the major steps of a protocol which was used to clone the cDNA for IFN-γ in 1982 by scientists at Genentech in the USA.

The cells were activated and total RNA extracted. Poly-A mRNA was recovered from this fraction using oligo-dT columns and this was then size fractionated on gels. The different size fractions were subsequently micro-injected into *Xenopus laevis* oocytes. To determine which mRNAs were translated into a cytokine product, the supernatants were assayed using an established biological activity or using an anti-cytokine antibody. The active mRNA fractions were then used to synthesise cDNAs which were ligated into plasmids or phage which were subsequently used to transform or infect *E. coli*. The resulting colonies or plaques were then replica plated onto nitrocellulose and hybridised using ^{32}P-cDNA probes prepared from the active mRNAs.

differential hybridisation

The fact that only stimulated cells possessed cytokine mRNAs was now used to advantage since the majority of the mRNA species in the cells would be the same ones as were found in unstimulated cells. By probing one replica membrane with ^{32}P-cDNAs prepared from mRNA from unstimulated cells and another with ^{32}P-cDNAs from stimulated cells, colonies or plaques were identified which hybridised strongly with only the ^{32}P-cDNAs from stimulated cells and at least some of these would be recombinants carrying the cytokine cDNA inserts. This process is called differential hybridisation. Some of these clones were then selected for cDNA sequencing and subsequent expression in *E. coli*, yeast or mammalian cells and the expressed product assayed for biological activity or identified by cytokine specific antibodies.

Π Can you think of a way by which these investigators (Figure 7.7) could have produced a much smaller library and thus saved themselves some work?

You will be able to check your ideas in the next section.

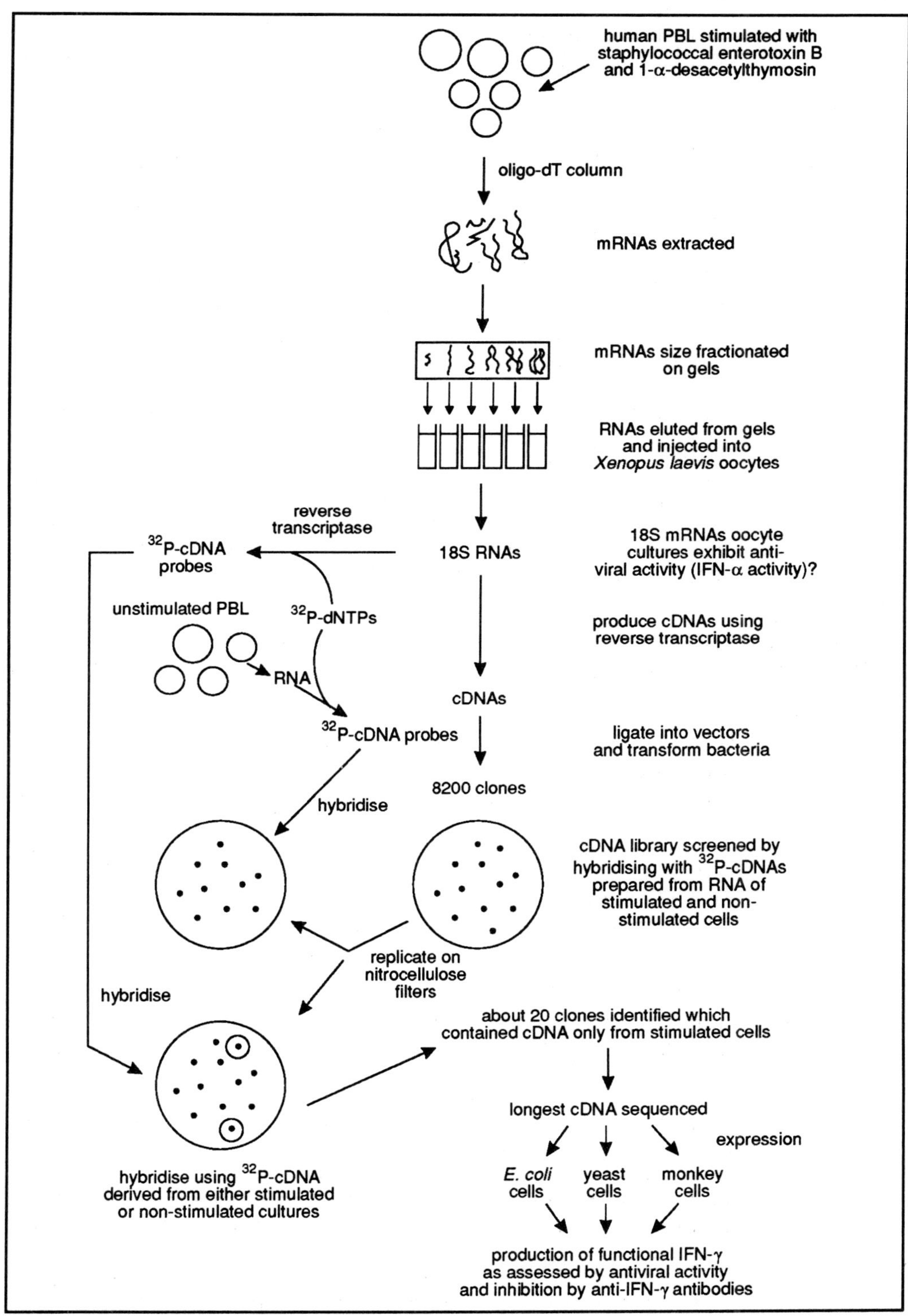

Figure 7.7 The cDNA cloning of IFN-γ using differential hybridisation.

7.5.2 Making subtracted libraries

subtracted library

They could have prepared what is called a subtracted library which would have resulted in much lower numbers of clones to screen but with a much higher percentage of recombinants carrying the IFN-γ cDNA insert. During their preparation of cDNAs from the 18S RNA fraction (see Figure 7.7) they would prepare single stranded cDNA using reverse transcriptase producing RNA-DNA hybrids. The RNA could then be removed by alkaline hydrolysis leaving single stranded cDNA. As we have already said, the majority of the mRNAs in stimulated cells are the same as in unstimulated cells so if we now add excess mRNAs prepared from unstimulated cells most of the cDNAs will anneal to these mRNAs. These can be separated from the few remaining cDNAs derived from mRNAs of the stimulated cells by passing the mixture down a hydroxylapatite column which binds the RNA-DNA hybrids. The cDNAs which pass through the column are now converted to double stranded cDNAs and cloned. Obviously, ^{32}P-cDNA probes could be prepared in a similar manner. The advantage of this method is that the investigators would not have had to screen 8200 clones but many times less.

RNA-DNA hybrids

SAQ 7.9

Without reference to Figure 7.7 set up a protocol from the following stages for the cloning of a cytokine gene (Stages not included in the list should be ignored).

1) Replicate on nitrocellulose filters.
2) Identify positive clones/plaques.
3) Express cDNA.
4) Gel electrophoresis of RNA.
5) Synthesise cDNAs.
6) Ligate into vectors.
7) Hybridise.
8) Prepare autoradiographs.
9) Stimulate cells.
10) Sequence cDNA.
11) Micro-inject oocytes.
12) Oligo-dT chromatography.

7.5.3 The hybridisation translation assay

A variation on the method just described for IFN-γ was used in the cloning of human IL-2 cDNA. To help you understand the protocol, follow the steps in Figure 7.8 while you are reading the text.

cDNA library from stimulated cells

hybridisation translation assay

groups of 24 clones

hybridised with mRNA

Following the establishment of a cDNA library from stimulated cells of the human T leukaemic cell line Jurkat III (cloned from Jurkat-FHCRC) using the same protocol as described for IFN-γ, Taniguchi's laboratory screened the library using the hybridisation translation assay to identify an IL-2 specific clone. This was done by randomly choosing 432 bacterial clones out of a total of 2000 and arranging them into 18 groups of 24 clones. Plasmids were prepared from these groups, digested with *Hind* III (cDNAs were in *Hind* III-*Hind* III restriction fragments) and the cDNAs from each plasmid group denatured and bound to individual nitrocellulose filters. Each filter was then hybridised with mRNA prepared from the active fraction of mRNA derived from stimulated Jurkat cells and the bound RNA was then eluted and injected into *Xenopus laevis* oocytes and the products assessed for IL-2 activity.

full length IL-2 cDNA

One group was found to be positive so plasmids from each individual clone from this group were then assessed in the same manner. This led to identification of a single clone p3-16 of about 650 bp which was IL-2 specific and this was used to screen another cDNA library with larger inserts in an attempt to obtain the full length IL-2 cDNA including the signal sequence. Such a clone was identified and subsequently sequenced. The IL-2 cDNA was expressed in both *E. coli* and mammalian cells.

Π Why were the 432 bacterial clones arranged into 18 groups of 24 clones? (Think about how many hybridisation translation assays need to be done).

Using the grouping procedure a total of 18 + 24 hybridisation translation assays need to be done to identify a single suitable clone amongst the original 432 bacterial clones. If the grouping procedure had not been adopted, then 432 such assays would be needed.

7.5.4 Screening using heterologous cDNA or oligonucleotide probes

Cloning of murine IL-2 cDNA using cloned human IL-2 cDNA

Π Assume that human IL-2 cDNA has been successfully cloned. Before you read on, can you think of a short cut investigators could take to clone murine IL-2 cDNA?

murine IL-2 cDNA

cross hybridisation with human IL-2 cDNA

Once a cytokine gene had been cloned from one species, then it is, in most cases, relatively straightforward to clone cDNAs for the same cytokine from cells of other species. Rather than going through the labour intensive procedures associated with translation assays, the radiolabelled cytokine cDNA could be used as a radiolabelled probe to identify IL-2 clones derived from cells of other species. This is possible because the nucleotide sequences of IL-2 from mouse and humans are very similar. A typical example is the cloning of murine IL-2 cDNA by Taniguchi's laboratory once they had the human IL-2 cDNA prepared as we described earlier. Polyadenylated RNA was isolated from PHA stimulated LBRM-33. After fractionating the mRNA, the active fraction was identified by cross hybridisation with human IL-2 cDNA on Northern blots. The cDNA library was then prepared from this fraction and the IL-2 clones identified by hybridisation to a human IL-2 cDNA sequence. Sequencing and expression of the murine cDNA in mammalian (COS) cells then followed.

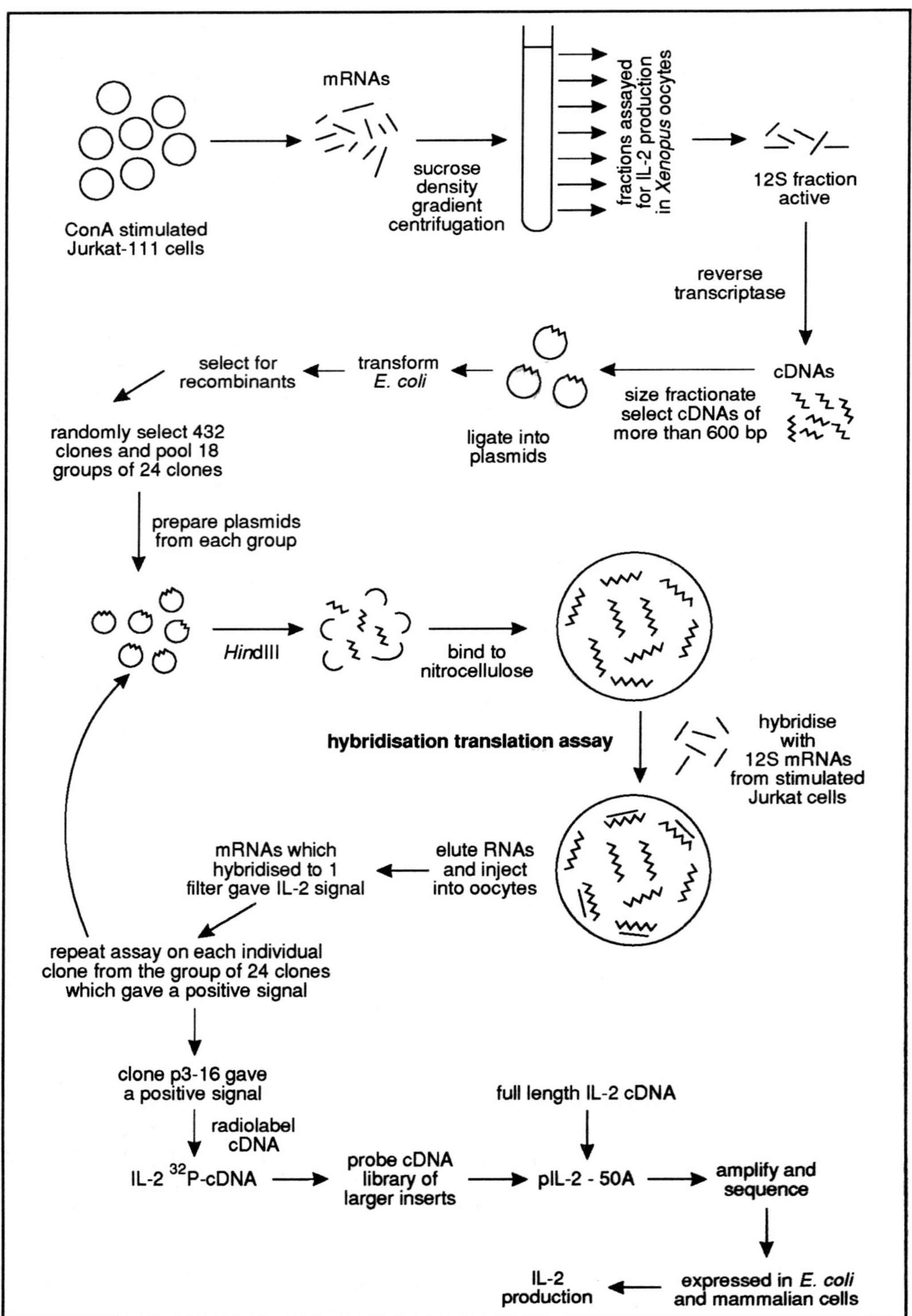

Figure 7.8 The cloning of human IL-2 cDNA by Taniguchi *et al* (1983) using the hybridisation translation screening assay.

bovine IL-2

34-mer oligonucleotide

Oligonucleotides based on the human IL-2 cDNA sequence were also used to clone bovine IL-2. Following extraction of mRNA from stimulated bovine lymphocytes, the size fractionated RNA that contained IL-2 mRNA was identified in Northern blots by hybridisation with a 34-mer oligonucleotide complementary to the carboxyl terminus of the human IL-2 cDNA gene. A cDNA library was then prepared which was screened with the carboxyl terminal oligonucleotide and a 40-mer oligonucleotides based on the amino terminal sequence of the human IL-2 cDNA gene.

Identifying cytokine clones using oligonucleotides based on the protein sequence

sequence tryptic peptides

In some cloning procedures, workers have been able to identify the cytokine cDNA using the peptide sequence of the purified cytokine. The cloning of human IL-6 from the human T cell line TCL-NA1 and human IL-12 from the EBV transformed cell line RPMI 8866 are typical examples of this approach (EBV = Epstein Barr Virus). This method requires that the cytokine has been purified to homogeneity. The cytokine is then subjected to digestion with proteolytic enzymes to produce tryptic peptides which can then be sequenced.

Π Can you generate single oligonucleotides based on these sequences which will hybridise to the cDNA of the cytokine thus identifying the clone?

DNA code is degenerate

multiple oligonucleotides

degenerate probe

Since the DNA code is degenerate, most amino acids are encoded by two or more codons. You cannot, therefore, use a single oligonucleotide to complementarily bind to the corresponding sequence in the cytokine cDNA but require multiple oligonucleotides which will include all the possible sequences for the individual codons to account for this degeneracy. This collection of oligonucleotides is often referred to as a degenerate probe. These probes are then radiolabelled and used to hybridise with the cDNA clones specific for the cytokine. One of them will hybridise to the cytokine cDNA clone in the library since it has the correct complementary sequence. The idea behind this approach is explained in Figure 7.9.

7.5.5 Cloning cytokine cDNAs using the polymerase chain reaction

We have already come across this technique when we were discussing the development of phage antibodies in Chapter 5 and it has proved very useful in the cloning of a number of cytokine cDNAs.

Π Can you remember what is the major difference between PCR and the other cloning methods we have discussed and what structural information do you need to perform PCR?

The advantage of PCR is that you do not need to develop a cDNA library so this technique is a short cut to cloning. Additionally, you do not need an abundant source of mRNA which is a requirement of the techniques so far discussed; in fact, PCR will selectively amplify just one or two molecules of cytokine mRNA to produce enough DNA to be seen on a gel with Southern hybridisation. In many cases, there is enough DNA to dispense with the initial cloning prior to sequencing as well.

However, there is a restriction to the use of PCR since you do need to know the sequences at the borders of your cytokine gene in order to prepare the PCR primers. This technique has, then, been applied to the cloning of cDNAs of cytokines of other species based on the sequence information gained from cloning of an original cytokine gene by conventional methods.

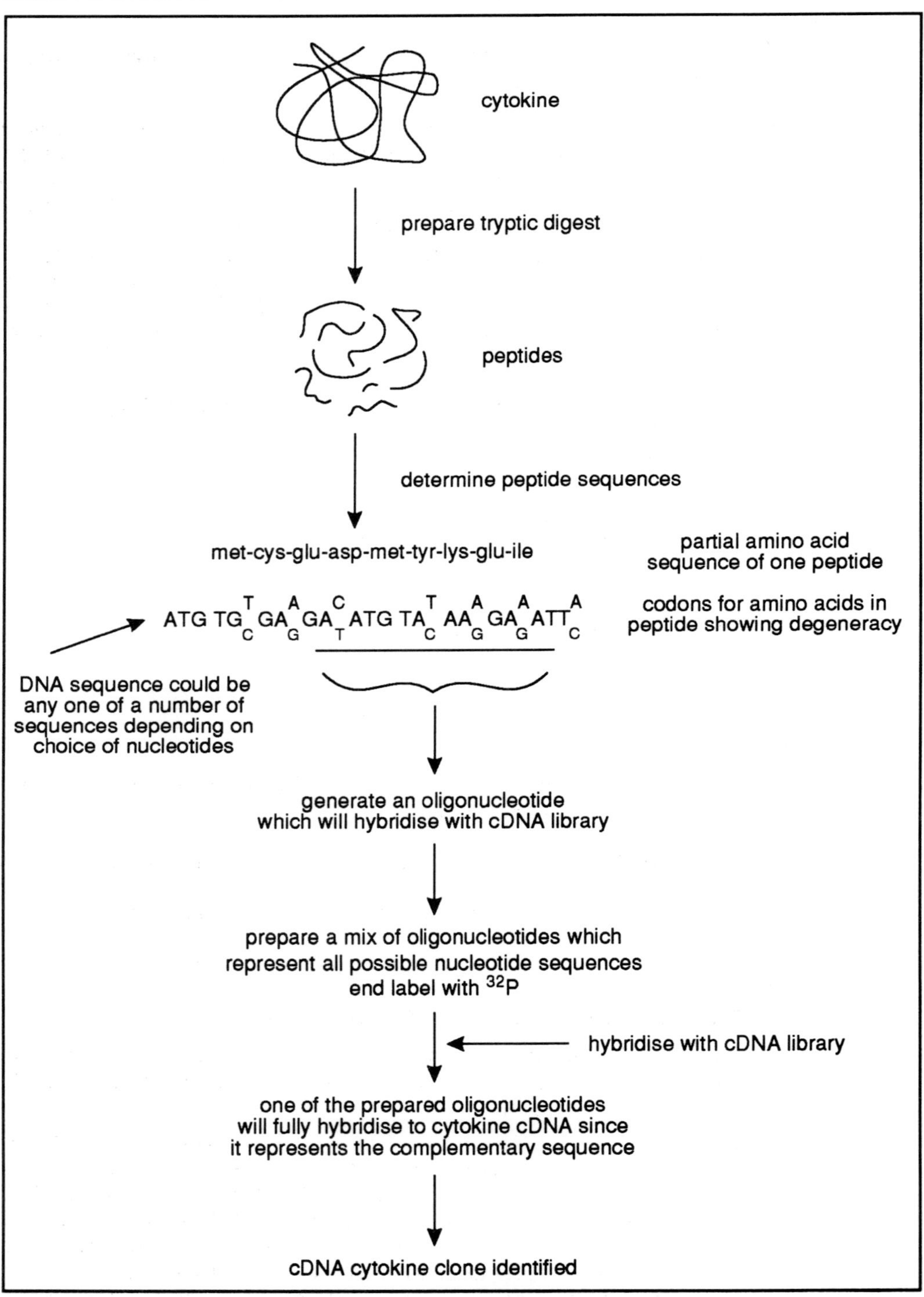

Figure 7.9 Identifying cytokine cDNA clones by screening with oligonucleotides. A typical choice of an oligomer is shown.

Π To remind yourself of this technique, read Section 5.5.2 and look at Figure 5.6 before reading on.

ovine and porcine IL-2 cDNAs

amplified using PCR

An example of its use was in the cloning of both ovine and porcine IL-2 cDNAs using PCR primers based on the bovine sequence. cDNAs were prepared from mRNAs extracted from mitogen stimulated PBL of either species and these were then amplified using PCR. The cDNAs were then either blunt end ligated into a poly linker site of a vector or restriction sites were introduced at both extreme ends of the cDNA during PCR amplification. The cDNAs were then ligated into a vector which was used to transform bacteria. The recombinants were identified based on antibiotic resistance. The cloning, sequencing and expression is then by conventional methods. The molecular details of these steps are described in the BIOTOL text 'Techniques for Engineering Genes'.

SAQ 7.10

Identify the major feature in the second list (a-e) of each of the procedures listed (1-5).

List 1

1) Hybridisation translation.
2) PCR.
3) Heterologous sequences.
4) Tryptic digest.

List 2

a) Cytokine cDNA from one species can be used as probe for the same cytokine in another species.
b) Oligonucleotide primers encoding borders of the gene are needed.
c) Oligonucleotides based on protein sequence of cytokine.
d) Assay based on detection of cytokine activity.

7.5.6 Expression screening in mammalian cells

Cloned cytokine genes can also be detected using a functional assay in mammalian cells. This was an approach taken by investigators at the Immunex Corporation, USA for cloning IL-7 cDNA.

positive pool

Briefly, the mRNA was isolated from a cell line producing high levels of IL-7 and a cDNA library was prepared in plasmid vectors which were used to transform *E. coli* which were then plated out at about 1000 colonies on each plate. Plasmid preparations from the colonies on each plate were pooled and then pools were separately transfected into COS cells and the supernatants subsequently assayed for the presence of IL-7 after a period of culture. The positive pool was then subcloned and the procedure repeated using smaller numbers of clones in each pool with testing of COS supernatants for the presence of IL-7. The process was repeated until an individual clone was identified which directed the synthesis of IL-7. The basis of this approach is depicted in Figure 7.10.

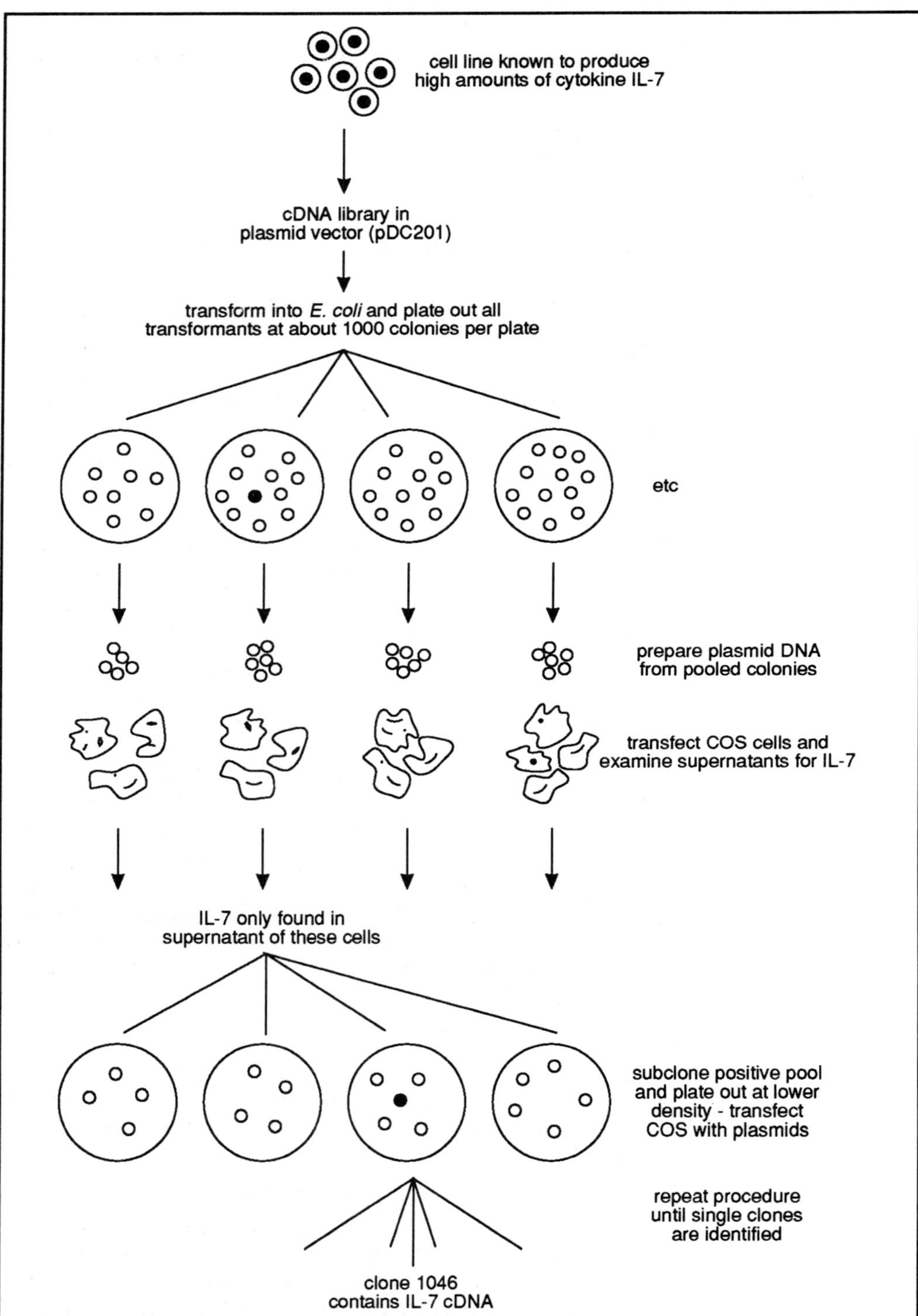

Figure 7.10 Expression screening in mammalian cells showing the cloning of IL-7. The filled circle represents the plasmid carrying the IL-7 gene.

Π How much of the cDNA message would you need for expression screening?

Since the cDNA is to be expressed in mammalian cells as a secreted functional protein you would need to clone full length cDNA including that encoding the leader peptide.

7.6 Cell surface receptors for cytokines

We have already intimated that cytokines act on target cells by binding to receptors which are specific for the particular cytokine. We are not going to examine the structure of the receptors for the individual cytokines since that would demand a chapter of its own but it would be useful for you to be aware of the sort of experimental approaches one adopts to investigate receptors and cytokine receptor interactions because they are applicable generally to many ligand receptor systems. The IL-2 receptor, which we shall keep referring to, has been described in detail in another text in this series ('Cellular Interactions and Immunobiology').

Π Before you read further, can you think of a very simple assay you could do to demonstrate the presence of receptors for a cytokine on a particular cell type for a cytokine if you had developed a bioassay for the cytokine?

7.6.1 A simple binding assay detects cytokine receptors on cells

A very simple technique was used to demonstrate that IL-2 receptors were present on cells responding to IL-2. Unstimulated and stimulated (PHA) T cells were incubated with a known amount of IL-2 in a culture dish for a short period. The supernatant was then harvested from the cells and assayed for the amount of IL-2 remaining in the supernatant. It was found that although levels of IL-2 remained unchanged in the supernatant from the unstimulated cells, the supernatant of the stimulated cells no longer supported the growth of T cells, ie the IL-2 had been removed by the cells.

This result suggested that IL-2 receptors (IL-2R) appeared on T cells following stimulation but they were not present on resting T cells. Similar assays could be adapted for use with other cytokines and their receptors. Obviously this is a rather simple approach and there are more definitive methods for characterising cytokine receptors. These include the development of antibodies to the receptors and the use of binding assays using radiolabelled cytokines.

7.6.2 Antibodies to cytokine receptors

The usual approach to developing antibodies to a particular cytokine receptor is to use a cell which expresses high levels of receptor as the immunogen. For instance, to generate anti-IL-2R antibodies, T cell leukaemia cells were used since these expressed many more IL-2R than stimulated PBL.

Bioassay for cytokine receptor antibodies

The developments of antibodies to a receptor will require an assay for the antibodies and this most commonly uses inhibition of the cytokine effects as the basis for the assay. For instance, again using IL-2 - IL-2R as the model, we know that IL-2 promotes the growth of T cells. If we add anti-receptor antibodies to the cultures, this should block the binding of IL-2 and thus prevent growth of the T cells. The assay, then, will involve dispensing aliquots of a T cell suspension into wells of a microculture plate together with a standard amount of IL-2 which is known to support the growth of the T cells.

The antibody solution is then serially diluted into a series of wells containing T cells plus IL-2 and the cultures are then kept in the incubator for 1-3 days. Radiolabelled thymidine is then added to each well and a few hours later, the cells are harvested and the incorporation of radiolabel determined. If the antibody is capable of blocking the receptor site, then wells containing antibodies will exhibit varying degrees of inhibition of growth of the T cells. A typical experiment and the results are shown in Figure 7.11.

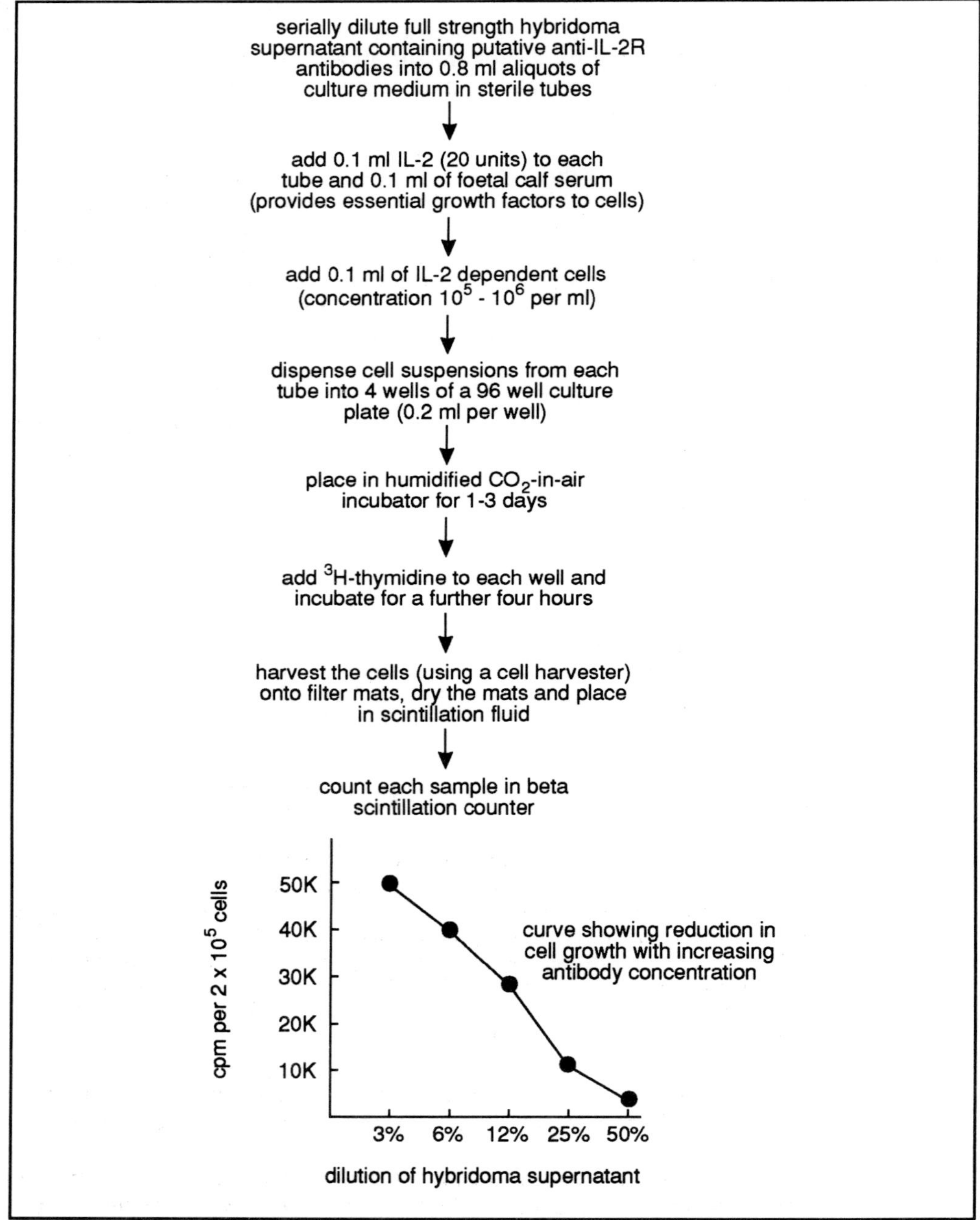

Figure 7.11 Bioassay for cytokine receptor antibodies showing the inhibitory effects of an anti-IL-2 receptor antibody on IL-2 supported growth of activated T cells.

7.6.3 A receptor binding assay using radiolabelled cytokine

A receptor binding assay is commonly used to investigate the binding of cytokines to their receptors and to determine their dissociation constants. The assay is relatively simple experimentally. Various concentrations of radiolabelled cytokine are added to aliquots of cells expressing the receptor and after a period of incubation, the cell suspension is layered on top of 10% sucrose in Eppendorf tubes. The tubes are then centrifuged pelleting the cell-bound cytokine together with the cells and leaving the unbound cytokine above the sucrose layer. The tubes are then frozen in liquid nitrogen, the tips containing the pellet are cut off and the amount of radiolabel attached to the cells counted. The remaining part of the tube is also counted, this representing the unbound cytokine.

cell bound cytokine

unbound cytokine

The counts in the pellets (bound cytokine) can be plotted against the amount of free cytokine contained in the gradient. From these data we can estimate the number of receptors for the cytokine per cell can be made. A typical result is shown in Figure 7.12. We can also transform the data in a similar fashion to that shown in Figure 2.1 and produce a Scatchard plot which leads to a determination of the dissociation constant which gives us information about the affinity of the binding of the cytokine to the receptor.

Scatchard plot

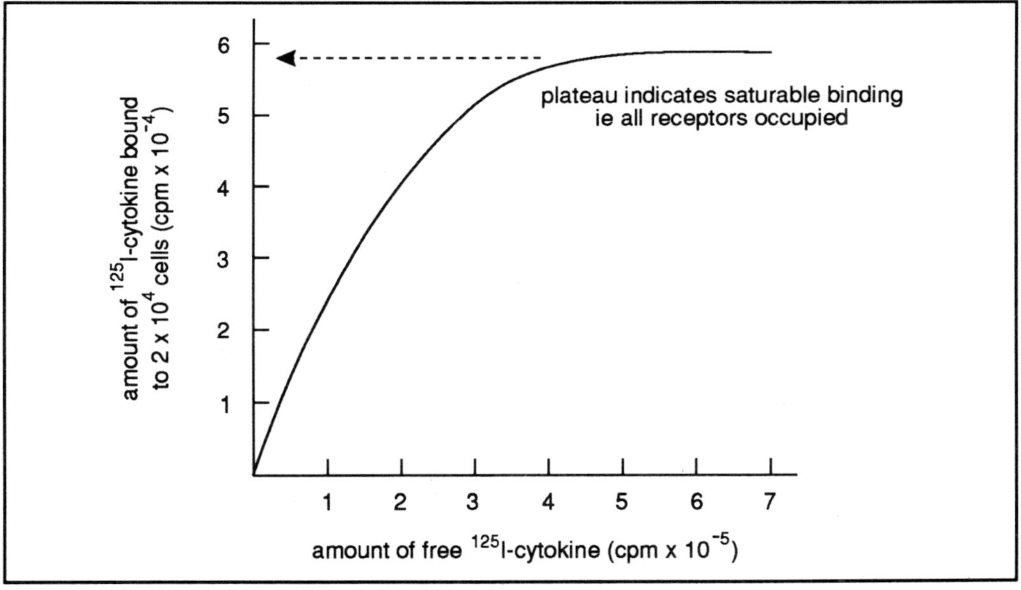

Figure 7.12 A typical curve for binding of radiolabelled cytokine by receptor expressing cells. Since the plateau represents saturable binding, this can be extrapolated back to the y axis to determine the amount of cytokine bound by a fixed number of cells. Knowing the specific activity of the radiolabel, you can then determine the number of cytokine molecules which have bound per cell and therefore the number of receptors expressed on each cell. This experiment can also be done using radiolabelled anti-receptor antibodies to obtain the same result.

Π If the cytokine gene had not been cloned you would not have had a pure cytokine to radiolabel. What alternative method could you have used?

The alternative approach to determining the number of receptors on cells is to used radiolabelled anti-receptor antibodies. However, you will realise that use of antibodies would not give you any information on the dissociation constants of the receptor - cytokine interactions.

SAQ 7.11

In one experiment, maximum (saturable) binding of 4500 dpm of ^{125}I-cytokine was achieved with 450 000 cells. Given that the specific activity of the radiolabelled cytokine to be 1.7 x 10^6 dpm per picomol (10^{-12} mol), and Avagadro's number is 6 x 10^{23} molecules mol^{-1}, calculate the number of cytokine receptors per cell.

7.6.4 Identification of cytokine receptors using affinity chromatography and affinity linking techniques

Π See if you can think of any method whereby you can isolate and identify cytokine receptors?

A cytokine Sepharose affinity column

cytokine Sepharose

One way is to prepare an affinity column of a cytokine attached to Sepharose beads and then pass a lysate of the receptor-bearing cells down the column. The receptor chains will attach to the column and they can be eluted off by changing the pH. The receptor fraction could be further characterised on SDS-PAGE. Alternatively the cells may be radiolabelled before lysis so that small amounts of receptor can be identified following SDS-PAGE and exposure to an X-ray film.

Affinity chromatography using anti-cytokine receptor antibodies

affinity column of anti-receptor antibodies

An alternative procedure is to prepare an affinity column of anti-receptor antibodies and use this to isolate the receptor chains from the cell lysates. Such purification methods can lead to the determination of the N-terminal amino acid sequence of the receptor chain for the preparation of oligonucleotide probes for cloning the receptor chain cDNA.

Affinity linking methods

affinity linking

cytokine receptor complexes (CRCs)

We have shown how this technique was used to identify two receptor chains for IL-2 in Figure 7.13. Cells expressing the receptor are incubated with radiolabelled cytokine and then the latter is covalently joined to the receptor by means of a linker - this is called affinity linking. The cells are then lysed using a detergent such as NP-40 and the cytokine receptor complexes (CRCs) are separated from the other cell surface components using anti-cytokine antibodies followed by precipitation using reagents such as protein A-Sepharose beads. The CRC antibody protein A-Sepharose complexes are then dissociated in SDS-PAGE sample buffer, the protein A-Sepharose beads are removed by centrifugation and the supernatant containing CRC and antibodies subjected to SDS-PAGE followed by exposure of the dried gel to an X-ray film. Knowing the molecular weight of the cytokine it is possible to determine the molecular weight of the receptor chains.

7.6.5 Characterisation of anti-receptor antibodies

Π Now that we have completed our discussion on cytokine receptors, make a list of the different techniques you would use to prove that a monoclonal antibody you had prepared was indeed specific for the cytokine receptor.

In Figure 7.14 we have described the various ways in which a monoclonal antibody against the p55 IL-2 receptor chain called anti-Tac was characterised. The approaches can obviously be used to characterise any antibody directed against a receptor if you have the ligand available. You should be able to check your list against the methods shown in Figure 7.14.

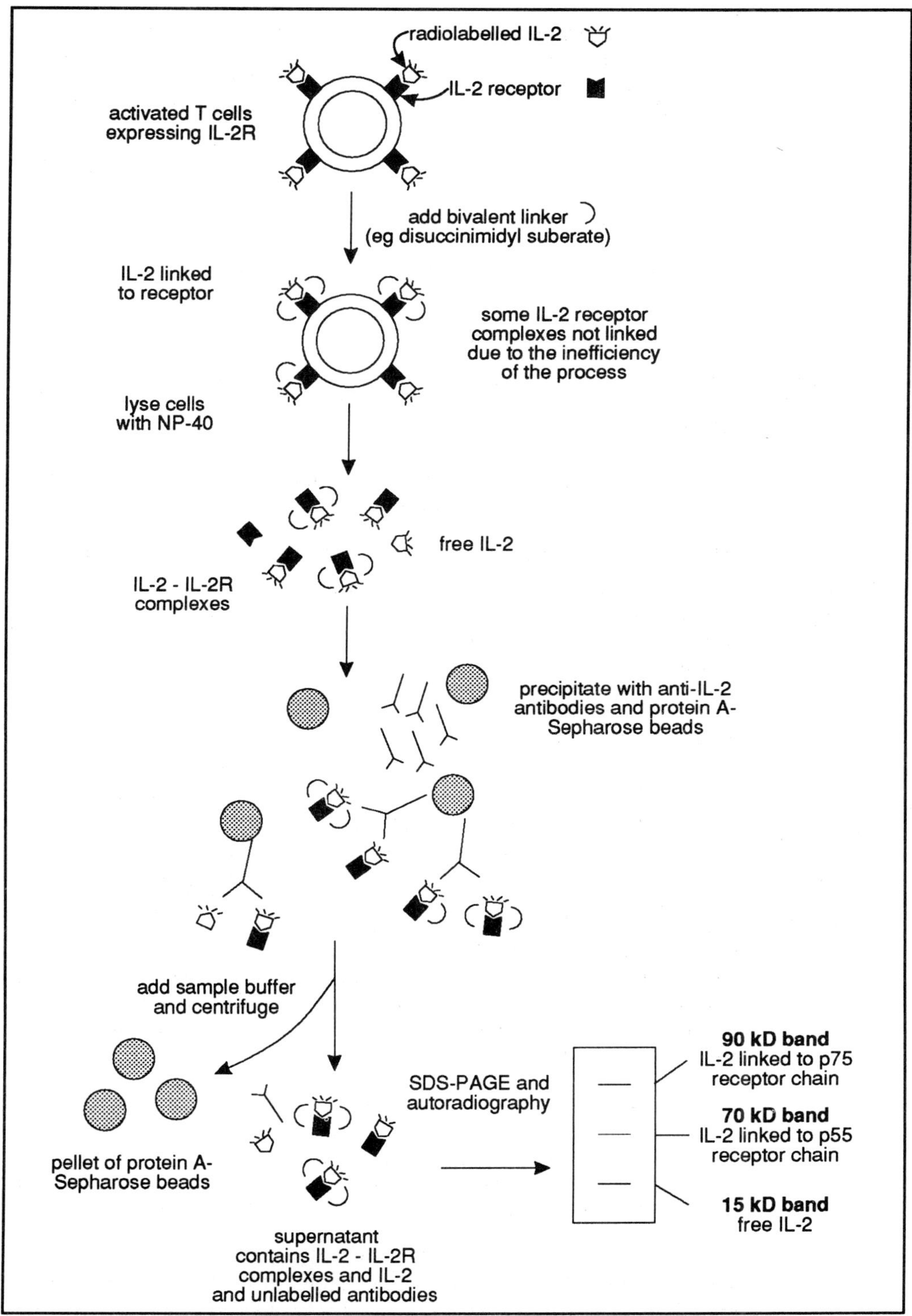

Figure 7.13 Identification of the p55 and p75 receptor chains of the IL-2 receptor using affinity linking techniques.

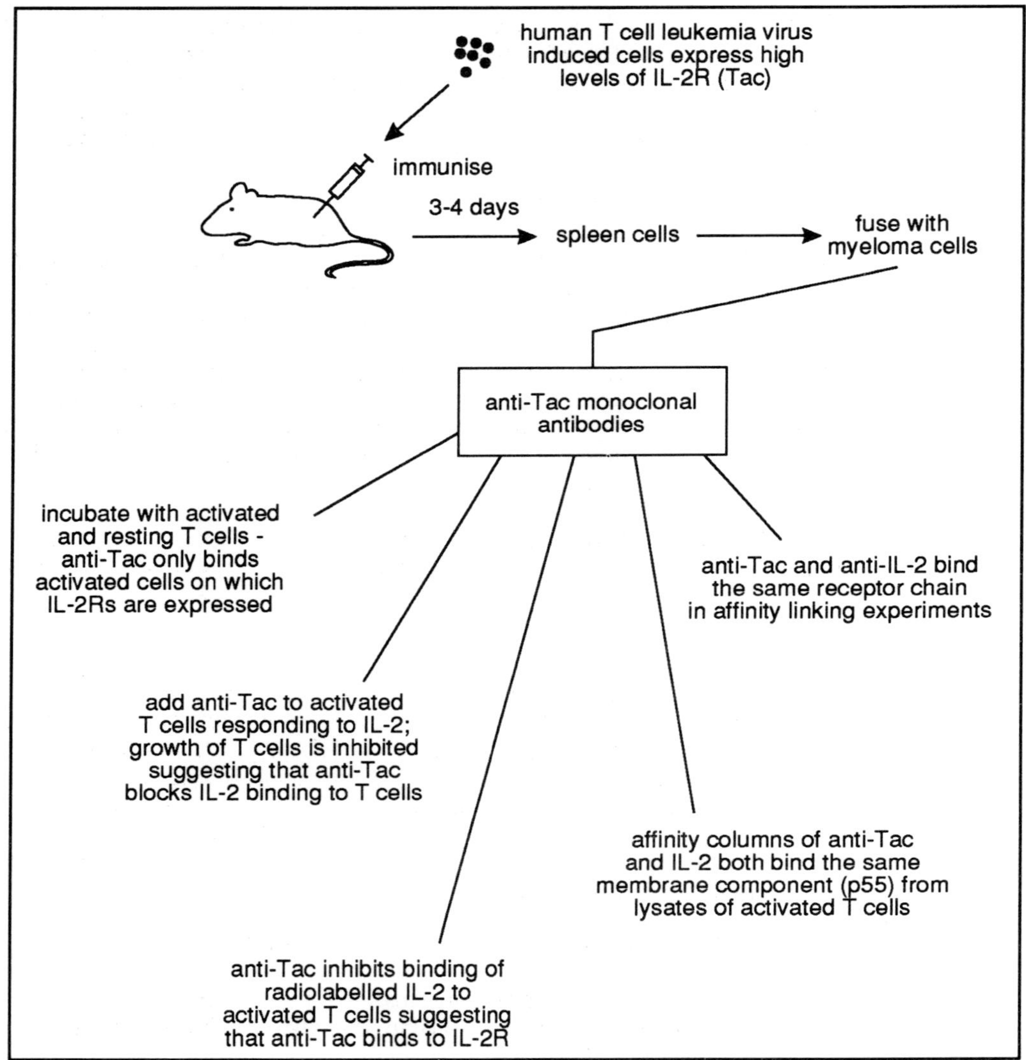

Figure 7.14 Experiments to demonstrate the specificity of an anti-cytokine receptor antibody such as anti-Tac.

7.7 Detection/measurement of cytokine production by individual cells or by limited numbers of cells

Having completed our discussions on how to produce cytokines and identify their receptors we shall now look briefly at just a few of the many ways by which we can detect cytokine production by individual cells or limited numbers of cells. There are two major approaches to do this, one is to identify the cytokine product by some immunolabelling technique such as variations of the ELISA and the other approach is to detect the mRNA in the individual cells.

Π In these sort of assays are we measuring biologically active cytokines?

Since the cells producing the cytokines are being placed in what must be considered abnormal conditions and in isolation away from other cells, the only conclusion one can make from these experiments is that the cell is capable of producing that particular cytokine. You cannot make any conclusion as to whether the cytokine is biologically active since you are not testing that parameter. We shall be examining bioassays later.

7.7.1 The Elispot assay

You may remember the use of this assay to detect antibodies; we mentioned it in Section 2.6.6. This technique involves coating of the wells of microculture plates with anti-cytokine antibodies followed by blocking with protein solution. The cells being assayed are then dispensed into the wells and incubated for various periods normally up to 24 hr. The cytokine secreted by individual cells will be trapped by the anti-cytokine antibodies and, after removal of the cells, the cytokines are identified using ELISA methodology. Since the cytokines on being secreted are trapped by antibodies in the immediate vicinity of the producer cells, this method results in a collection of coloured spots in the well which identify the individual cells which have secreted the cytokine.

coloured spots

This technique could be useful in examining the kinetics of production of different cytokines by a homogenous population of cells such as T cell clones stimulated by various reagents.

7.7.2 The cell blot assay

This represents a slight variation on the Elispot assay and also assesses cytokine production by individual cells. In this method, the cells are bound to nitrocellulose or nylon membrane and the membrane is submerged in culture medium containing anti-cytokine antibodies. When the individual cells secrete cytokines they become attached to the antibodies so these are localised next to the producer cells. The cells are then fixed and stained using ELISA methods. After colour development, the cytokine producing cells are identified by a coloured zone, the size and colour intensity of which relates to the amount of cytokine produced.

SAQ 7.12

Which of the following statements are correct?

1) Antibodies directed against the p55 IL-2R could be substituted for the anti-IL-2 antibody used in the affinity linking experiment described in Figure 7.13.

2) The molecular weight of the larger IL-2R chain characterised in Figure 7.13 is 90 kD.

3) An antibody generated against a cytokine receptor should bind the same membrane component as the cytokine itself.

4) The Elispot assay detects cytokine production by single cells.

5) The cell blot assay detects biologically active cytokines.

7.7.3 Northern analysis of cytokine mRNA

The alternative method is to look for cytokine mRNAs in the cells since if these are present then the cell is producing the cytokine. We shall briefly describe the principles behind the two main assays which are Northern blot analysis and *in situ* hybridisation.

Northern analysis

In Northern analysis, the mRNA is extracted from the cells of interest and size separated on agarose gels. It is then transferred to nitrocellulose or nylon membranes to which the RNA becomes bound. The RNA blots are then probed with radiolabelled complementary oligonucleotides or cytokine cDNAs which are detected using autoradiography. Using this method, it is possible to make semi-quantitative assessments of the amount of RNA in the original cell samples. Obviously in this method you are measuring levels of mRNA in a cell population.

7.7.4 *In situ* hybridisation

in situ hybridisation

In situ hybridisation involves the analysis of messenger RNA in individual cells. In this method, tissue sections or other cell preparations are probed with radiolabelled oligonucleotides based on cytokine cDNA sequences which hybridise with the mRNAs fixed in the tissue section. The slides are then covered in radiation sensitive emulsion or film and after a suitable exposure time, they are developed. The tissue sections are then observed under the microscope and the cells containing hybridised RNA identified by the presence of silver grains.

7.7.5 Cytokine message amplification phenotyping or MAPPing

amplification of cytokine mRNAs by PCR

This rather complex title simply means the amplification of cytokine mRNAs by PCR and their detection. This method enables the simultaneous assessment of different cytokine mRNAs in a single cell or a few cells.

The mRNAs are isolated from the cells by a microtechnique and this is reverse transcribed using oligo-dT as primer. Following first strand synthesis, samples of the cDNA are aliquoted into separate reaction tubes and cytokine primers are added to individual tubes and amplified by PCR. This is followed by electrophoretic analysis to determine which cytokine mRNAs were present in the cells. If you include an internal cDNA standard the mRNAs present in the cell(s) can be quantified. Examine Figure 7.15 which illustrates a typical experiment.

7.7.6 Other methods for measuring cytokine production

We have already met the other major methods for measuring the levels of cytokines in supernatants or other samples. These include the ELISA and the methods which use radiolabels (see Chapter 2).

ELISA

The major ELISA method is a cytokine trapping method where the wells of the microtitre plate are coated in anti-cytokine antibodies. Serial dilutions of a cytokine standard and unknown cytokine test samples are then added. Following incubation and suitable washing steps, the second enzyme linked anti-cytokine antibody is added followed by substrate and colour development is determined. There are various modifications to this basic scheme as described in Chapter 2.

Π To remind you of the details of the ELISA re-read Section 2.6.2 substituting cytokine for antigen.

radioimmuno-assay (RIA)

The radioimmunoassay (RIA) can also be used to measure cytokine levels. Serial dilutions of cytokine standards or test samples are incubated with aliquots of anti-cytokine antibodies. A standard amount of radiolabelled cytokine is then added to each tube and this will be bound by any antibodies which have empty binding sites. The cytokines (radiolabelled and non radiolabelled) bound by antibodies are now recovered from the reaction mixture using an anti-globulin attached to beads. A standard curve is produced from the standards from which you can estimate the amount of cytokine in your samples.

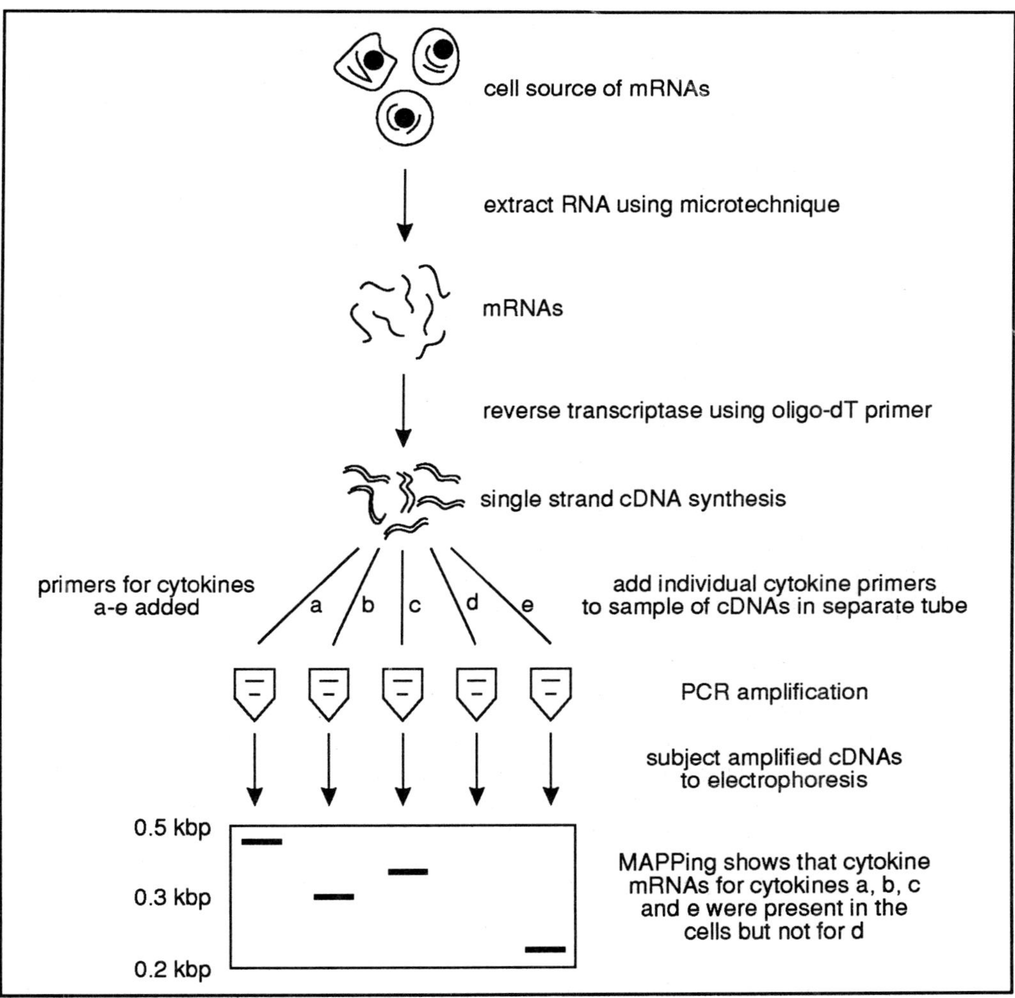

Figure 7.15 The MAPPing technique using PCR to determine levels of cytokine mRNAs in single cells or limited numbers of cells.

immunoradiometric assays (IRMA)

In immunoradiometric assays (IRMA), the anti-cytokine antibody is attached to beads. Serial dilutions of cytokines and samples are prepared and the beads are added to these solutions. After an incubation period and washing steps, a ^{125}I-anti-cytokine antibody is added to all the tubes. This binds to the cytokine molecules bound by the beads. The beads are then recovered and washed and counted in a gamma counter. The counts per minute will be directly related to the amount of cytokine which was initially bound by the antibody on the beads.

SAQ 7.13

Match the items in the left column with those in the right column using each item only once.

1) Northern analysis	a) PCR
2) MAPPing	b) Electrophoresis
3) RIA	c) Tissue section
4) ELISA	d) Radiolabelled cytokines
5) *In situ* hybridisation	e) Enzyme anti-cytokine antibodies

7.8 Bioassays for detecting and measuring cytokines

immunoreactive cytokines

The assays we have discussed so far actually measure what we call immunoreactive cytokines which basically means that the actual protein has been detected or measured. However, there is no indication from these tests that the cytokine is biologically active since this requires a bioassay which measures the effect of a cytokine on a particular cell type or sub population of cells.

7.8.1 Problems with bioassays

These are obviously very important assays which cannot be substituted by any of the tests mentioned earlier. However, they are far from perfect, particularly because the tests are rarely cytokine specific. Let us give you a couple of examples. When IL-2 was originally discovered, it was thought to be the only T cell growth factor. However, we now know that IL-4 and IL-7 also promote growth of some T cells. Not only that, but yet other cytokines are involved in the response to IL-2. IL-1 induces IL-2 production and quite a few cytokines, including IL-6, have been implicated in inducing IL-2R expression.

The original assay for IL-1, the thymocyte co-stimulation assay, was thought to be specific for this cytokine. It is now known that it can also detect IL-2, IL-4, IL-6 and the tumour necrosis factors among others.

Π If you had established a test for a cytokine and discovered it was affected by another cytokine, how could you make the test specific for the one cytokine?

neutralising antibodies

Sometimes bioassays can be made specific for a single cytokine by the use of neutralising antibodies. For example, the IL-2 T cell growth assay is known to detect IL-4 using some cell lines. The effects of this cytokine can be neutralised by adding anti-IL-4 antibodies to the test system. You could also use this test to measure IL-4 by adding anti-IL-2 antibodies.

For many of these bioassays, cell lines have been established which are fairly specific for particular cytokines. Examples of these are murine cytotoxic T cell lines (CTLL) used to measure IL-2 and growth factor dependent cell lines which measure some of the haematopoietic cytokines.

We obviously cannot describe each bioassay which is available for each cytokine so we shall give typical examples of the sort of assays which are performed to assess various functions mediated by cytokines.

7.8.2 Cell growth assays

T cell growth assay

Use Figure 7.16 to follow each step involved in this assay. CTLL are maintained in culture in the presence of IL-2. The cells are harvested by centrifugation, resuspended in medium and dispensed into wells of a microculture plate at a concentration of 10^4 to 10^5 per ml. Serial dilutions of standard IL-2 and unknown IL-2 are then added to the wells and the cells are cultured for about 18 hours. Radiolabelled thymidine is then added to each well and after further incubation, the cells are harvested and assessed for uptake of the radiolabel. The cells exposed to IL-2 will exhibit dose dependent uptake thus indicating IL-2 dependent growth.

reference IL-2 standards

50% maximal stimulation

By titrating reference IL-2 standards as well you can determine the number of units of biological activity present in your test samples. As shown in Figure 7.16, the data derived from both standard and test IL-2 samples are plotted. The amount of IL-2 which produces 50% stimulation can be determined from the standard curve. In Figure 7.16, this is a $\frac{1}{16}$ dilution of the standard. Since the full strength standard has a concentration of 40 units ml^{-1}, we can calculate that 50% stimulation is achieved by an IL-2 concentration of $\frac{1}{16} \times 40 = 2.5$ units ml^{-1}. From the test curve shown in Figure 7.16, we can determine that a $\frac{1}{6}$ dilution of the unknown gives 50% stimulation. Thus $\frac{1}{6}$ dilution contains 2.5 units ml^{-1}. Therefore the original contains $6 \times 2.5 = 15$ units ml^{-1}.

Π What is the purpose of the two controls in Figure 7.16?

medium control

cell control

The medium control consists of wells containing medium only with no added cells or IL-2. These wells should show minimal radioactivity unless there is microbial contamination. Thus significant radioactivity in these wells indicates contamination and you have to repeat the whole assay. The cell control consists of medium and cells with no added IL-2. Since the CTLL are IL-2 dependent, these cells will die during the incubation period and will incorporate a minimal amount of radio-isotope. The counts per minute (cpm) values of the cell control should be subtracted from the cpm of the cells exposed to IL-2; this will yield data reflecting true IL-2 supported cell growth.

Other proliferative assays

IL-5

IL-6

IL-7

IL-4

There are many other proliferative assays which are used to quantitate individual cytokines. For instance, murine IL-5 is assayed by its proliferative effects on the murine B cell line BCL-1, human IL-5 on the TF-1 cell line and IL-6 on various hybridoma cell lines such as B9. IL-7 is assayed for its proliferative effects on a pre-B cell line. IL-4 was originally assayed for its proliferative effects on B cells activated with anti-immunoglobulin, *Staphylococcus aureus* or phorbol esters. It is now usually assayed on a number of cell lines including B cell lines such as BALM-4 and L4. Since it also has a proliferative effect on T cells you can assay its effects in the IL-2 assay just described by assessing the additional growth effects of serial dilutions of IL-4 on cells exposed to a suboptimal amount of IL-2.

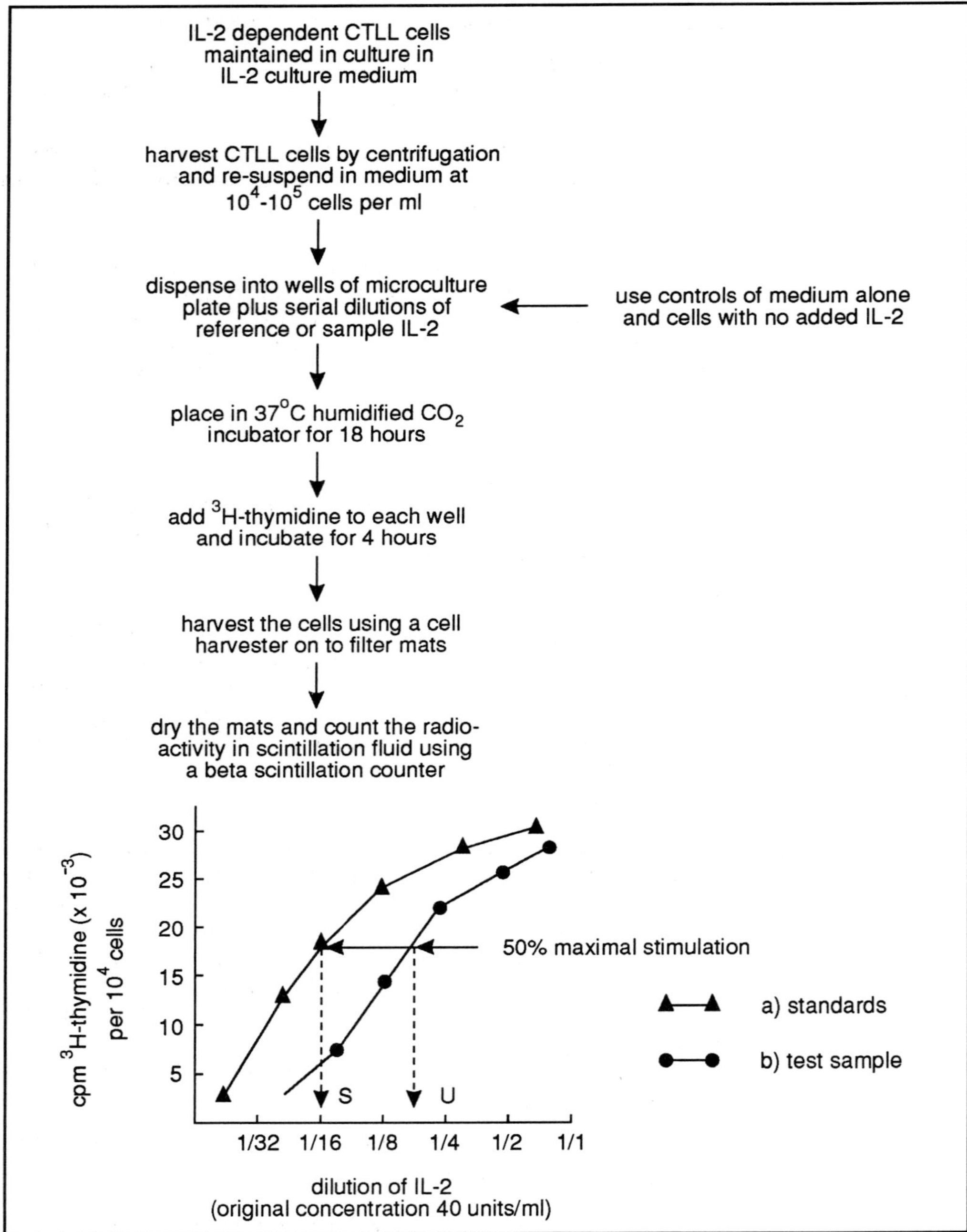

Figure 7.16 The T cell growth assay. In this example a reference sample of IL-2 - a) and a sample of unknown IL-2 concentration - b) was used. By comparing the 50% maximal stimulation of both parallel parts of the curve, the concentration of the unknown sample can be determined. 50% maximal stimulation was observed with 1/16 standard IL-2 (S) and 1/6 (U) unknown IL-2. Since the reference IL-2 concentration was 40 units per ml, the unknown sample had a IL-2 concentration of 6/16 x 40 = 15 units ml^{-1}.

Special assays for haematopoietic growth factors

There are two principal methods for assaying these factors, the first employs freshly isolated haematopoietic cells and the second uses established growth factor dependent cell lines.

bone marrow

In the first method, cells are recovered from the bone marrow of mice or Man and these are suspended in serum free medium in agar supplemented with the cytokine. After a few days of growth, the number of colonies which have developed reflect the presence of the cytokine.

growth factor dependent cell lines

When using growth factor dependent cell lines, the cells are cultured until they attain an exponential growth rate and they are then resuspended in culture medium in the presence of the cytokine. Growth can be assessed after 1-2 days using incorporation of tritiated thymidine as for the IL-2 assay. Examples of cell lines used for these assays include FDCP-2 which is IL-3 dependent, DA1 (GM-CSF dependent) and M1 (G-CSF dependent).

7.8.3 An assay involving induction of cytokine release

Since one of the properties of IL-1 is to induce the production of IL-2 by T cells, this is the basis of a bioassay for this cytokine. For instance, LBRM-33 cells which we met earlier, can be induced to produce IL-2 by exposing the cells to the IL-1 sample and the IL-2 can then be measured using the CTLL assay.

7.8.4 A chemotactic assay

chemokinesis

You may remember that IL-8 is a chemotactic factor and was detected by an assay for chemotaxis. The most appropriate assay for this cytokine which is relatively straightforward is one which involves measuring chemokinesis which is an enhancement of random movement. This bioassay involves separation of neutrophils from peripheral blood and then estimating their migration from solidified agarose droplets in response to the cytokine.

7.8.5 Other assays

There are many assays which we have not been able to describe due to lack of space. These include assays involving the stimulation of antibody synthesis by B cells, the detection of interferons assayed by their ability to reduce viral killing of cells and of tumour necrosis factors using a cytotoxic assay. There are very few laboratories, if any, which can perform a wide spectrum of assays since many of them involve the maintenance of special lines which is labour intensive and costly. Most laboratories therefore specialise in just a few assays such as those for B cells or T cells or those involving a single cytokine and know very little about the assays for other cells or cytokines. You should not then feel disadvantaged in not knowing the bioassays for all the cytokines.

SAQ 7.14

Examine the following statements and decide which best fit mRNA assays, binding assays or bioassays.

1) Assays detecting cytokine production by single cells.
2) Examination of cytokine production by pathological tissue specimens.
3) Examination of body fluids for cytokines.
4) Measurement of T cell growth.
5) Measurement of immunoreactive cytokines.

Summary and objectives

Having studied this chapter you should now have at least a passing acquaintance with most of the known cytokines and their major functions and you will understand the complexities of cytokine biology. We described in some detail the methods by which cytokines were produced from supernatants of activated cells, cell lines, T hybridomas and T cell clones and showed how these methods have been replaced by cytokine gene cloning and expression. We then discussed methods used for the identification and characterisation of cytokine receptors. We described various approaches to detect and measure cytokine production including bioassays in the final sections of the chapter.

Now that you have completed this chapter you should be able to:

- describe and interpret the first experiments which reported the existence of cytokines;
- list the major cytokines involved in various immune activities;
- list the major properties of cytokines;
- explain what is meant by interleukin, monokine, lymphokine and cytokine;
- calculate the recovery, specific activity and purification factor of a cytokine in a purification protocol;
- plan suitable cytokine purification protocols and interpret the results;
- discuss the advantages and disadvantages of using activated cells, cell lines, T cell hybridomas and T cell clones as sources of cytokines;
- describe, in detail, the cloning of cytokine genes using the following screening methods - differential hybridisation, subtracted libraries, hybridisation translation and the use of cDNA or oligonucleotide probes;
- explain the advantages of the polymerase chain reaction in the cloning of cytokine genes;
- describe the technique of expression screening in mammalian cells;
- calculate the numbers of cytokine receptors on a cell from cytokine binding assay data;
- describe affinity linking experiments to identify cytokine receptors and interpret the results of such experiments;
- explain the principals of the Elispot assay, the cell blot assay, Northern analysis, *in situ* hybridisation and MAPPing techniques;
- describe various bioassays for detection and measurement of cytokines.

Antibodies and cytokines as immunoreagents

Antibodies and cytokines as immunoreagents

8.1 Introduction

The major part of this chapter will examine the potential of antibodies and cytokines as immunotherapeutic agents with quite a lot of emphasis on the potential treatment of cancer patients since this is where a large part of the research activity is directed. We are going to describe to you the generation of new antibody reagents such as recombinant globulins, immunotoxins and bispecific antibodies which may prove effective in targeting and killing *in vivo* various cells, including tumours cells, which have been implicated in the disease process. An alternative approach we shall discuss is the potential use of cytokines as novel biological drugs in immunotherapeutic approaches to cancer and infectious diseases.

Although there have been considerable successes already using some of these novel antibody or cytokine-based therapeutics, most of their applications lie in the future. However, irrespective of whether the research on these novel drugs is successful or not, it is important to realise that antibodies have for many years played a central role in the clinical laboratory as diagnostic tools and also in the treatment of various diseases and we shall briefly look at a few ways in which antibodies fulfil this role. Additionally, antibodies are extremely important tools to the research scientist and we shall also briefly review this area.

8.2 Antibodies and the clinician

8.2.1 Antibodies as diagnostic tools to detect and measure antigens

diagnostic tools

Because of the unique specificities of antibodies, they occupy a very special place in clinical laboratories where they are used as diagnostic tools to detect and measure a whole range of different antigens of importance to the clinician and, in addition, are used as immunotyping reagents to identify the HLA antigen specificities and all the blood group antigens. It is difficult to imagine how we ever got along without them since they are probably the most important tools the clinical laboratory possesses and new antibody specificities are being developed continually to meet the demands of modern diagnostic medicine.

In Figure 8.1 we show some of the current uses of antibodies in the clinical laboratory. The methods used to detect the many different antigens obviously depends on whether the antigen is in soluble form, is expressed on cells or is present in pathological tissue specimens. However, these methods have been covered for the most part in Chapter 2 since the examples we quoted then do not only refer to antigen-antibody reactions but demonstrate how antibodies can be used to identify antigens.

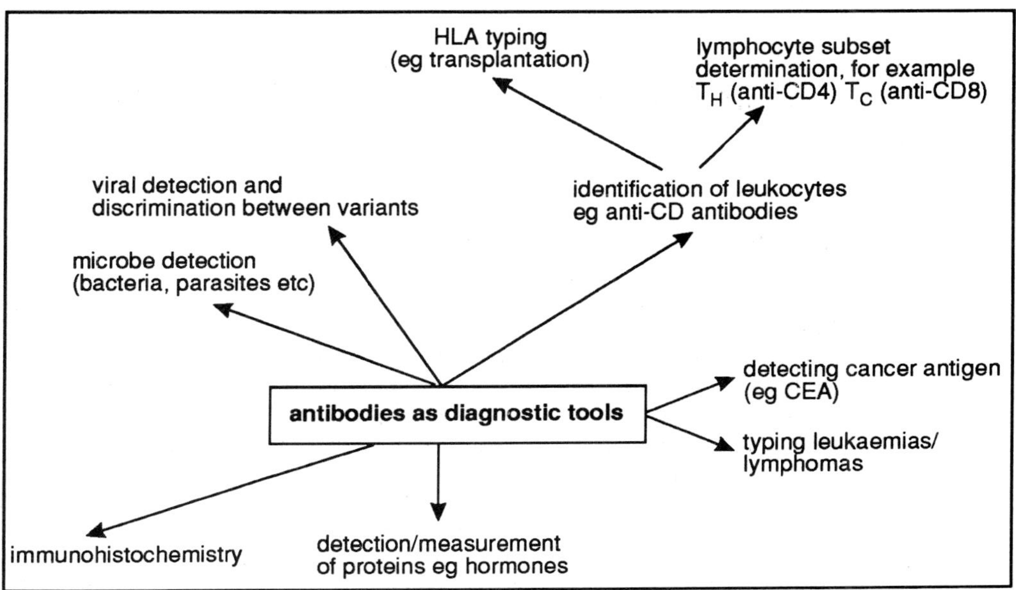

Figure 8.1 Some of the uses of antibodies in the clinical laboratory.

soluble antigens

Thus if you wish to detect or measure soluble antigens such as drugs, steroid hormones, peptide hormones, soluble cancer antigens (eg carcinoembryonic antigen CEA found at elevated levels in patients with gastrointestinal cancers) or IgE levels in serum you would do it using an antibody reagent in either radio-immunoassays or ELISA-based assays. If you wish to measure IgG, IgA or IgM levels in serum you would use a less sensitive immunoprecipitation technique.

In a laboratory supporting the transplant unit in a hospital, there is a battery of antibodies available which can be used to distinguish between the many HLA Class I and Class II specificities and all the different members of the blood groups thus enabling cross matching of the cells of donor and recipient. There are other T cell specific antibodies, for example, anti-CD3, anti-CD4 and anti-CD8, which are used to monitor the T cell population during immunosuppressive regimens following transplantation of the organ.

In the microbiology laboratory, there are diagnostic kits which use antibodies to detect individual viruses and distinguish between viral subtypes such as the influenza variants. There are other antibodies which can be used to identify parasites, bacteria and fungi.

In the histopathology laboratory there is a wide range of antibody reagents which can detect antigens in tissues using enzyme labelling and immunofluorescent techniques and other antibodies which can be used to type leukaemias and lymphomas. There are also a whole range of antibodies to the CD antigens on different types of lymphoid cells which have many diagnostic uses.

8.2.2 Antibodies are indicative of many diseases

As well as antibodies which have been generated in the laboratory as specific reagents to identify antigens, we also find that in many diseases antibodies are found in patients which give the clinician strong indications as to the nature of the disease or of the disease-causing organism. They can be considered to be nature's own contribution to the diagnosis of disease and these antibodies can be detected by a variety of methods, many of them referred to in Chapter 2.

Antibodies are often found in the sera of patients with infectious diseases. An example of current interest is the detection of antibodies in a majority of patients with HIV infection. These are normally detected using enzyme immunoassays eg ELISA assays using HIV or components of it as the antigen.

Clinicians can also be aided in their diagnosis of immune disorders by identifying antibodies in the sera. These include anti-thyroglobulin antibodies in a majority of patients with Hasimoto's thyroiditis, anti-treponemal antibodies in patients with syphills (these are also used to identify the microbe), antibodies to pancreatic islet cells in patients with insulin dependent diabetes mellitus, IgE antibodies in patients with allergies, antinuclear antibodies in patients with the autoimmune disease systemic lupus erythematosus and so on.

8.2.3 Therapeutic uses of antibodies

We shall be discussing various potential uses of new antibody reagents later but we will give you a few examples of how antibodies are used in therapy now.

anti-digitoxin antibody

OKT3

If a patient is suffering from drug toxicity say from the heart drug digitoxin, then the drug can be removed from the circulation by injection of an anti-digitoxin antibody. In transplant patients, the antibody OKT3 which has a anti-CD3 specificity (CD3 is a common antigen on all T cells) is used to remove the T cells. We shall be looking at more selective ways of doing this later. Thus OKT3 acts as an immunosuppressor.

antibodies to treat cancer

anti-idiotypic antibodies

There have, of course, been many attempts to use antibodies to treat cancer since the first tumours were treated about 65 years ago with hyperimmune sera from rabbits immunised with tumours. Some of the most successful treatments have been against B cell cancers such as follicular lymphomas and Burkitts lymphoma where anti-idiotypic antibodies were used and some regressions were observed and one complete remission was reported.

Π Can you explain why anti-idiotypic antibodies should have this effect?

A B cell cancer arises from a single cell and because of this monoclonality, all the cells will express the same idiotype in the antibody receptor. An antibody directed against this idiotype could bind to the tumour cell and promote killing of the tumour cells by natural killer cells, macrophages and other cells bearing Fc receptors.

However, although there are many reports of successes in antibody treatment of tumours in experimental animals, this is not reflected in the clinical setting where very few successful treatments have been reported.

Passive antibody therapy is much more successful in the treatment of infectious disease where we saw in Chapter 3 the possible uses of immune globulin both to prevent and to combat infectious disease. One of the outstanding successes of antibody therapy is the use of anti-Rh antibodies to prevent haemolytic disease of the newborn (refer back to Figure 3.9).

SAQ 8.1

Which of the following statements is/are incorrect?

1) Anti-idiotypic antibodies could be useful to treat T and B cell proliferative diseases.

2) Anti-CD3 antibodies (for example OKT3) are ideal for prevention of transplant rejection.

3) ELISA assays are commonly used to detect and measure antibodies in patients' sera.

4) IgG and IgE antibodies in sera are measured using immunoprecipitation techniques.

5) Antibodies in serum always reflects past history of disease in the patient.

8.3 Antibodies in the research laboratory

In the earlier chapters we introduced you to many uses of antibodies which are pivotal to the success of many research programmes. Thus antibodies are used to detect and quantify antigens and we have discussed this in some detail in Chapter 2. We have also described the use of antibodies to affinity purify antigens, we have described how antibodies are used to distinguish between different cell types and how they can be used to characterise cell specific membrane markers on these cells. We have shown how they have proved to be extremely useful in identifying cytokines and their receptors and in purifying these molecules. They are also used for epitope analysis of proteins which is essential for identifying B and T cell epitopes in the generation of new vaccines and antibodies have played a central role in elucidating the mechanisms involved in cell interactions. Their uses have extended beyond immunology. For instance, they are used to identify expressed products in molecular genetics and are used more and more in plant biology to identify and purify molecules of interest.

In other words, they are universal research tools due to their specificity. This property of antibodies has recently led to a further application of antibodies - as new enzymes.

8.3.1 Catalytic antibodies - abzymes

Using hybridoma technology complemented by molecular biology, it is possible to generate highly specific antibodies to virtually any molecule of interest and this could, in principle, allow development of antibodies which possess enzyme activities ie antibodies that can catalyze a chemical reaction.

Antibodies were first reported to be capable of catalysing chemical reactions in 1986 and, since that time, many reactions have been demonstrated to be catalysed by antibodies thus indicating that antibodies can be used as alternatives to enzymes and, perhaps, more importantly, to generate new enzymes where no natural enzyme is available. These catalytic antibodies have been called abzymes and have been generated by immunisation of mice with the enzyme substrate coupled to a protein carrier such as keyhole limpet haemocyanin (KLH). Many enzymes, as you know, have metal or cofactor requirements for activity and antibodies have been produced which incorporate additional sites for these. We will not go into the complexities of the chemical reactions catalysed by these antibodies as they are beyond the scope of this text but you should be aware of this important development representing yet another major use for antibodies. These reagents will not only be of use in research laboratories but will be used in industry for the generation of a variety of novel products using abzyme-catalysed reactions. If you wish to learn more about abymes, we recommend the BIOTOL text 'Technological Applications of Biocatalysts'.

8.4 Antibodies as immunoreagents

8.4.1 Bispecific or hybrid antibodies

Antibodies bearing two receptor sites of differing specificities are called bifunctional antibodies, hybrid antibodies or bispecific antibodies and the main aim of using these is to direct cytotoxic cells to kill tumours and pathogens.

Targeting of T cells to tumours using hybrid antibodies

anti-tumour and anti-TCR antibodies

One way to demonstrate the potential of bispecific antibodies is to prepare two monoclonal antibodies one with specificity for tumour cells and the other with specificity for the T cell receptor (anti-TCR). When these antibodies are covalently linked together and mixed with tumour cells and T cells, the tumour cells are killed.

Π You know, of course, that T cells are antigen specific and act in a MHC-restricted fashion. Before you read on, can you explain this phenomenon?

It is true that all T cells, including cytotoxic T cells, are antigen specific and MHC restricted but you can bypass these requirements by using an antibody to the TCR which activates the cell in the same way. It is known that anti-TCR antibodies when linked to Sepharose beads induce T cells to release various cytokines after binding and cytotoxic T cells will lyse cells which have anti-TCR bound to the surface.

Consequently, if a hybrid antibody consisting of anti-TCR and anti-tumour antibodies are incubated with the tumour cells, the antibodies will attach to the cells and the anti-TCR antibodies on the surface will be available to bind to and activate any cytotoxic T cell, irrespective of its antigen specificity, to kill the cell to which the bispecific antibody is bound. These ideas are illustrated in Figure 8.2.

In Figure 8.2, A shows the normal activity of cytotoxic T cells where they will only kill cells expressing the peptide bound to MHC I for which the particular T cell is specific. B shows how this mechanism can be bypassed using antibody directed against the T cell receptor. If this antibody is covalently linked to an anti-tumour cell antibody then binding of the bispecific antibody to its respective ligands will activate the T cell and it will kill the cell to which it is attached. The T cell cannot distinguish between the antibody linkage and the MHC I-peptide linkage to the target cell. Using this method all cytotoxic T cells, irrespective of their specificity, can be recruited to kill the tumour cells.

anti-CD3 antibodies

CD2

CD28

There are alternative ways of activating cytotoxic T cells other than with anti-TCR antibodies. You will remember that all TCR molecules are closely linked to the CD3 complex and anti-CD3 antibodies will also mediate the same effects. Additionally, antibodies to CD2, another trigger molecule present on all T cells and the CD28 activation molecule which normally interacts with B7 on B cells, will augment cytotoxic activity.

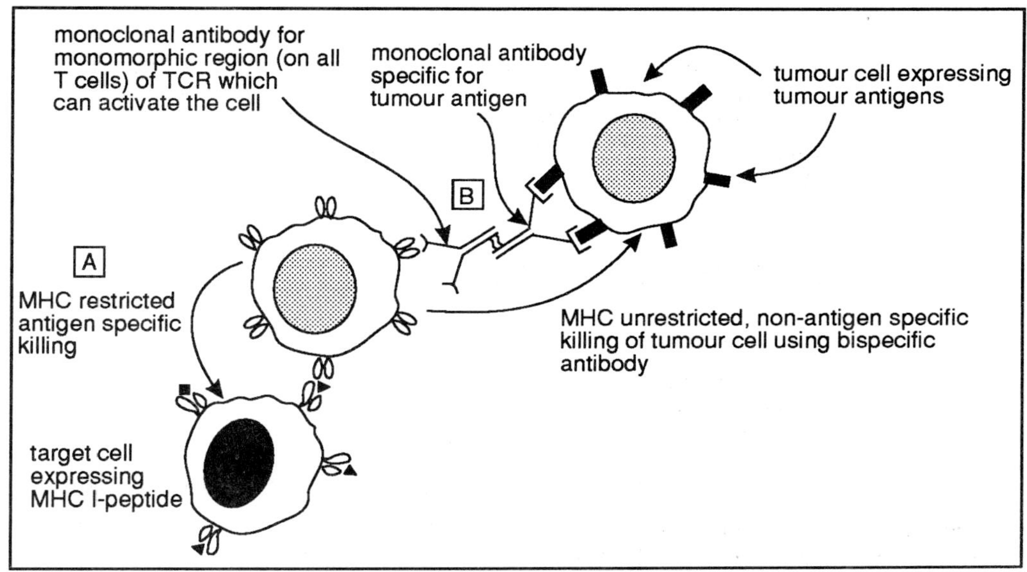

Figure 8.2 The use of bispecific antibodies to promote MHC unrestricted, antigen nonspecific killing by cytotoxic T cells. See the text for an explanation of A and B.

Presumably, if you wished to treat a cancer patient with a bispecific antibody, you would inject the antibody which hopefully would target and bind to tumour cells in metastases in the patient using the tumour specific component of the heteroconjugate. Any cytotoxic T cell which came into contact with these tumour cells would then be bound by the anti-T cell antibodies and would lyse the tumour cells. Obviously, since the activation molecules identified by the anti-T cell antibodies are also expressed on other T cells such as T helper cells, these would also become bound and would release cytokines into the microenvironment of the tumour and might augment the killing effects. One way this could be achieved is by the release of IL-8 which could attract other cytotoxic cells to the tumour.

Targeting other effector cells to tumours and pathogens

antibody dependent cellular cytotoxicity

Fc receptors

Macrophages, neutrophils, natural killer cells and eosinophils can all perform antibody dependent cellular cytotoxicity (ADCC) if they are presented with antibody coated target cells including tumours. Macrophages and neutrophils can also phagocytose antibody coated pathogens. These functions are dependent on the expression of Fc receptors of various kinds (FcγR1, FcγR11, FcγR111) for IgG.

Π Before you read on, can you think how you could use bispecific antibodies to promote tumour killing by these cells?

The Fc receptors on these circulating cells are probably occupied by antibodies of a wide variety of specificities and they will not kill tumour cells in the patient unless the latter are coated with IgG antibodies. However, once again, you could use bispecific antibodies to target the tumour cells as before; this time the second specificity will be anti-FcγR and this will bind to the receptors on the effector cells thus bringing them close to the tumour cell surface. The binding of the anti-FcγR to FcγR will also activate the effector cells resulting in ADCC. Similar approaches could be used to promote either effective phagocytosis or cytotoxicity of pathogens by macrophages, neutrophils or eosinophils. Examine Figure 8.3 which explains this mechanism.

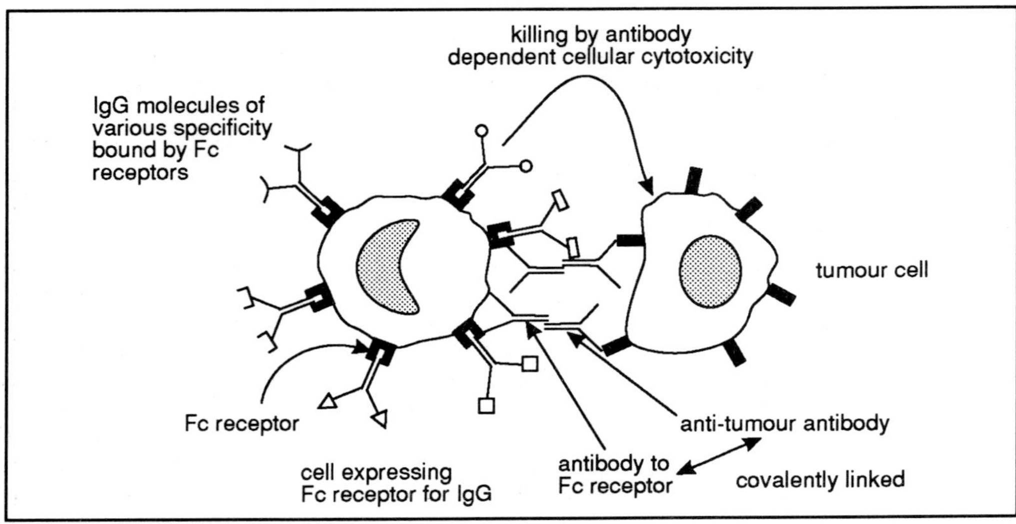

Figure 8.3 The potential use of bispecific antibodies to promote killing of tumour cells by Fc receptor bearing cells. This use can be extended to the killing of pathogens by macrophages, neutrophils and eosinophils. In the examples shown, the biospecific antibody binds to tumour antigen and to Fc receptor.

SAQ 8.2

Complete the table below indicating your choice with a + or a -. (To answer this you need to know quite a lot about the surface components of T cells, macrophages and natural killer (NK) cells and also quite a lot about their biological activities).

	Type of cell being targeted to tumour		
Component/process involved	**T cell**	**Macrophage**	**NK**
1) Anti-CD3			
2) ADCC			
3) Anti-CD2			
4) Anti-TCR			
5) Anti-tumour cell			
6) Anti-FcR			
7) MHC restricted			
8) Cytotoxicity			
9) Phagocytosis			
10) Cytokine production			

Preparation of bispecific antibodies

The first bispecific antibodies were prepared by chemically linking the antibodies together with such agents as SPDP which we met in chapter 2. The alternative approach is to fuse two hybridoma cells making the different antibodies resulting in what is called a quadroma; some of the resulting antibodies will be truly bispecific, each V_H V_L pair having a different specificity. Sometime in the near future F(ab')$_2$, genetically engineered in *E. coli* will probably replace these methods.

Π Can you think of an advantage of using $F(ab')_2$ over that of whole bispecific antibodies in the targeting of tumour cells in the patient?

The fact that the antibody fragment is smaller means that it may have better access to the tumour outside the blood stream.

Clinical trials

There have been some successes reported using bispecific antibodies to treat various cancers. In one trial, T cells from patients were first treated with an anti-CD3-anti-tumour antibody and the cells were then infused back into patients with carcinoma of the breast, lung or ovaries. Considerable destruction of tumour cells was observed. In another study, patients with brain tumours were treated with IL-2 activated and expanded natural killer cells (obtained from the patient) together with bispecific antibodies containing anti-tumour specificity following surgical excision of the primary growth. This treatment resulted in disappearance of the tumours in a majority of the patients and was more effective than giving natural killer cells alone.

SAQ 8.3

Assume you have two clones of hybridoma cells, one produces anti-T cell antibodies, the other produces anti-tumour antibodies. You are attempting to produce bispecific antibodies with anti-T cell anti-tumour specificities. You propose to fuse the two types hybridomas.

Which of the following statements are correct with regard to the preparation of these bispecific antibodies using hybridoma fusions?

1) A majority of the primary hybridomas will successfully fuse.

2) Of the fused cells, a majority will be making the correct bispecific antibody.

3) It will be difficult to purify the bispecific antibody from the supernatant.

4) Many of the antibodies produced will not have specificity for the tumour.

5) Many of the antibodies will not have specificity for the T cells.

8.4.2 Recombinant globulins

recombinant globulins

Recent advances in recombinant DNA technology have facilitated the production of recombinant genes which encode novel proteins. We have already discussed production of novel recombinant antibodies in Chapter 6 but recombinant globulins are significantly different from chimaeric or humanised antibodies since it is the antigen binding end of the antibody which has been replaced. Recombinant globulins then consist of the Ig constant domain to which is attached at the amino end some other protein. Thus you can study the properties of non-antibody proteins using all the reagents which react with the constant domains and the fusion proteins can be purified from supernatants of mammalian cells by affinity chromatography using protein A/G or anti-globulins. The usefulness of such reagents is shown in Figure 8.4.

There is a lot of information in Figure 8.4, so read through this figure before continuing with the text.

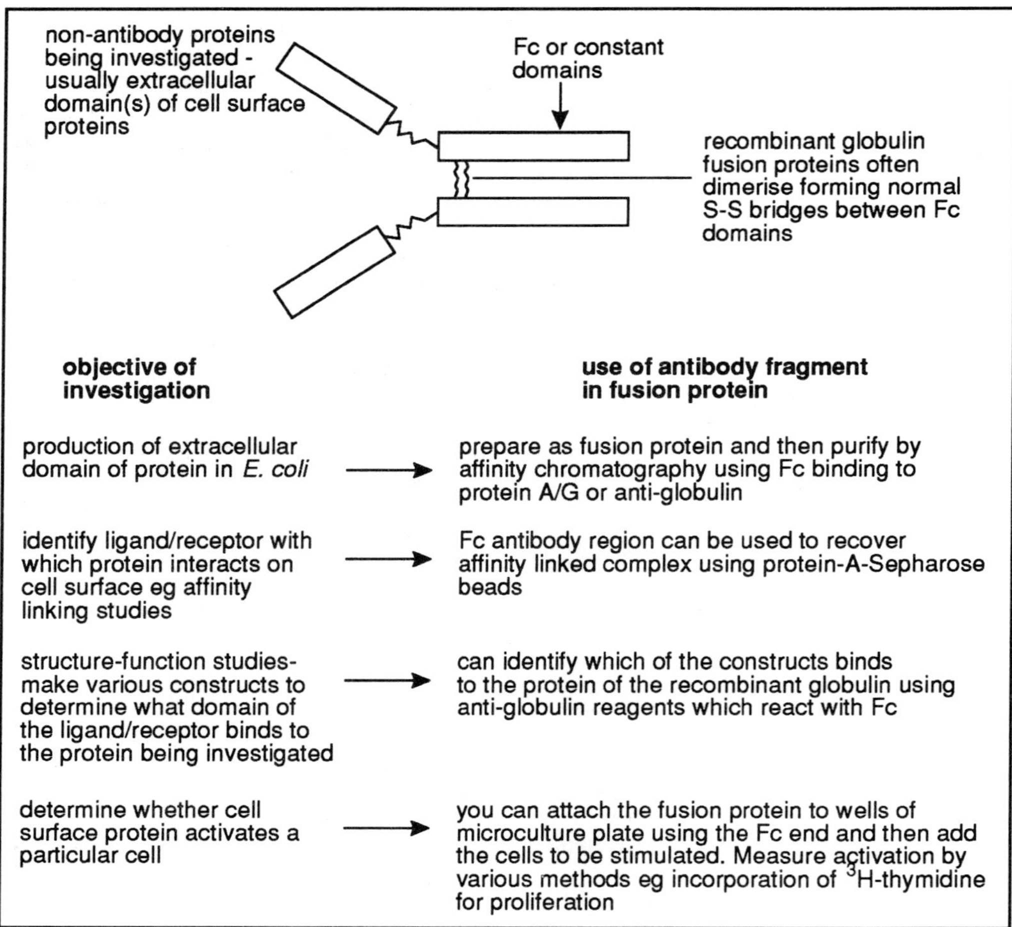

Figure 8.4 Some of the uses of recombinant globulins. The protein, which in most cases has been derived from an extracellular domain(s) of a cell surface protein is attached to C_H1, 2 and 3 or C_H2 and 3 of IgG heavy chains using recombinant DNA techniques. Some of the potential uses to which the Fc end can be put to characterise the protein are shown. See also Figure 8.5 and the text.

These novel reagents have been used to study the T cell receptor, ligand-receptor interactions particularly involving adhesion molecules and protein structure. The particular application we will discuss is the possible use of recombinant globulins as novel pharmaceutical reagents in AIDS and malaria patients. These particular recombinant globulins have been called immunoadhesins.

immunoadhesins

CD4-immunoadhesin for the treatment of AIDS patients

The entry of HIV into T helper cells and other cells is mediated by the CD4 molecule which is expressed in large numbers on all T helper cells and macrophages. A surface component of the HIV virus, gp120, binds to CD4 resulting in infection of the cell. The reduction of, and final elimination of, the T helper cell population renders the patient immunodeficient and susceptible to infection and cancer finally leading to death.

soluble CD4 molecules

One approach to prevent infection of the T helper cells is to use soluble CD4 molecules which would compete with the membrane bound CD4's on T helper cells for the virus.

Recombinant DNA technology has resulted in a C-terminally truncated, water soluble form of CD4 which lacks the domains which anchor it to the cell membrane. This novel reagent actually prevents HIV infection of cells *in vitro*. However, it has a relatively short half life of about 1-2 hours and does not have any effect on the virally infected cells or the virus. Thus the AIDS patient would have to receive a continual infusion of CD4 to prevent further infection of his/her cells. However, preliminary studies have shown a reduction of circulating virus in patients treated with high doses of this reagent.

CD4 immunoadhesins consist of the truncated CD4 molecules attached to the amino ends of an IgG Fc region. The cDNAs for both components are spliced together and expressed in mammalian cells. The fusion product has the capacity to bind to the gp120 and to block infection of CD4 expressing cells. The Fc portion is needed to give the CD4 immunoadhesion a long half life and it also facilitates the binding of the immunoadhesion by the Fc receptors of phagocytic cells.

Consequently, the CD4 immunoadhesin does not need to be continually infused and it can also mediate antibody dependent cellular cytotoxicity towards HIV-infected cells. If the CD4 is linked to the appropriate IgG subclass such as IgG1 it can also be transported across the placenta thus possibly affording protection to the unborn infant. The mechanism is depicted in Figure 8.5.

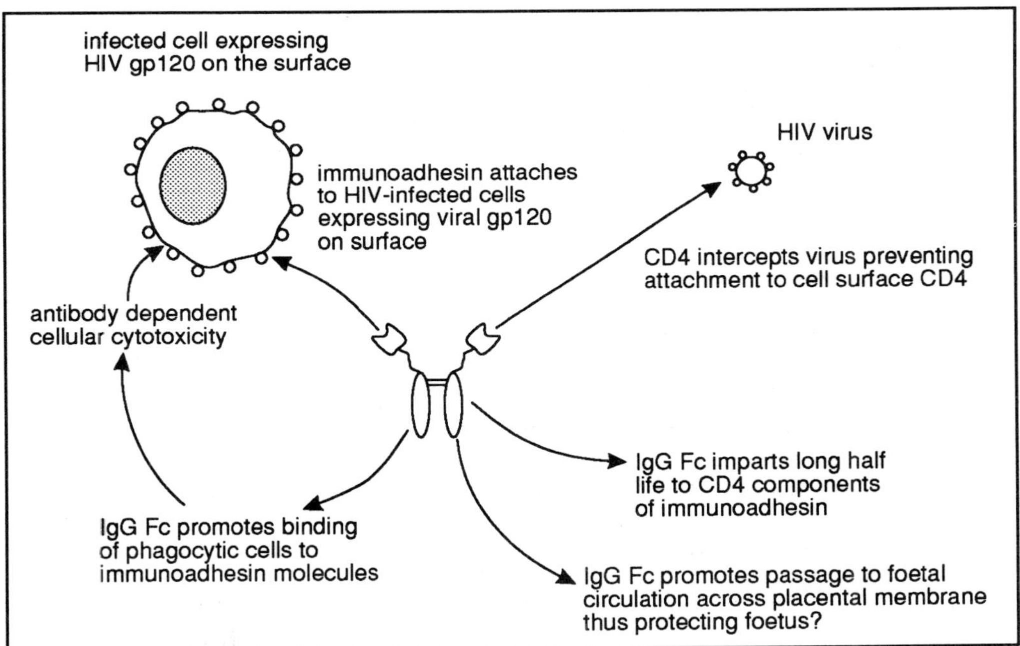

Figure 8.5 The potential roles of CD4-immunoadhesin in AIDS patients.

Another immunoadhesin has been developed as a potential treatment for individuals infected with the malaria parasite. This immunoadhesion consists of the intercellular adhesion molecule (ICAM-1) attached to the CH2 and CH3 domains of IgG1. It has been shown that red cells infected with the mature form of the malaria parasite bind to ICAM-1 expressed on endothelium in small blood vessels and this action may protect the parasite from destruction in the spleen and may also contribute to the pathology of malaria.

Treatment of patients with this immunoadhesin, it is hoped, will prevent the adhesion of the infected red cells to the endothelium since they will become attached to the immunoadhesin and this will promote the killing of the parasite by Fc mediated phagocytosis of the infected red cells. We await further developments in this interesting area.

SAQ 8.4

Complete the following statements using the word list below.

Recombinant globulins are molecules composed of [ ] [ ] of antibody molecules attached to other proteins. [ ] are one form of recombinant globulins which target cells for [ ] or [ ]. CD4-immunoadhesin extends the [ ] [ ] of CD4 and also promotes the [ ] of cells infected with virus which enter the cell through the CD4 molecule. ICAM-1 immunoadhesin prevents the binding of [ ] [ ] infected erythrocytes to venule [ ] through the [ ] molecule and promotes [ ] of the infected red cells and the parasite by [ ] [ ] [ ] cells such as [ ].

Word list: malaria parasite; immunoadhesins; phagocytosis; Fc receptor bearing; half life; endothelium; ICAM-1; cytotoxicity; destruction; macrophages; constant domains; killing.

8.4.3 The magic bullet - immunotoxins

Immunotoxins consist of toxic molecules attached to whole antibody molecules or antibody fragments. The rationale behind this approach is that the antibody has specificity for the target cell such as a tumour cell and binds to its surface. The whole immunotoxin is then taken up into the cell and will kill the cell.

Types of toxins

holotoxins, diphtheria toxin, ricin, *Pseudomonas* exotoxin A, ribosome inactivating proteins, pokeweed antiviral protein, gelonin and saporin

The most popular toxins used in the earlier studies are called holotoxins and consist of disulphide-linked A and B chains. They include diphtheria toxin secreted by some strains of *Corynebacterium diptheriae* and ricin from the seeds of the castor bean *Ricinus communis*. *Pseudomonas* exotoxin A is a toxin which has been used recently for the preparation of recombinant immunotoxins. This toxin is a single protein consisting of 3 domains, one for cell binding, one with enzyme activity and the third for movement of the whole molecule across the membrane into the cytoplasm. Another class of plant toxins inactivates ribosomes. This class is called RIPs or ribosome inactivating proteins. These single chain molecules are very similar to the A chains of the holotoxins and do not possess affinity for most cells since they lack a B chain. These toxins include pokeweed antiviral protein, gelonin and saporin.

Action of holotoxins

The B chain is responsible for binding of the toxin to the target cell and is usually galactose or lactose specific whereas the A chain possesses enzyme activity. After binding of the toxin to the cell membrane, it is taken up into the cell by endocytosis. The resulting endosome becomes increasingly acid, this causes the cleavage of the disulphide link between the two chains and the A chain passes through the endosomal membrane into the cytoplasm where it blocks protein synthesis and leads to cell death (Figure 8.6).

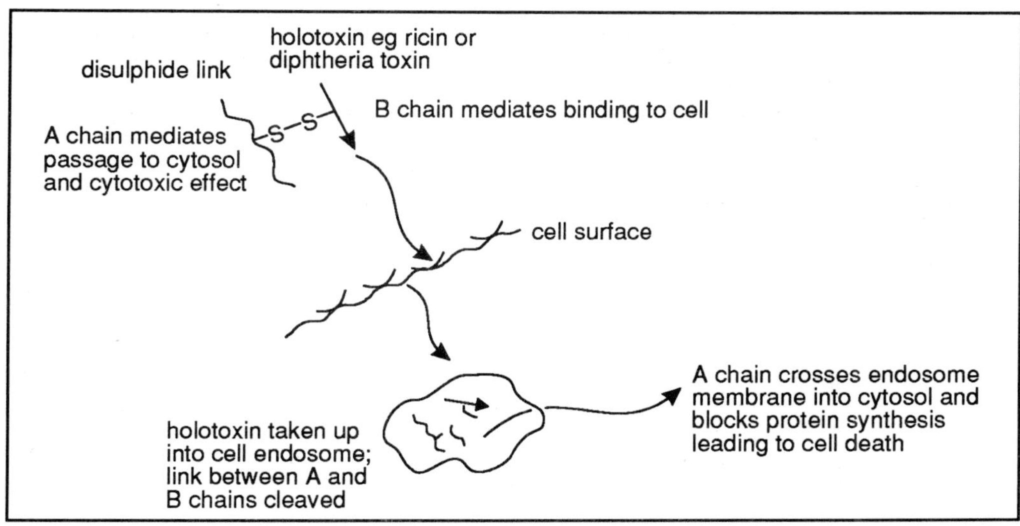

Figure 8.6 Basic mechanism of a holotoxin. Some toxins eg *Pseudomonas* exotoxin consist of one chain with 3 domains. Domain 1 mediates cell binding (equivalent to chain B of holotoxin), domain 2 mediates translocation to the cytosol and domain 3 possesses the cytotoxic activity (the activities of domain 2 and 3 equivalent to the A chain of holotoxin).

Π How would you prepare a tumour cell specific immunotoxin?

Using a holotoxin, it would obviously be preferable to remove the B chain since its presence may lead to destruction of non-target cells. However, although removal of the B chain and its replacement by an antibody specific for the tumour cell leads to better targeting, the cytotoxicity is often much weaker than the whole toxin molecule.

Π Make a list of the approaches you could adopt to deal with this problem.

excess of free B chains

There are a number of approaches you could adopt to get over this problem of reduced A chain cytotoxicity although there is little evidence that they would work *in vivo*. You could inject an excess of free B chains along with the A chain immunotoxin (A-IM) or with the holotoxin (A and B chain)-immunotoxin (AB-IM). This would prevent non-specific binding of the immunotoxin. Alternatively you could use the AB-IM with an excess of galactose or lactose (B is galactose or lactose specific). However, although this would be effective in a culture dish, it is unlikely to work *in vivo* since the sugar would be rapidly metabolised or excreted.

Another approach would be to prepare separate A and B immunotoxins and inject them together. Both immunotoxins would target the same cells and the presence of the B chain in the endosome may enhance the cytotoxicity of the A chain. Yet another approach could be to engineer out the lectin (sugar binding) activity of the B chain and use the whole holotoxin or inject B chains linked to anti-A antibodies along with A-IM. This latter approach would of course mean that the B chains would accumulate in the same areas as A. Since A is also attached to an antibody, it will only accumulate in areas in which the ligand (antigen) for this antibody is found. Thus both A and B accumulations are mainly confined to the target areas.

SAQ 8.5

Which of the following strategies would cause killing of tumour cells in culture? Assume that the immunotoxin has specificity for the cancer cell.

1) Ricin alone.

2) A-IM along with B chains linked to anti-A antibodies.

3) Bispecific antibodies for A chains and tumour cells + free A chains.

4) A chain, anti-A chain antibodies and anti-tumour cell antibodies.

5) AB-IM + anti-B chain antibodies.

Limitations when using immunotoxins

Π Make a list of what you consider will be the limitations of immunotoxins in the treatment of cancer.

Although these immunotoxins show unique specificity and cytotoxicity for cancer cells *in vitro* there are problems which will have to be addressed for them to be successful in the treatment of various tumours in patients. These problems include the following:

- accessibility of the tumour for immunotoxins. If the immunotoxin is of high molecular weight it may have difficulty in targeting the tumour or permeating the tumour mass. However, with the advent of recombinant DNA techniques, most immunotoxins will consist of single chain Fv's linked to toxin fragments which mediate the cytotoxic activity. This approach will increase the accessibility of the cells within the tumour;

- immunotoxins can induce antibodies in the patient. Even though the Ig part of the molecule is human, the actual toxin component will induce an immune response;

- tumour antigens shed from the surface of tumour cells will be bound by the immunotoxin thus preventing any action on the actual tumour. The complexes formed from the interaction of shed antigens and immunotoxins will be removed by cells expressing Fc receptors thus causing damage to non-target cells. Again, using scFv immunotoxins, you would eliminate the Fc receptor-mediated problems;

- cells not expressing the tumour antigen will not be affected by the immunotoxin. It is obviously essential that all the tumour cells express a sufficient density of antigen for binding of the immunotoxin to occur;

- the immunotoxin must be cleaved in the endosome to release the A chain or its equivalent to the cytoplasm. In some cases it has been difficult to induce endosomal uptake of the immunotoxins; instead they have been sequestered into lysosomes where they have been degraded.

In spite of these potential problems, immunotoxins appear to have a future in immunotherapy and some successes have already been reported (see later).

Genetically engineered immunotoxins

genetic engineering

The early immunotoxins were produced by chemically linking the chains by disulphide bridges and there were then problems with purification since heterogeneous products were formed. The present approach is to prepare them using genetic engineering and the most popular toxin incorporated into the reagents is *Pseudomonas* exotoxin.

Π Diphtheria toxin (DT) has also been used to make a recombinant IL-2-DT immunotoxin for targeting activated T cells in T cell leukaemias or transplant patients. Would this be effective?

The problem with using diphtheria toxin is that most individuals have antibodies to DT due to being immunised with the triple vaccine diphtheria, pertussis and tetanus. You would therefore expect that these antibodies would neutralise the effects of the toxin.

To give you an example of an immunotoxin, one group at the National Institute of Health in the USA made a construct encoding the amino acids 253-613 of *Pseudomonas* exotoxin (essentially domains 2 and 3) and fused it to cDNAs encoding a single chain Fv of the variable regions of the anti-Tac antibody as shown in Figure 8.7. The immunotoxin was expressed in *E. coli* and was found in the inclusion bodies from which it was extracted and purified. In this construct, the exotoxin is attached to the carboxyl end of the light chain variable segment. In a subsequent study, the same group prepared a second immunotoxin in which the first 388 amino acids of DT were attached to the amino end of the heavy chain variable segment of the scFv.

These immunotoxins were shown to be very effective in killing activated T cells which express the p55 receptor chain of the IL-2R complex but have no effect on resting T cells which do not express this receptor chain.

Π Can you think of uses for this immunotoxin?

You could use this immunotoxin to selectively kill T cells which have become activated and are expressing their p55 IL-2 receptor chains. You would expect to find activated T cells in patients with autoimmune disease, those who have received a transplant and patients with T cell based tumours.

Π Can you suggest an alternative immunotoxin which could serve the same purpose?

IL-6 immunotoxin

Instead of using cDNAs based on anti-receptor antibodies you could also use the cytokine itself ie IL-2 linked to an exotoxin or DT. An IL-6 immunotoxin has also been developed to target IL-6 receptors on myelomas and hepatomas and epidermal growth factor immunotoxins have been produced to target various cancers which express high levels of the epidermal growth factor receptor.

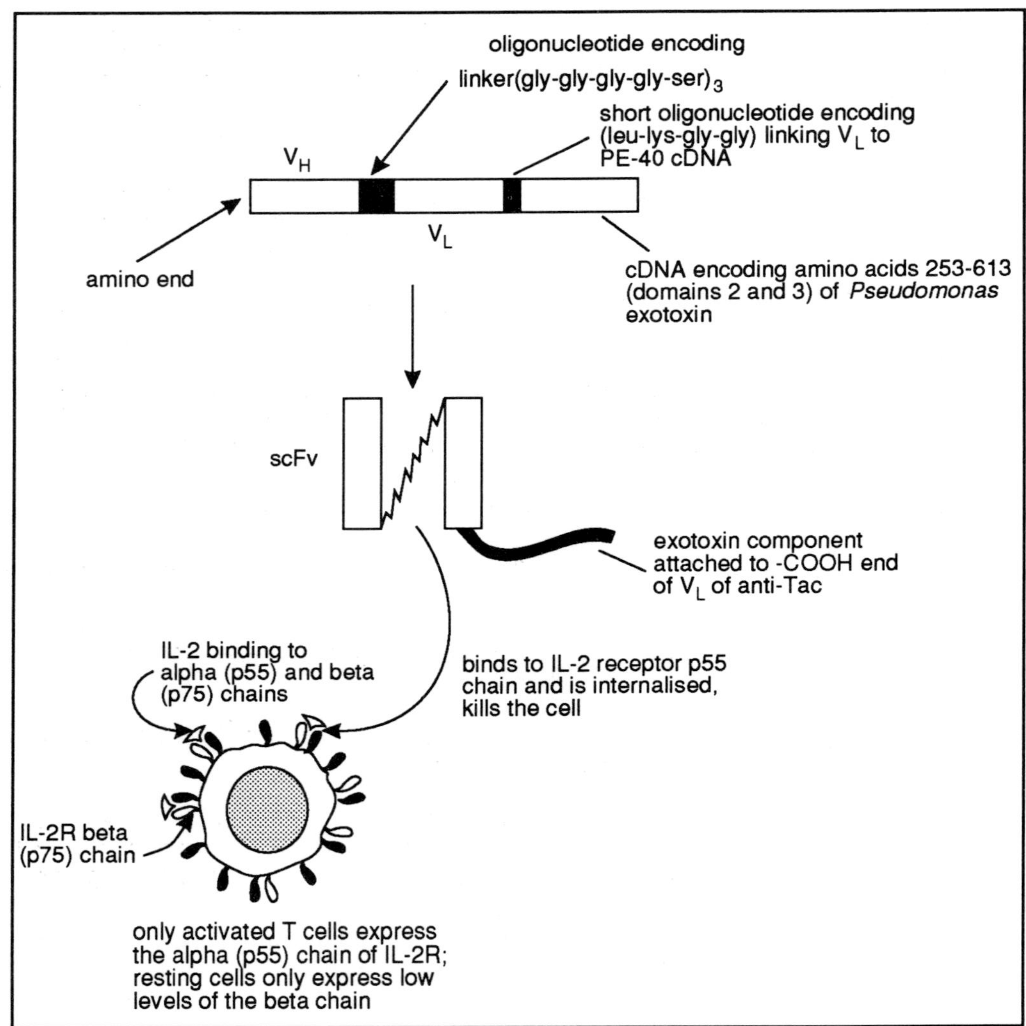

Figure 8.7 An example of an immunotoxin showing an scFv with specificity for the alpha (p55) chain of the IL-2 receptor linked to a truncated form of *Pseudomonas* exotoxin. This immunotoxin targets activated T cells which express the IL-2R alpha chain.

Trials in mice and Man

Some immunotoxins have been assessed for their effectiveness in immunodeficient mice prior to clinical trials in patients. SCID mice are immunoincompetent since they lack functioning T and B cells and these mice will accept, without rejection episodes, various cells from Man. Some SCID mice were injected with Daudi cells (B cell cancer) and these produced a disseminated B cell lymphoma in the animals similar to what would be produced in Man. Treatment of the mice with ricin A-anti-CD22 resulted in a major reduction in the number of cancer cells in the mice, demonstrating that the immunotoxin had succeeded in killing many of the cancer B cells. CD22, by the way, is a molecule present on all B cells and so all B cells would be targeted.

ricin A-anti-CD22

anti-CD7-ricin A

anti-CD19-pokeweed antiviral protein

In a similar experiment, anti-CD7-ricin A reduced, by 200-fold, the number of human leukaemic T cells in mice and in another study, anti-CD19-pokeweed antiviral protein reduced the death rate of mice injected with a human B cell acute lymphatic leukaemia cell line by about 70%. As before, CD7 is present on all T cells and CD19 on all B cells and the immunotoxins would kill non-cancer cells as well.

In human trials, an anti-CD19-ricin immunotoxin was given to 23 patients with B cell lymphoma. All signs of the cancer disappeared in one patient and there were major improvements in two other patients. Using a Fab anti-CD22-ricin A immunotoxin on 15 patients with B cell lymphoma, major reductions in tumour size were noted in 6 patients. Some patients made antibodies to either the mouse antibody or the toxin.

So you can see that some progress is being made using immunotoxins in cancer patients and we can expect exciting developments in the next few years. The major problem appears to be the development of antibodies to the toxin which limits the administration of multiple doses of immunotoxin over an extended period.

SAQ 8.6

Which of the following statements are correct?

1) Anti-CD19 immunotoxins are not useful immunoreagents since they kill all B cells.

2) Immunotoxins which incorporate whole human IgG antibodies have no major side effects.

3) Generally, immunotoxins must target cell surface molecules which naturally enter the cell by endocytosis.

4) If immunotoxins target cell surface molecules which are expressed on both cancer cells and normal cells, then immunotoxintherapy will only be possible when the cell surface molecules are not expressed on stem cells.

5) Anti-IL-2R-toxin is a better immunotoxin than anti-CD3-toxin to prevent transplant rejection.

Bone marrow rescue using immunotoxins

bone marrow rescue

In some leukaemia's, such as acute lymphocytic leukaemia (ALL), a considerable success rate has been achieved using bone marrow rescue. In this procedure, a bone marrow sample containing haematopoietic stem cells is removed from the patient during a remission period. In ALL a majority of the blood lymphocytes are of leukaemia origin derived from some of the stem cells which have undergone carcinogenesis. The patient is now aggressively treated using high dose radiotherapy and/or chemotherapy which destroys all the blood cells and the remaining stem cells in the bone marrow.

Π What would be the immune status of this patient?

Since all the blood cells including the lymphocytes have been destroyed, this patient is totally immunodeficient and must be kept in germ free conditions to prevent infection. The bone marrow stem cells which were withdrawn from the patient before treatment are treated with the immunotoxin which selectively destroys the cancer stem cells. The remaining healthy stem cells are then injected back into the patient and after a period of time will reconstitute the haematopoietic cells lost during radiotherapy.

Use of radioimmunoconjugates

You can use radioimmunoconjugates as an alternative to immunotoxins to target cancer cells. These are antibodies with specificity for the cancer cells to which a radioactive isotope has been attached. However, in many instances, the level of radioactivity needed to achieve significant clinical responses is very toxic to other cells. For instance, in a recent clinical trial, patients with B cell lymphomas were treated with anti-B cell specific antibodies labelled with radioactive iodine or yttrium. The patients then had to be infused with autologous bone marrow which had been removed prior to treatment to reconstitute the patient with haematopoietic cells. So you can see this approach is not as specifically targeted as that using immunotoxins.

SAQ 8.7

Complete the table below indicating your choice with a + or -.

Characteristic	Recombinant globulins	Immunotoxins
1) Single chain constructs		
2) Many are antigen specific		
3) Immunoadhesins		
4) Fc receptor involved		
5) Protein A purification		
6) Variable regions		
7) Genetically engineered		
8) Needs endocytosis		
9) May be of non-antibody construction		

8.5 Cytokines as immunotherapeutic agents

An alternative approach to the use of antibody targeting in cancer treatment is to try to use cytokines to boost the activities of the patient's immune system which will hopefully result in the destruction of the cancer cells. In most cancer patients, there is evidence to suggest the presence of immunosuppressive mechanisms arising from the cancer or the patient's own immune system which prevent the clonal expansion of the patient's T cells which are specific for the cancer cells and would normally be capable of destroying the cancer cells. Thus the use of cytokines is an attempt to overcome these suppressive effects and to reactivate the immune system to destroy the cancer cells.

immuno-suppressive mechanisms

This is a very attractive approach since natural products of the immune system are used which should be less traumatic than the very aggressive treatments currently used following excision of the primary tumour, namely radiotherapy and chemotherapy. Since cytokines injected into the patient with cancer or an infectious disease can modify the response, they have been called biological immune response modifiers. To date, there is only very limited success using this approach and an alternative strategy which is being actively pursued at the present time is cell-based immunotherapy.

biological immune response modifiers

cell based immunotherapy

adoptive immunotherapy

active immunotherapy

The basis of this strategy is shown in Figure 8.8; it involves generating an anti-cancer response using two distinct approaches. In the first method, called adoptive immunotherapy, limited numbers of anti-cancer cells are harvested from the patient and these are expanded *in vitro* using cytokines before re-infusing them back into the patient. In the second method, called active immunotherapy, cytokine genes are introduced into either tumour cells or anti-tumour T cells followed by injection of these cells into the patient in an attempt to produce high levels of cytokine *in vivo*. Let us now look in more detail at each of these strategies.

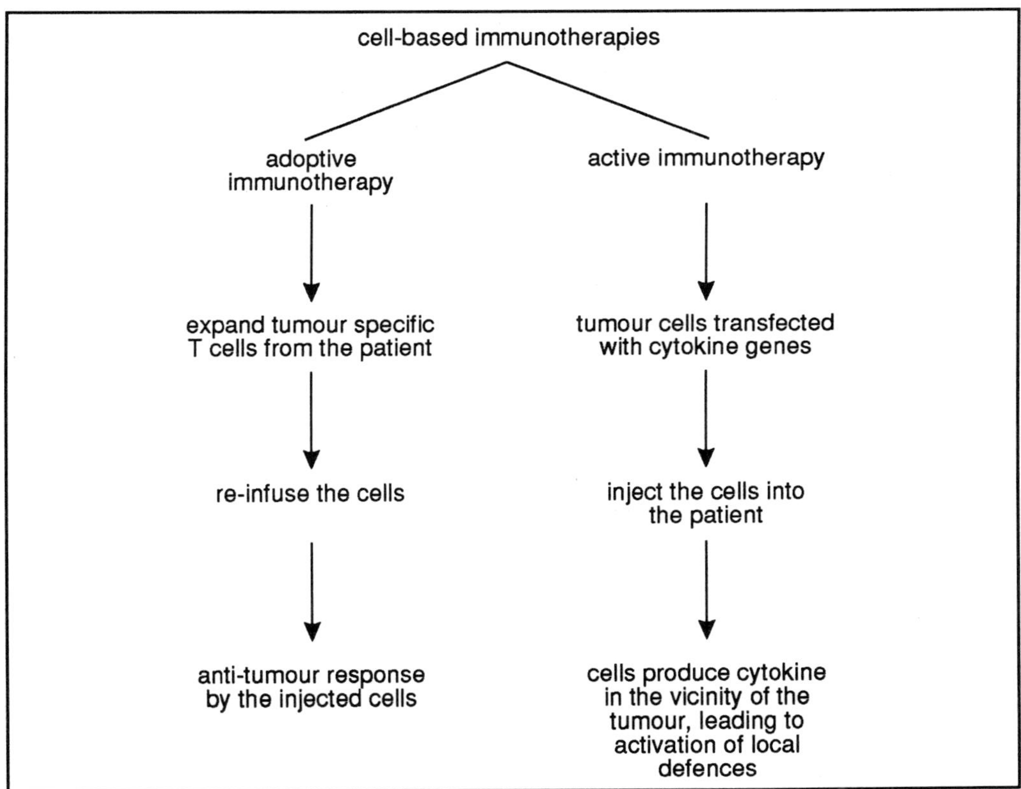

Figure 8.8 Two methods being investigated for the immunotherapy of cancer.

8.5.1 Cytokine therapy in cancer

hairy cell leukaemia

interferon-alpha

AIDS related Karposi's sarcoma

This has usually involved multiple injections of high amounts of a particular cytokine into the patient and many different cytokines have been used in preliminary studies. There have been some successes but they are rather few in number. For instance, treatment of patients with hairy cell leukaemia with interferon-alpha has resulted in a complete response (absence of disease) in about 14% of patients and improvements in another 54% of patients. However, it is expected that most patients will suffer a relapse in the future. Treatment of AIDS related Karposi's sarcoma with the same cytokine also resulted in some apparent 'cures' in about 12% of patients. There are also some reports of success using extremely high doses of IL-2 to treat some cancers such as metastatic melanomas and renal cancers.

Apart from these reports, there is little evidence at the present time that treatments of cancer patients with individual cytokines have any major curative effects and high doses of cytokines have considerable toxic effects on patients. The problem with this strategy is that we do not know how much cytokine is reaching the metastases to influence the local T cells and other lymphoid cells at the site, and therefore the emphasis has switched to cell-based therapies in an attempt to more specifically target the cancer cells.

8.5.2 Cytokine genes in tumour cells

If you prepare a cell suspension from a resected primary tumour, you will find that about 95% of the cells are tumour cells and the remainder consist chiefly of various lymphoid cells including tumour specific cytotoxic T cells. Since these are found in the tumour mass they have been called tumour infiltrating lymphocytes (TIL). We shall discuss these later. The important point we are making at this stage is that some cytotoxic T cells specific for the tumour do succeed in searching out the tumour but do not succeed in destroying it. We have already suggested that this may be due to suppression of these cells by the tumour cells.

tumour infiltrating lymphocytes (TIL)

The question now being asked by investigators is: 'can these cells be activated if they are supplied with a local source of cytokines'. There are two ways of supplying these cytokines, you could make the cancer cells supply them or you could make cancer specific T cells supply them. The major progress has been made using the former method and we shall concentrate on the sort of experiments which have been done in mice.

The idea is to transfect tumour cells with cytokine gene(s) and inject these into mice and then examine whether the tumour is rejected and whether the mice have developed long term tumour specific immunity. A typical experiment showing the effects of treatment of mice with IL-2 gene-transfected tumour cells is shown in Figure 8.9. You will notice that if you first inject the transfected cells into the mice and a few weeks later challenge the mice with the non-transfected tumour cells, no tumour develops. This demonstrates that the treatment was effective in generating tumour immunity, both the transfected cells and the untransfected cells must have been destroyed.

IL-2 gene-transfected tumour cells

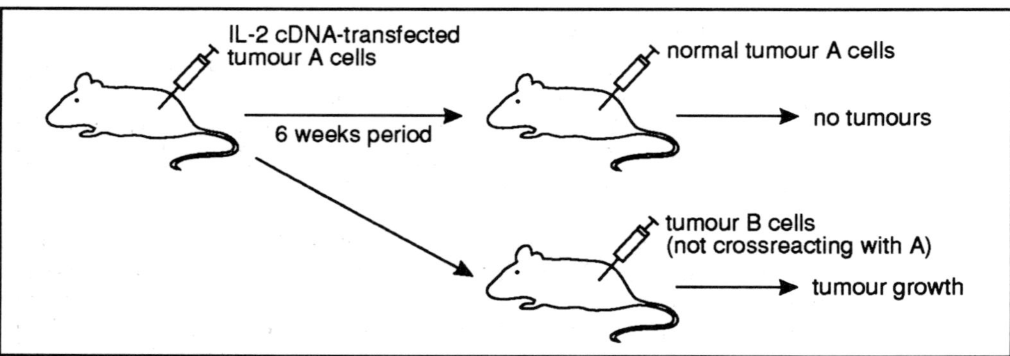

Figure 8.9 Experiment to demonstrate the effectiveness of pretreating mice with IL-2 cDNA-transfected tumour cells in producing tumour immunity.

Similar successes have been reported using tumour cells transfected with other cytokines including IL-4 and tumour necrosis factor and we await further development in this exciting area of tumour therapy.

SAQ 8.8

Which of the following statements are incorrect?

1) Adoptive immunotherapy involves clonal expansion of tumour specific T cells in the patient.

2) The experiment described in Figure 8.9 shows that IL-2 acts in an antigen specific manner in inducing tumour immunity.

3) The idea of active immunotherapy is to produce large amounts of cytokine at the tumour site.

4) IL-2 alone is ineffective as an immunotherapeutic reagent.

5) It is reasonable to assume that TIL need large amounts of IL-2 to be effective in tumour immunity.

8.6 Adoptive immunotherapy using LAK and TIL

adoptive immunotherapy

lymphokine activated killer cells (LAKs)

Adoptive immunotherapy pre-dates active immunotherapy by a few years and is principally due to the work of one investigator, Steven Rosenberg and his group at the National Cancer Institute in the USA. Adoptive immunotherapy is so-called since the host (patient) accepts (adopts) cells derived from the host. In this method, cells are removed from the patient, activated and expanded with cytokines and then re-infused back into the patient. Two types of cells have been used in this approach, lymphokine activated killer cells (LAKs) and the tumour infiltrating lymphocytes we mentioned earlier. There are two major differences between these two cell types. Firstly, LAKs are obtained from the peripheral blood of cancer patients or the spleens of mice whereas TIL are obtained from the actual tumour mass itself. Secondly, TIL are tumour specific whereas LAK are not.

8.6.1 Lymphokine activated killer cells

LAK cells were first described in 1980 when the addition of IL-2 to human peripheral blood cells caused the proliferation of non-T cells. These were called lymphokine activated killer cells since they were found to kill some tumour cell lines *in vitro*. They were finally identified as activated natural killer (NK) cells and although they can kill some tumour cells they are not antigen specific.

Rosenberg and his coworkers examined the ability of these cells to cause the regression of tumours in mice. To do this they made cell suspensions from spleens of normal mice, cultured these cells with IL-2 for 3 days in culture dishes and then transferred the cells along with IL-2 into tumour bearing mice where there was metastatic spread. The found that many tumours regressed with this treatment (Figure 8.10).

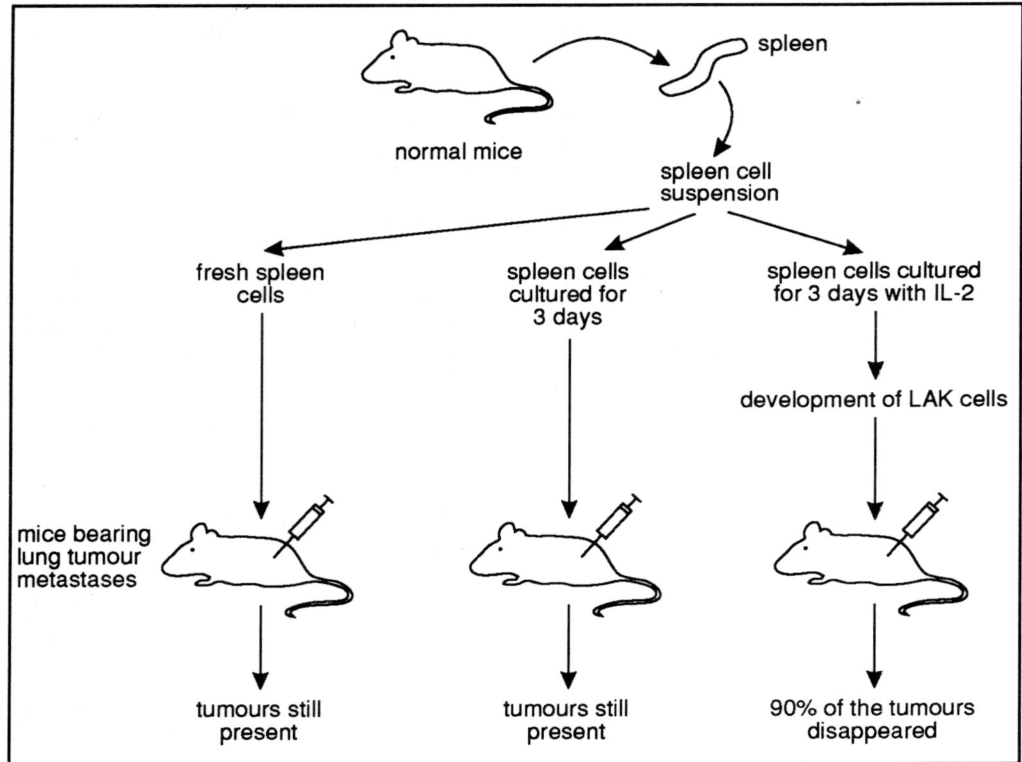

Figure 8.10 The development and use of lymphokine activated killer cells in tumour bearing mice. In Man, LAK cells are obtained by leukophoresis from the blood stream.

They then applied this method to terminally ill patients who had already undergone all the traditional treatments (cancer excision, radiotherapy, chemotherapy) without success. Millions of cells were removed from the patient by leukophoresis and the cells were then incubated with IL-2. The LAK cells which developed in culture were found to kill a variety of cancer cells including those from melanomas (skin cancers), cancer of the colon and some sarcomas (cancers of the connective tissues). Subsequently, the LAK cells were infused into cancer patients along with IL-2 and by 1990, 300 patients had been subjected to this type of immunotherapy with impressive results. 35% of patients with cancers of the kidney, 21% of patients with melanoma and 17% of patients with colorectal cancer showed either complete or partial regression of the cancers. However, the treatment had no effect on patients with sarcoma, lung cancer or breast cancer.

Π You will have noticed that IL-2 is always administered along with the LAK cells. Can you suggest why?

In the treatment of cancer patients, huge numbers of LAK cells (often 50 billion) were injected. Many of these would not succeed in targeting the metastatic growths, and this is one of the reasons why so many cells are injected. Additionally, it has been estimated that many thousands of cells are needed to kill a 1 cm wide tumour so many cells need to accumulate at each tumour site. The IL-2 was found to induce proliferation of these cells at the tumour site; this did not occur in the absence of IL-2 and the tumours did not regress unless IL-2 was also administered.

Based on this work, this approach is now accepted USA-wide as a treatment for patients with advanced melanoma and kidney cancers. However, there is a more effective approach using tumour specific T cells.

8.6.2 Tumour infiltrating lymphocytes

The discovery of TIL was based on the idea that if these cells are specific for the tumour then the tumour mass itself should be a rich source of such cells. Rosenberg's team, therefore, removed a tumour from a mouse, chopped it up into small pieces and then treated it with enzymes to separate the cells (Figure 8.11). The cells were then placed in cell culture with IL-2 for a period of about a week.

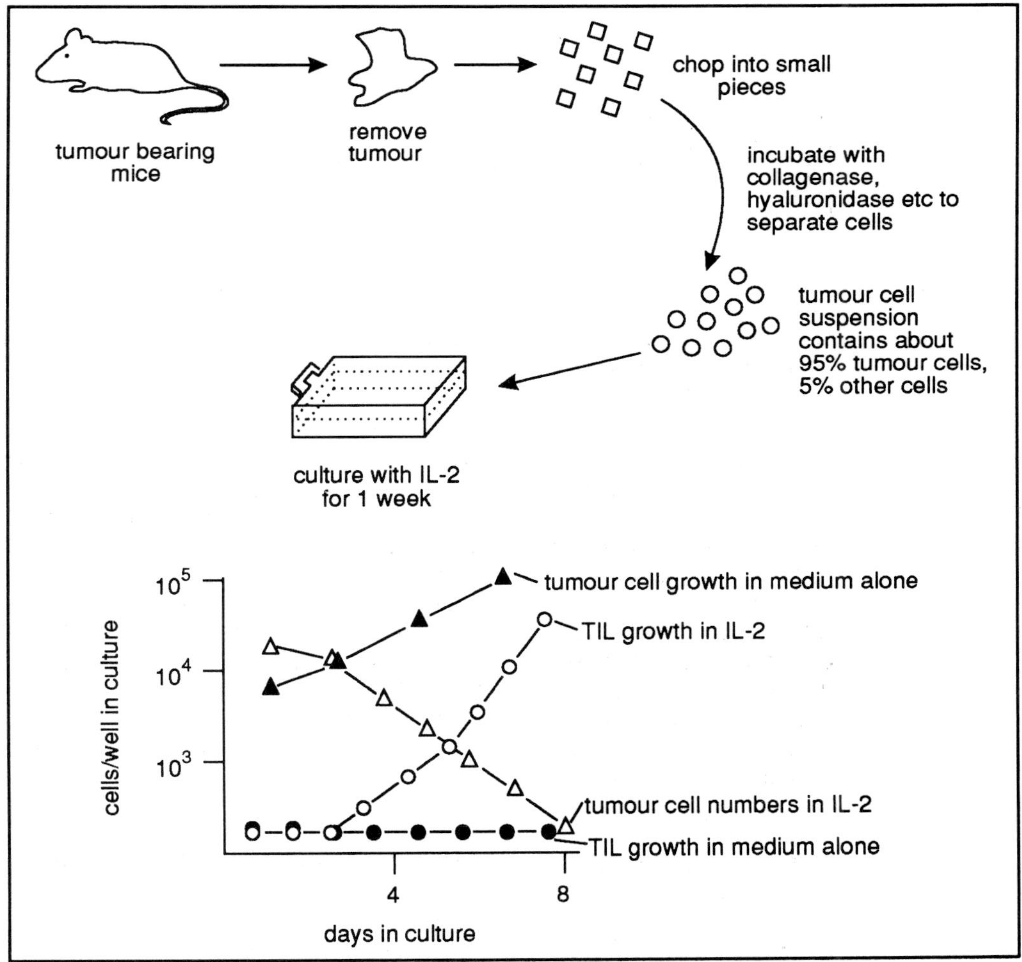

Figure 8.11 Production of tumour infiltrating lymphocytes from murine tumours showing the death of tumour cells and expansion of TIL in the presence of IL-2 (see text for details).

Π What would you expect to happen over this period?

When the cultures were examined after this culture period, it was found that almost all the remaining living cells were lymphocytes and the majority of these were cytotoxic T cells with specificity for the tumour. So the T cells, under the influence of IL-2, had proliferated in culture and had progressively killed all the tumour cells.

The TIL were then expanded further in IL-2 over a period of another 2 weeks and they were then injected into mice with established lung metastases. In many mice, these tumour growths were eliminated. The interesting point was that the number of TIL which resulted in effective treatment was about 100 times less than the number of LAK cells needed to obtain a similar result. When the TIL were subsequently examined for their anti-tumour effects on large metastatic growths in mice, it was found that they were not effective. However, if the drug cyclophosphamide was also given, the TILs successfully eliminated the tumours. It is thought that this drug eliminates suppressor T cells which would block the activity of the TIL. As with LAK cells, TIL did not appear to be effective against all cancers.

Rosenberg's group then obtained TILs from cancer patients, expanded the TILs in IL-2 and refused the patients with these expanded TILs and with IL-2. Treatment of the first 20 patients with melanoma was reported on in 1988. Over half the patients responded well to this therapy showing complete or partial remission. Thus, it would seem that cell based therapy using TIL's will probably be effective with further development. This will include the introduction of cytokine genes into the TIL to make them more effective.

SAQ 8.9

Complete the table below indicating your choice with a + or -.

Characteristic	TIL	LAK
1) T cells		
2) Antigen specific		
3) Expanded in IL-2		
4) Natural killer cells		
5) Kills more than one type of tumour cell		
6) MHC restricted		
7) Effective against all cancer		

Summary and objectives

Having completed this chapter you will now be familiar with most of the recent developments in the use of antibodies as immunoreagents. We first described their immense importance in clinical laboratories and in research and then looked at their potential uses mainly as immunotherapeutic reagents for the treatment of cancer and infectious diseases and you will now be familiar with such products as bispecific antibodies, recombinant globulins and immunotoxins. We also examined the usefulness of cytokines in immunotherapy, the potential use of engineered cells transfected with cytokine genes and the progress made in adoptive immunotherapy using lymphokine activated killer cells and tumour infiltrating lymphocytes.

Now that you have completed this chapter you should be able to:

- list the diagnostic uses of antibodies in the clinical and research laboratories;
- give examples of antibodies used in disease therapies;
- describe the potential uses of bispecific antibodies in cancer and infectious diseases and describe the mechanisms involved;
- explain the term 'recombinant globulin' and list the uses of such products;
- explain how immunoadhesins can improve the potential of molecules such as CD4 and ICAM-1 as immunotherapeutic agents;
- describe the mechanism of action of holotoxins and the different approaches taken to produce viable immunotoxins from these molecules;
- list the most useful features of immunotoxins and the difficulties encountered in their use as immunotherapeutic reagents;
- compare the major structural features and properties of recombinant globulins and immunotoxins;
- distinguish between adoptive and active cellular immunotherapy;
- describe the production and uses of cytokine transfected tumour cells, LAK and TIL in the treatment of cancer.

Responses to SAQs

Responses to Chapter 1 SAQs

1.1 Items 2, 3, 4 and 6 would all be immunogenic in Man. Human albumin (item 1) and host lymphocytes (item 5) are self and not immunogenic. An ABO-compatible transfusion is not immunogenic because of the red cells since they are of the same blood group in donor and recipient, hence the phrase ABO-compatible. However, whole blood also contains white cells and, as we shall find out later, they carry on their surfaces molecules which would be immunogenic in another individual. *Streptococcus pneumoniae* (item 3), being a bacterium, is obviously immunogenic. The fact that the virus is killed (item 4) does not alter the fact that it possesses many proteins which are non-self to the host and which will induce the production of antibodies but no infection. Donor cells from another individual (item 6) will be immunogenic for the same reason as for item 2.

1.2 A conformational epitope 1) arises in an antigen due to molecular folding E). MHC I- peptides 2) are recognised by cytotoxic T cells D) and MHC II-peptides 4) are recognised by T helper cells B). A linear determinant 3) or epitope is represented by an amino acid sequence A) and an antibody 5) is antigen specific C).

1.3 The only statement which is incorrect is statement 3. Opsonisation which means 'make ready to eat' is the coating of antigen by C3b particles and not the subsequent uptake and killing.

1.4

Characteristic	MHC I	**MHC II**
1) Associates with β_2-microglobulin	+	-
2) 3 domains/chain	+	-
3) Polymorphism	+	+
4) Activates cytotoxic T cells	+	-
5) Binds peptides	+	+
6) Activates T helper cells	-	+
7) Humoral immunity connection	-	+ (a)
8) $\alpha\beta$ heterodimer	-	+
9) Expressed on APCs	+	+ (b)
10) Related to immunoglobulins	+	+ (c)

a) The connection is that T helper cells are required for the majority of B cell responses and these recognise MHC Class II-peptide complexes. b) One of the major characteristics of APCs is the expression of MHC II but since these cells are nucleated they also express MHC I molecules. c) From Figure 1.7 both MHC I and II molecules show homology with immunoglobulin constant domains and are part of the immunoglobulin superfamily.

1.5

1) Incorrect. The peptides would appear on MHC Class I molecules since the peptides are derived from endogenous antigens ie the mouse serum albumin would be produced inside the cell.

2) Correct.

3) Incorrect. There may not be a T cell with a receptor specific for the particular MHC-peptide combination. This proposal has led to the 'hole in the repertoire theory' which states that nobody possesses a full spectrum of T cells expressing receptors for every MHC-peptide combination.

4) Incorrect. Specificity is not a general characteristic of antigen presenting cells. Although B cells take up the antigen in an antigen specific manner through antibody receptors, other APCs use nonspecific methods such as through C3b and Fc receptors.

5) Incorrect. Since the same peptide may bind to more than one MHC II allelic product it can activate more than one T helper cell.

1.6 Did you manage to work this out? The principal is that if the antibody has two binding sites it can link antigen molecules together and if the complex is large enough it will precipitate out of solution. Only one of the 5 answers fulfils this requirement, that of IgG treated with pepsin 2) since the resulting $F(ab')_2$ fragment retains both binding sites. This can act like a whole antibody molecule and link antigens together.

1.7 The hinge region 1) contains many proline D) residues. The antigen combining site 2) consists of the variable domains of a heavy and light chain, V_H V_L C). A domain 3) is about 110 amino acids B) in length. $F(ab')_2$ consists of two antigen binding sites, 2 (V_H V_L E) linked together. The heavy chain 5) of each antibody class is an isotype A).

1.8

Characteristic	IgG	IgM	sIgA	IgD	IgE
1) No of domains/molecule	12	70	24	12	14
2) Possess allotypes (+/-)	+	-	+	-	- (a)
3) Isotype (Greek letter)	γ	μ	α	δ	ε
4) Monomers/molecule	1	5	2	1	1
5) Molecular weight (kD)	146	900	360	190	180 (b)
6) J chain ? (+/-)	-	+	+	-	-
7) Extended COO^- end (+/-)	-	+	+	+	- (c)

a) This obviously refers to heavy chain allotypes. If you put + on all five due to the possible presence of kappa light chains which possess allotypes, you would have been correct. b) We expected you to work out the molecular weight for sIgA which could have been done by assuming a monomeric molecular weight similar to IgG + J chain and secretory piece. c) As well as IgM and IgA, you will remember that there is an octapeptide on the end of the IgD heavy chains.

1.9

1) False. The greatest heterogeneity by far is due to idiotypes as they represent differences in antigen binding sites.

2) False. IgM and IgE are not known to possess allotypes but all 3 antibody classes are different isotypes.

3) True. An individual produces anti-Ids to his/her own antibodies, but they can also be generated in other individuals or other species.

4) False. The CDRs are within the paratope (binding site) which interacts with the epitope.

5) False. Allotypes have not been found on lambda light chains and therefore the last part of the statement is inaccurate.

1.10 Fab 1) matches with virus-receptor inhibition D) since this part of the antibody blocks this interaction.

IgE 2) matches with mast cell E) since the latter possesses high affinity receptors for IgE.

Opsonisation 3) matches with antigen-antibody complex B) since this process involves coating of antigen by antibodies.

IgG1 4) matches with neutrophil Fc receptor C) since neutrophils possess receptors which bind the Fc regions particularly of IgG1 and IgG3.

ADCC 5) matches with cell lysis A) since antibody dependent cellular cytotoxicity (ADCC) results in lysis of cells coated with antibodies.

1.11

1) False. As you can see from Figure 1.19 there are other ways of activating complement apart from antibodies. As we saw early in the chapter, some micro-organisms and C-reactive protein are two examples of other factors which can activate complement.

2) False. C3a does indeed activate mast cells but is not a chemotactic factor itself. This is a property of C5a.

3) True.

4) False. This process may also involve coating of antigens by antibodies.

5) True.

1.12

We are sure you managed to work this out. The question was to demonstrate that because of DNA rearrangement of gene segments a large number of antibodies can be encoded for by a small part of the genome. You will know from other biochemistry studies that each amino acid is encoded by 3 bases. It is then relatively simple to work out. You multiply the number of segments by the amino acids encoded by the segment multiplied by 3 (triplet codon).

Kappa chain genes:	V segments:	250 x 98 x 3 =	73 500
	J segments:	5 x 13 x 3 =	195
			Total 73 695
Heavy chain genes:	V segments:	1000 x 98 x 3 =	294 000
	D segments:	12 x 8 x 3 =	288
	J segments:	4 x 13 x 3 =	156
			Total 294 444

Grand total for both light and heavy chain genes = 368 139.

This figure as a % of the total genome (3.5×10^9 bp) is 0.0105.

Answer 4) 0.01% is the correct answer.

1.13

The one correct answer is 5). Let us look at the other responses.

1) It is the exons, not the introns, which encode the protein products.

2) VDJ joining only occurs in heavy chain genes, there being only a VJ joining in light chain genes.

3) The whole leader peptide is not encoded by the leader exon since the last 4 carboxyl terminal amino acids are encoded by the V segments.

4) From Figure 1.20 you can see that J segments encode longer sequences in heavy chains not light chains.

1.14

Somatic mutation 1) matches with immune response B) since this occurs in B cells which encounter antigen and not before. Junctional diversity 2) matches with CDR3 C) since it represents imprecise joining of the segments resulting in the formation of CDR3. N regions 3) are oligonucleotides D) inserted into the CDR3 region. $V\lambda2J\lambda4C\lambda4$ 4) is a nonfunctional rearrangement A) as $C\lambda4$ is a pseudogene.

1.15

1) Incorrect. Cytokines are produced in all responses not just primary responses.

2) Incorrect. As indicated in Figure 1.23 somatic mutation occurs later in the primary response and some of the B cells which have undergone somatic mutation then differentiate into plasma cells and produce antibodies of a different quality other than the germline antibodies.

3) Incorrect. Class switched progeny often undergo somatic mutation and the higher affinity cells are selected by antigen. Hence the specificity of the parent B cell is altered.

4) Correct.

5) Correct.

Responses to Chapter 2 SAQs

2.1 When you plotted the graph of r/c against r (Scatchard plot) the straight line would intercept the x axis (r) at 2.0 giving an antibody valency of 2.

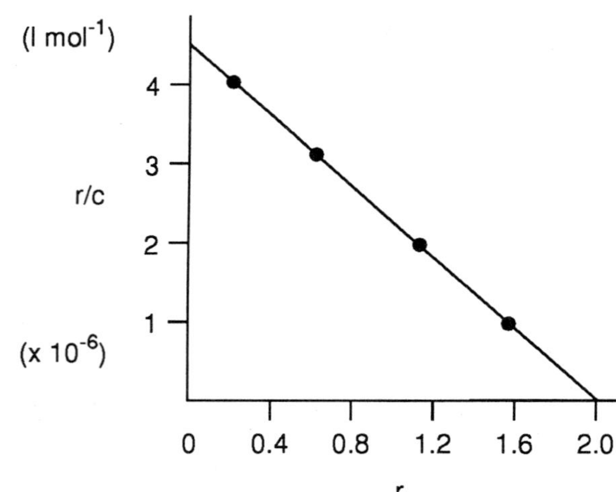

You then should have calculated the value of the slope ie $\Delta y/\Delta x$. This is most easily done by simply using the intercepts on the y and x axis. The value of r/c is about 4.4×10^6 and r = 2. K is then 2.2×10^6 1/mole.

2.2

1) Incorrect. The higher the K value, the higher the affinity.

2) Incorrect. Antibody affinity describes the binding strength of each binding site. Antibody avidity is the binding strength of the whole antibody. For IgG this would be the combined strength of both binding sites which, you will remember may be 1000 times higher than that of monovalent binding.

3) Incorrect. Since the ligand must be dialysable, this method is not applicable for large molecules.

4) Correct.

5) Incorrect. With high affinity antibodies, all the binding sites are occupied at low concentrations of ligand.

2.3 Since, at equivalence, all the antibodies and antigen are in the precipitate we can calculate the molar ratio quite easily. Using the figures given in the text, 0.393mg antibodies and 85μg albumin are in the precipitate. The molar ratio is calculated by dividing these weights by the molecular weights and directly comparing the results.

Molar ratio is $\frac{0.393 \text{ mg IgG}}{150\,000} : \frac{0.085 \text{ mg albumin}}{68\,000}$

2.62×10^{-6} IgG : 1.25×10^{-6} albumin

giving a molar ratio of 2.1 IgG molecules per molecule of albumin

2.4 After squaring the diameters of the precipitin rings, you should have plotted these against the concentrations producing a graph similar to that in Figure 2.7. Thus:

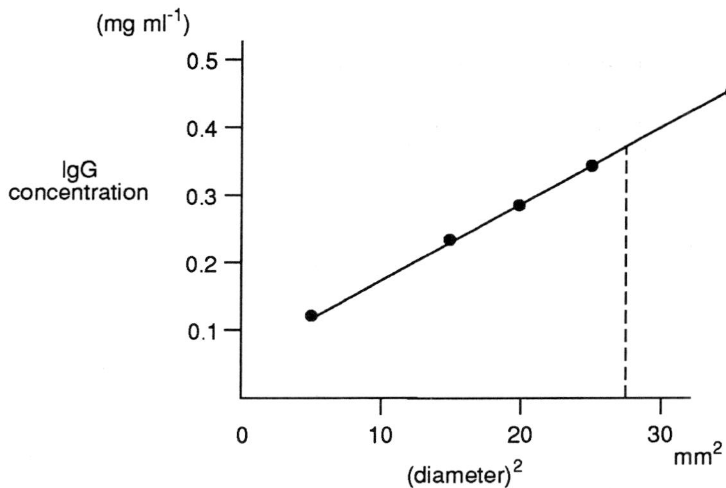

Since the unknown sample produced a ring of diameter 5.24, the square of this would be 27.46. From the graph this would be produced by an antigen concentration of 0.37mg ml^{-1}. Since the sample was diluted once in saline the original concentration was 0.74mg ml^{-1}.

2.5 This question would have helped you interpret Ouchterlony precipitin patterns.

1) Incorrect. The precipitin line is formed by the anti-DNP antibodies in the anti-DNP-IgG$_{hu}$ antiserum reacting with DNP epitopes on DNP-BSA. The antibodies are not specific for BSA.

2) Correct. The spur on this precipitin line pointing towards the DNP-IgGSh indicates epitopes present on human IgG which are not present on sheep IgG.

3) Correct. Antibodies directed against DNP epitopes form precipitin lines with both DNP-IgG$_{sh}$ and DNP-BSA because of the common DNP epitopes. The spur on line 4 shows that antibodies to human IgG in the antiserum and crossreactive with sheep IgG are precipitating the sheep IgG. This of course is nonidentical to BSA in line 5, hence a reaction of partial identity.

4) Correct, Since the lines are crossed and not fused this indicates no common epitopes on the two antigens.

5) Incorrect. The antiserum does not contain antibodies to BSA and there are no epitopes on human IgG which are common to BSA. We therefore observe no precipitation. Of course, if the BSA is extremely dilute then even if antibodies which react with BSA were present, a visible precipitate may not be produced.

2.6

Technique	Quantitative	Qualitative
1) Immunoelectrophoresis		+
2) Rocket IEP	+	
3) Mancini	+	
4) Ouchterlony		+
5) Counter immunoelectrophoresis		+
6) 2 dimensional IEP	+	
7) Nephelometry	+	

We should point out that all these methods can be used in a qualitative manner but some of the above methods are used principally for detecting antibodies or antigens (qualitative) while others measure the amounts of antigens or, in some cases, antibodies (quantitative).

2.7 Based on the information given in the question, the principal approach would be to mix the serum from one of the surviving mice with a suspension of the bacteria. If the serum contained antibodies to the bacteria then you would observe an agglutination reaction. This observation would not, of course, prove that anti-bacterial antibodies were directly responsible for the survival of the mice. To prove this you would have to inject some of the serum into fresh mice and then inject the bacteria. Survival of the mice would suggest that the serum was protective against the bacteria.

2.8

1) Wells A1-3 show no haemagglutination even though wells containing more dilute antibodies do. This is likely to be due to the prozone effect e) where excess antibodies prevent links between the red cells. Thus wells A1-3 match with e).

2) Wells D1-4 are indicative of a low titre anti-SRC antiserum c). You can calculate the titre as follows. The antiserum was diluted 1:10 initially and then serially diluted for 4 wells where haemagglutination was observed. The dilution in each of the 4 wells is 1:20, 1:40, 1:80 and 1:160. Thus the titre is 160.

3) Wells A1-8 are indicative of a high titre antiserum against SRC d) since it can be serially diluted from an initial 1:10 dilution 8 times. The dilutions would result in wells of 1:20, 1:40, 1:80, 1:160, 1:320, 1:640, 1:1280 and 1:2560. The titre of this antiserum is 2560 - a high titre antiserum.

4) Wells of C1-2. Based on the choice you have, this could represent the effects of a heterophile antiserum a). Many normal sera possess low levels of anti-SRBC antibodies which would be seen in 1 or 2 wells when serially diluted.

5) Wells B1-9 represents normal rabbit serum b) which contains no anti-SRC antibodies.

2.9 Dividing each experimental O.D value by the O.D for the 100% haemolysis tube would give you a set of figures for % haemolysis. If you had plotted these against the quantity of serum (complement) added you would have obtained a curve similar to that in Figure 2.15. Thus:

Tube number	% haemolysis	
1	100	100
2	$\frac{0.62}{0.69} \times 100$	89.8
3	$\frac{0.46}{0.69} \times 100$	66.7
4	$\frac{0.21}{0.69} \times 100$	30.4
5	$\frac{0.04}{0.69} \times 100$	5.8
6	$\frac{0.01}{0.69} \times 100$	1.4

Thus graphically:

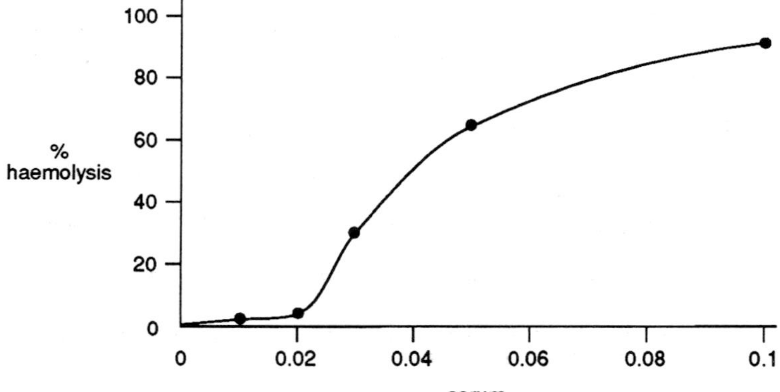

You could then determine the CH_{50} value by determining the amount of complement added which resulted in 50% haemolysis. You should have obtained a figure of about 0.04ml. Since the CH_{50} value can be reported as the amount of complement in one ml, the figure is about 25 units ml^{-1} (2).

2.10

1) This matches with complement fixation d).

2) IgG is measured in serum by the radial immunodiffusion method a).

3) Antibodies to influenza virus can be detected by haemagglutination inhibition e). This is because influenza virus expresses haemagglutinins on its surface which agglutinates some red cells. Antibodies specific for the haemagglutinins block (inhibit) this reaction.

4) Anti-albumin antibodies can be detected rapidly by counter immunoelectrophoresis b) because the albumin is negatively charged and migrates to the anode.

5) Antibodies to red cells are detected by the haemagglutination technique c).

2.11

1) Incorrect. Antibody is bound to the microtitre plate which then binds antigen and a separate antibody is then used to quantitate the antigen.

2) Incorrect. The competition for binding to antigen is between test antibodies and the antibody-enzyme conjugates for binding to antigen.

3) Incorrect. In the indirect antigen ELISA there is an extra layer of anti-globulin-enzyme conjugate which quantitates the antigen specific antibodies and therefore does not contain these antibodies.

4) Correct.

2.12

Dot-ELISA 1) matches with insoluble product d) as the reaction is carried out on a membrane such as nitrocellulose. Plate ELISA 2) matches with soluble product e) as the colour is read on a spectrophotometer. Elispot 3) matches with plasma cells a) since this is an ELISA adapted for detection of products such as antibodies secreted by plasma cells. Antibody capture 4) matches with solid phase antigen b) as the antibody in the sample becomes bound to the antigen attached to the solid phase and is then quantitated. Anti-globulin-enzyme 5) binds to the Fc regions c) of the antigen specific antibodies used in the enzyme immunoassays.

2.13

1) Incorrect. These assays only use radiolabelled antibodies.

2) Incorrect. Not all IRMA's use antibodies attached to the solid phase. For example, the antigen inhibition assay uses antigen fixed to the solid phase to capture antibodies in competition with free test antigen.

3) Correct.

4) Incorrect. Lactoperoxidase catalyses the iodination of tyrosine residues.

5) Correct. This reagent includes a tyrosine residue as part of its structure and this is attached to lysine residues or amino terminal residues in the protein thus increasing the number of tyrosines.

2.14

Avidin 1) matches with signal enhancement d) due to its multiple binding sites for biotin. For example, each avidin molecule can bind more than one biotin-enzyme conjugate thus increasing the colour intensity of the reaction.

IRMA 2) involves the use of radiolabelled antibodies a).

FITC 3) absorbs UV light e) and emits light of a different wavelength.

Immunoferritin 4) is an electron dense reagent c) used in electron microscopy.

FACS 5) separates cells b) based on their immunofluorescence.

2.15

All except 3) can be used in Western blotting. The substrate quoted in 3) results in a soluble product which could not be seen on a membrane.

Responses to Chapter 3 SAQs

3.1 The incorrect statement is 4). Members of an inbred stain can produce slightly different antibodies to a single antigen. It is useful to comment on 3). Not all antisera are monospecific; indeed some are not intended to be so. For instance, there are antisera to whole human serum which contain antibodies to a majority of the proteins present in human serum. This is useful since it can be used to detect protein deficiencies in pathological samples or major changes in the levels of some proteins detected by immunoelectrophoresis.

3.2 You may have thought that statement 1) might be correct or that antibodies produced in the rabbit against sheep IgG may cross react to some extent with sheep and human albumin. In fact, statement 1) is incorrect since there are no common epitopes between IgG and albumins. The multiple precipitin lines are most likely due to albumin impurities in the IgG fraction used to immunise the rabbit, so statement 2) is correct. Statement 3) is correct since the impurity in the sheep IgG would be sheep albumin not human albumin. The rabbit has produced antibodies to sheep albumin epitopes, some of which are common to human albumin. Statement 4) is obviously incorrect since the rabbit has not been immunised with human albumin. Statement 5) is likely to be incorrect since we have already implied that the anti-sheep albumin antibodies may not all cross react with human albumin. This statement would only be correct if all the sheep albumin epitopes were also present on human albumin.

3.3 Note that the question only asks you how to separate, the antibodies, you are not asked to purify them. Based on the text thus far, you could incubate the antisera mixture with sheep red cells and then centrifuge out the red cells. This would remove the anti-red cell antibodies. You could then prepare an affinity column of either DNP-KLH or human IgG and pass the red cell absorbed antisera mixture over the column which would result in retention of one set of antibodies leaving the remaining monospecific antibodies in the eluate.

3.4 MDP 1) matches with cytokines b) since this substance has been found to be the major stimulant of the inflammatory response.

Marginal ear vein 2) matches with test bleed d).

Mineral oils 3) matches with antigen depot e) since they cause slow release of antigen from a local deposit of immunogen following immunisation.

ISCOM 4) is an immunostimulatory complex made of saponin and antigen a).

Hyperimmunisation 5) results from multiple injections of an immunogen f).

Conjugate 6) matches with hapten and carrier c).

3.5

1) Correct. Since the absorption does not include elution of antibodies from a column using acid conditions it is a less harsh method.

2) Correct. It is good practice to test the mouse serum for specific antibodies before proceeding with a costly and laborious hybridisation.

3) Correct. This is because all the monoclonal antibodies are specific for the antigen whereas the globulin fraction of an antiserum invariably contains contaminating non specific antibodies and other non antibody proteins which will compete with the specific antibodies in the fraction for sites on the affinity column. This will lower the numbers of affinity sites able to capture the antigen.

4) Correct. This is because monoclonal antibodies are specific for one epitope which may have a limited distribution on the cells whereas antisera contain many different antibodies which will be able to bind many different epitopes and thus promote agglutination.

5) Incorrect. Plasma contains the clotting proteins and these are no longer present in serum.

3.6

1) Incorrect. Since you are giving antibodies these have a limited half life and therefore only give protection over a limited period.

2) Incorrect. Antibodies in xeno-antisera normally have shorter half lives than human IgG antibodies .

3) Incorrect. Quite a few of the older individuals in the population are probably already immune to Hepatitis A.

4) Incorrect. IgA antibodies have a rather short half life. In a breast fed infant, however, this is compensated by the supply of milk containing fresh IgA over the first year or so of life.

5) Correct.

3.7

1) This is not the ideal treatment since the patient should have been immediately given rabies immune globulin to destroy the virus. The post exposure treatment also involves a multiple injection regimen.

2) This treatment would induce long term resistance to polio. Some countries such as Sweden have almost entirely relied on the parenteral adminstration of the inactivated polio vaccine for protection. As we saw earlier, the intranasal administration of this vaccine does not provide extended protection but this vaccine induces circulating antibodies which block the spread of the virus from the point of entry in the gut to the central nervous system. On the other hand, the Sabin vaccine actually blocks viral entry in the gut through IgA.

3) This will not induce long term protection if the immunocompromised patients cannot make antibodies against the mumps virus or if there is a deficiency in the cytotoxic T cell response to the virus.

4) If administered at the same site the immune globulin may block the activation of B cells and T cells by covering the epitopes on the virions.

5) This treatment is the one routinely administered to children in developed countries resulting in long term protection.

3.8

1) Toxoid should be matched with inactivated neurotoxin c).

2) Antigenic drift represents minor sequence differences e) in H and N antigens of influenza virus.

3) Subunit vaccine represents an immunogenic sub component of a whole organism b).

4) Recombinant live vaccine which does not cause the symptoms of disease is an attenuated vector a).

5) Vaccinia is a viral immunogen g).

3.9

Statements 1) and 3) describe advantages of peptides as vaccines. Statement 2) is not correct as many peptides do not induce antibody responses without the help of carriers or special adjuvants. Statement 4) is not correct as many peptides either do not bind the MHC II or are not recognised by T helper cells. Statement 5) is not correct as antibodies induced by the peptides do not always bind to the native antigen due to conformational differences between peptide and native protein.

Responses to Chapter 4 SAQs

4.1

Chimaeric antibodies 1) match with e) since they consist of mouse variable domains and human constant domains. Monoclonal antibodies 2) are derived from hybridomas d). Reshaped antibodies 3) are human antibodies with grafted murine CDRs a). Polyclonal antibodies 4) are contained in antisera c). Phage antibodies 5) are composed of antibody fragments b), namely the variable domains.

4.2

The incorrect statement is 4). The HAT sensitive (HGPRT) mutants were made years ago and dispensed to many laboratories.

4.3

Examine Figure 4.4. T cells in the presence of anti-transferrin antibodies would grow in the presence of IL-2 since this is an experimental control. Thus the line on the graph represents a normal response of T cells to IL-2. Since T cells in the presence of antibody 1 exhibit similar growth characteristics, it is apparent that antibody 1 has no effect on the interaction of IL-2 with the IL-2 receptors and hence has no effect on the growth of the T cells. However, this does not mean that antibody 1 is not specific for IL-2 since it could bind an epitope not involved in binding of IL-2 to the receptor.

In contrast, antibody 2 totally inhibits the growth of the T cells and therefore blocks the effect of IL-2 and is likely to be specific for the part of IL-2 which binds to the receptor and is obviously IL-2 specific.

Neither antibody is likely to be specific for the receptor since the immunising antigen was IL-2.

Based on these arguments, the responses should be as follows:

	Antibody 1	**Antibody 2**
1)	-	+
2)	-	+
3)	- (?)	- (?)
4)	-	+
5)	?	+

4.4 A cell count of 200 in 4 squares means 50 cells per square. Since the volume of cell suspension on each square is 0.1 mm^3, the number of cells per cm^3 of culture fluid is 50 x 10^4. In 20 ml of medium the total number of cells is 10^7. To obtain 2 x 10^6 cells you would need 4 cm^3 of this cell suspension. The answer is 2)

4.5 From the data given you should have obtained the following figures for the mean IL-2 concentrations in μg ml^{-1}: 1/50 - 7.4; 1/100 - 3.7; 1/250 - 1.48; 1/500 - 0.74; 1/1000 - 0.37; 1/2500 - 0.148; 1/5000 - 0.074 and 1/10 000 - 0.037.

The mean optical densities of each row (a-g) are: row 2 - 2.111; row 3 - 2.178; row 4 - 1.943; row 5 - 1.439; row 6 - 0.906; row 7 - 0.434; row 8 - 0.233; row 9 - 0.167.

These concentrations and optical densities resulted in the following graph:

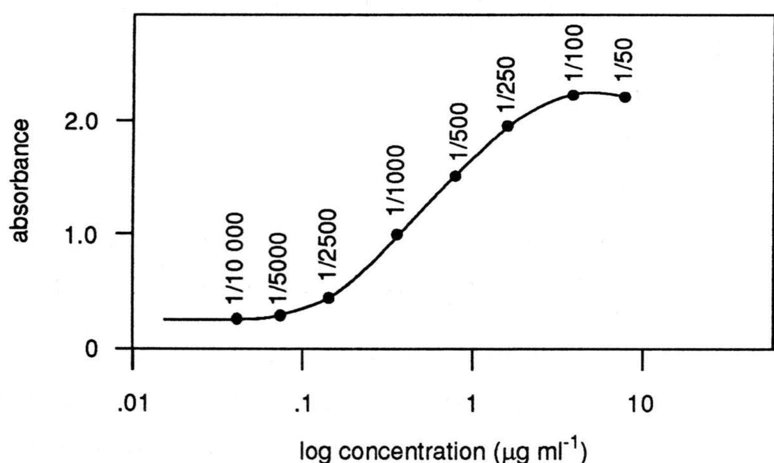

1) From the graph, since the optical density of X is 1.5 the IL-2 concentration of sample X is about 0.8 μg ml^{-1} - answer 3).

2) For the second calculation, the lower limit of sensitivity is 0.074 μg ml^{-1}. This is 0.074 x 10^{-6} g ml^{-1}. This figure divided by the molecular weight of IL-2 gives the molar concentration in mol ml^{-1}.

$$\frac{0.074 \times 10^{-6}}{15\,000}$$

that is 4.93 x 10^{-12}, ie 4.93 picomols ml^{-1} - answer 3).

4.6 The correct answer is 4). If the same mouse strain is not used, the hybridoma cells may be rejected by the host. The reasons for the other statements being incorrect are as follows.

1) Hybridomas can lose chromosomes and become non-producers.

2) The cloning procedure does not select for high affinity B cells. The presence of high affinity cells in the splenic population selected for fusion is dependent on the maturation of the immune response due to somatic hypermutation in the animal prior to the fusion procedure. The cloning procedure does not alter the relative numbers of low and high affinity hybrids generated during the fusion.

3) Ascites fluid contains much higher amounts of specific antibodies than serum in an immunised mouse since there are more cells producing the antibodies, and since they are immortal, they produce antibodies continually. In contrast, plasma cells in a normal spleen only produce antibodies for a relatively short time.

5) Feeder cells do not need to be from the strain of mouse - they can even be from a different species. For instance murine hybridoma clones can be supported by rat splenocytes.

4.7 The steps should have been arranged in the following order:

3) Immunisation.

8) Test serum for antibody.

9) Grow myeloma cells.

11) Pre-infusion immunisation.

1) Fusion.

2) Selection in HAT medium.

5) Observe development of hybrids using microscope.

10) Test supernatants.

7) Cloning by limiting dilution.

4) Expand the cultures.

6) Isotype the antibody.

4.8 The only correct statement is statement 4). The reasons why the other statements are incorrect are as follows.

1) Human hybrids generally produce more than one antibody as the fusion cell partners are antibody secretors.

2) Many EBV transformed B cells only produce antibodies for short periods.

3) Since most hybrids produce IgM it suggests that the B cells have not class switched.

5) Heterohybridomas are so called because they consist of cells from more than one species.

Responses to Chapter 5 SAQs

5.1 1) The best answer is non-covalent association, d) since Fv consists of non-covalently linked heavy and light chain variable domains.

2) The answer is e) since a chimaeric antibody is about two parts human and one part mouse since there are 4 variable domains (murine) and 8 constant domains (human) in a whole IgG molecule.

3) The match is a) since in single chain Fv (scFv) the variable domains of light and heavy chains are joined by a linker peptide.

4) dAb consists of a single type of domain, b), the variable domain.

5) Reshaped antibody matches with c) since it contains murine CDR's.

5.2 All the answers are correct except 5) since myeloma cells would not survive in the presence of these drugs unless they had been transfected with the pSV2 plasmids which confer resistance.

5.3 This question will have helped you to sort out the major features of producing reshaped antibodies. The list should be as follows:

Immunise mouse, 3).

Production of hybridomas, 9).

Extract murine mRNAs, 4).

Sequence murine V domain cDNAs, 8).

Synthesise oligonucleotides, 10).

Insert human V_H and V_L into M13 vectors, 5).

Site directed mutagenesis, 1).

Ligation into pSV2 vector, 2).

Transfect myeloma cells, 6).

Analyse reshaped antibodies, 7).

5.4

1) Incorrect. The main disadvantage is the loss of a long half life for Fab fragments derived from IgG antibodies.

2) Incorrect. The combinatorial approach espoused by Lerner generated random combinations of heavy and light chains in Fab fragments.

3) Incorrect. We have seen that repertoire cloning can be done from genomic DNA which does not require immunisation. It is difficult to conclude that the majority of the clones derived by reverse transcribing mRNAs are due to previous encounters with antigen since mRNA is necessary to synthesise IgM membrane receptors in the absence of antigen stimulation. Counter to this argument is the relatively recent dogma that not many B cells survive on exit from the bone marrow unless they contact antigen and clonally expand. So if you said this was correct based on similar arguments, you could be right. Having said that, multiple choice questions are always based on the best answer and there is definitely a correct statement (see 5)).

4) Incorrect. You would have to go back to your studies on antibody genes for the answer to this one. You will remember that a successful rearrangement only occurs on one of the two chromosomes.

5) Correct.

5.5 This is an open ended question. Here we will suggest some of the answers to these questions. It may well be that you have other correct answers which we have not thought of.

1) Based on our discussions on repertoire cloning, we mentioned successful experiments which had demonstrated the production of antibodies to self components by harvesting variable genes from B cells of un-immunised donors. So some B cells appear to be autoreactive although some of these specificities may have been generated due to the random pairing of heavy and light chain variable genes. However, we know that such autoreactive B cells do exist, they are unreactive to antigen possibly due to lack of T cell help or the presence of suppressor T cells. Repertoire cloning may therefore provide us with unique reagents directed against human components which were not available heretofore.

2) The point of this question is that germ line V segments in early B cells are separated by introns from both diversity and joining segments. It may indeed be interesting to try to generate antibody fragments encoded by these V segments and containing CDR1 and 2 only. It may be that we would find some of them still bind to some common antigens which we regularly encounter. You could also prepare oligonucleotides to replace the CDR3-framework 4 region resulting in newly engineered antibodies with unique specificities!

3) One of the attractions of repertoire cloning and phage display is that you will detect a number of antibodies to a single epitope. Mixing of these antibodies may result in an extremely effective polyclonal antibody reagent!

Responses to Chapter 6 SAQs

6.1

1) Incorrect. Since hybridoma supernatants contain only antibodies derived from a single clone, ie monoclonal antibodies, there is no need for such a purification step.

2) Incorrect. Protein A affinity purification separates total IgG and cannot distinguish between one antibody and another with regard to antigen specificity.

3) Incorrect. Since the fragments consist of the amino ends of the antibodies, the Fc region will be missing and it is the Fc region which binds protein A.

4) Correct.

5) Incorrect. This was difficult. Although this is true for antiserum and ascites fluid, it is not true for hybridoma supernatants containing foetal calf serum as a supplement since this does not contain any antibodies at all. Hence the only antibodies in hybridoma supernatants are the monoclonal antibodies.

6.2

This question was to help you identify the steps for obtaining various antibody fractions.

Crude serum antibody fraction, 1) results from ammonium sulphate precipitation, c).

Pure serum IgG, 2) is obtained from serum by ammonium sulphate precipitation followed by protein A or G affinity chromatograph, a).

IgM 3) is prepared by dialysis against water to produce the euglobulin fraction followed by gel filtration, d).

IgG with minor contaminants, 4) is prepared by ammonium sulphate precipitation and ion exchange chromatography, b).

6.3

The answer is 2) = 3.27 ml.

Let X = the volume of saturated ammonium sulphate that is used. This will contain X x 100 units of ammonium sulphate.

The final solution will have a volume of X + 4 ml and ammonium sulphate concentration of 45% saturation.

Thus this will contain (4 + X) x 45 units of ammonium sulphate. But since this ammonium sulphate has been derived from the saturated ammonium sulphate solution.

$(4 + X) \times 45 = X \times 100$. Thus X = 3.27 ml.

In other words 3.27 ml of saturated ammonium sulphate needs to be added to 4 ml of mouse serum to produce a 45% saturated solution.

6.4

The paragraph should read like this:

Anion exchange is a popular method for the purification of **IgG** from serum. In this method, **positively charged** groups covalently linked to an insoluble matrix have negatively charged **counter ions** which can be exchanged for **negatively charged** proteins which are being separated. If the protein, however, is **positively charged**, it will not bind to the matrix; this is the basis of the **batch** method for separation of IgG. In the **chromatographic** method, the IgG, along with all the other proteins in the sample, are **negatively charged** and adsorb to the column. By **increasing** the **ionic strength** or **lowering** the pH of the eluting buffer, you can selectively elute IgG from the column before the other proteins since antibodies possess more basic **isoelectric points** than the other serum proteins.

6.5

There are several following possibilities for producing pure IgM. Here we give two examples.

If you have an IgM producing hybridoma then this is the best source of pure IgM. You obviously cannot purify it on a protein A or G column as these methods are specific for IgG. However, you could purify it on an anti-IgM affinity column or even an antigen affinity column since you would know the specificity of the antibody. A cheaper method is to precipitate the IgM from serum free hybridoma supernatant with ammonium sulphate and then purify it by gel filtration.

6.6

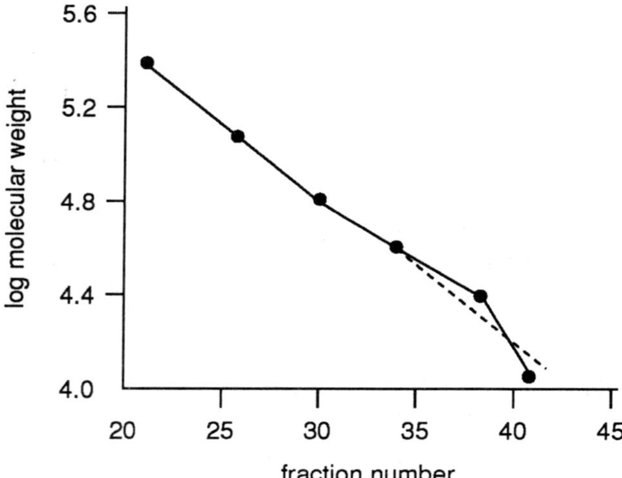

Having plotted the data (log molecular weight against fraction number) you would obtain a more-or-less straight line suggesting a linear relationship.

You can then see that a fraction number of 25 for the unknown antibody corresponds to log molecular weight 5.18 giving an apparent molecular weight of 150 000 D which is the molecular weight of IgG.

Calculated data:

Protein	Molecular weight	Log M. Wt	Fraction number
cytochrome C	12 400	4.09	41
chymotrypsinogen	25 000	4.40	38
ovalbumin	43 000	4.63	34
bovine serum albumin	66 000	4.82	30
lactic dehydrogenase	135 000	5.13	26
catalase	240 000	5.38	21
antibody protein			25

6.7

1) Incorrect. Although you would have been quite right in stating from Table 6.3 that human IgG3 does not bind to protein A, this fact is the basis for a purification protocol. Since the purified IgG fraction only contains the 4 subclasses, if you apply them to a column of protein A Sepharose, human IgG3 will not bind and can be recovered in the non binding fraction.

2) Incorrect. They bind with medium affinity to protein G.

3) Correct

4) Correct.

5) Incorrect. You have been provided with the capacity for protein A Sepharose in the text; this is about 10 - 20 mg antibody per 1 ml column. A column volume of 0.2 ml would have a maximum capacity of 4 mg. Although this would indeed bind the 1 mg of specific antibodies if they had been separated from the

background antibodies in serum, they would not all be bound on this column due to competition by the additional background antibodies which form at least 90% of the total IgG in the mouse serum.

You may also have concluded that this statement was incorrect since mouse IgG1 does not bind well to protein A. This would be an incorrect conclusion for the same reason as for statement 1).

6.8 This was a useful exercise for you to become familiar with the practical aspects and aims of affinity chromatography. Your list should read as follows:

6)	Prepare antiserum	g)	Produce anti-isotype Abs
10)	Check cross reactions	b)	Are the Abs monospecific?
5)	Affinity purification of anti-isotype Abs	i)	Eliminate crossreactivity
4)	Prepare affinity matrix	k)	Conjugate Abs to beads
1)	Precycle with normal serum	f)	Block high affinity Abs
11)	Apply sample to column	a)	Adsorption step
8)	Wash with neutral buffer	e)	Remove non specific binding
9)	Elute with acid buffer	c)	Recover first lot of Abs
2)	Elute with alkaline buffer	j)	Recover second lot of antibodies
7)	Dialyse and concentrate	d)	Make ready for storage

6.9

1) Incorrect. Since the IgG peak from gel filtration would contain some IgA (they are of similar molecular weight) and protein A binds human IgA, IgA would be a contaminant.

2) Incorrect. Antigen affinity chromatography is sometimes used especially when antigen is freely available. Purification of IgM monoclonals using this method is very effective since it is fairly difficult to totally purify IgM any other way and takes multiple steps.

3) Incorrect. Pre-cycling of antigen affinity columns is not done with normal serum but with the acid eluting buffer such as glycine HCl to remove any proteins which are loosely bound or which have leached off the column during storage. Treatment with normal serum, you will remember, is performed to block high affinity antibodies used in affinity purification of isotypes.

4) Correct. This procedure usually results in more than 95% pure Mabs.

5) Incorrect. Sometimes antibodies can be eluted using less extreme pH conditions. In other instances, buffers of extreme acid and alkaline pH fail to remove the antibodies and you may have to use chaotropic agents or agents such as high molar urea or guanidine HCl.

6.10 Having done this question, you should be quite familiar with current attempts to induce the bacteria to produce functional antibody fragments instead of having to assemble them *in vitro*. All the statements are supported by the evidence in the figure, but let us go through each one to emphasise the reasons for such conclusions.

1) Functional Fv's are not created in the cytoplasm since it is a reducing environment thus preventing the formation of most disulphide bridges in the products.

2) Since the periplasm is an oxidising environment, the conditions are right for assemblage of antibody fragments.

3) Since the antibody fragments are already assembled there is no need for the denaturing reagents which promote disassemblage of component chains so they can be reassembled properly in a subsequent step.

4) You should expect this, since firstly, many fragments in inclusion bodies are never recovered as functional antibodies and secondly, the multi-step processing *in vitro* will result in losses as well.

5) This is correct so there has to be a suitable cleavage site between the haemolysin signal peptide and the antibody chain which can be used to recover the functional antibodies *in vitro*.

6.11 We often use recipes for various reagents in the laboratory without much thought as to why the individual components were added. We though it might be useful for you to think of reasons, most of which are straightforward, for some of these additions since it will help when you develop new recipes for reagents.

1) Urea is a solubilisation reagent which helps to prevent aggregation and non covalent associations between molecules so that the individual proteins move separately into the resolving gel.

2) Glycerol is added to the sample buffer to increase the density of the sample so that when the sample is pipetted into the well under the electrode buffer, it settles at the bottom of the well and is not dispersed.

3) Bromophenol blue is a tracker dye which is negatively charged and of lower molecular weight than any of the proteins being separated. It therefore moves through the gel slightly ahead of the protein of lowest molecular weight and marks the so called dye front; when it reaches the bottom of the resolving gel, the electrophoresis is complete.

4) The glycine is essential for the stacking process since glycinate ions help, with chloride ions, to stack the proteins into a highly dense narrow band so that all proteins enter the resolving gel together.

Responses to Chapter 7 SAQs

7.1

1) You cannot make this conclusion based on the evidence presented. The cells from un-immunised guinea pigs were treated with supernatant from immune cells and under these conditions responded to both MIF and MF. The control experiment for such a conclusion would be to put the lymphocytes from the un-immunised guinea pig in cell culture and then test the supernatant for the presence of mitogenic factor. This experiment is not shown and no conclusion can be drawn. Obviously, this experiment was reported in the paper from which the data were taken and no mitogenic factor was found to be produced by the un-stimulated lymphocytes which is what you would expect.

2) You could not make this conclusion since the data suggest that the cells need to be re-stimulated with antigen.

3) This is correct.

4) This is an incorrect conclusion since this experiment only demonstrates that a mitogenic effect appears to be exerted by some cells in the PEC. There is no identification of a cytokine being released by the cells which could be responsible for the MIF effects. This is demonstrated by the experiment using the supernatant.

5) This is not strictly correct, although you may well have been led to believe that T cells were involved based on the labelling of the diagram. However, the data does not identify T cells as the producer of MIF and MF since there was no attempt to separate the T cells from the other cells.

7.2 Although you may well have little knowledge of individual cytokines this early in the chapter, this question would help you to appreciate firstly, the wide spectrum of activities cytokines are involved in, secondly, the likelihood that different cytokines may have some common properties and thirdly, that most immune activities involve multiple cytokines.

IL-3, IL-7 and IL-11, 1) matches with d) since these are involved in haematopoiesis. IFN-γ and IL-4, 2) matches with c) since both these cytokines up-regulate MHC II expression. Although this combination could also match other items, there is no other group of cytokines which possesses this ability. IL-2, IL-4 and IL-6, 3) are all involved in the growth of T cells, e). IL-4, IL-5, IL-6 and IL-13, 4) all mediate effects on antibody synthesis, a), most of them being involved in class switching mechanisms. IL-1, IL-6 and TNF-α, 5) all have major roles in inflammatory responses which also involves tissue trauma, b).

7.3

1) Incorrect. Since many cytokines are produced by non T cells, this statement cannot be correct.

2) Incorrect. Pleiotropism means that a cytokine acts on many cell types.

3) Incorrect. Few cytokines act in this way. Most act in an autocrine fashion, that is, on the producer cell, or in a paracrine fashion, that is, on neighbouring cells.

4) Correct.

5) Incorrect. It is not sufficient to have purified a cytokine to homogeneity, it must have been cloned and sequenced, the nucleotide and inferred amino acid sequences must be different to other interleukins.

7.4

1) TGF-β since it is generally anti-proliferative and causes class switching to IgA which is the antibody found on mucosal surfaces.

2) The 5 interleukins produced by T cells which promote B cell responses are IL-2, IL-4, IL-5, IL-6 and IL-13 although we still know little about the role of the last one.

3) Interleukins which exert their major effects during haematopoiesis are IL-3, IL-7 IL-9 (not very well documented yet) and IL-11.

4) Interleukin 2 is the cytokine which supports the long term growth of T cells but IL-1 is also needed for its production.

5) There are two cytokines which up-regulate MHC II, namely IL-4 and IFN-γ. However, only IL-4 acts on B cells whereas IFN-γ actually inhibits MHC II expression on these cells.

7.5 The table should have been completed as follows:

Activity	IL-1	IL-2	IL-3	IL-4	IL-5	IL-6	IL-8	IL-10	IFN-γ
1) MHC expression				+					+
2) T cell activation	+	+	+			+		+	
3) Chemotactic factor							+		
4) T/NK product		+							+
5) Ab production	+	+	+	+	+	+		+	
6) Macrophage product	+					+	+		

7.6 The specific activity calculation is straightforward since you only have to determine the units of biological activity per μg of material.

Specific activity of crude material: $\dfrac{3.9 \times 10^6}{38\ 000} = 102.63$ units μg^{-1}

Specific activity of purified material: $\dfrac{4.4 \times 10^6}{40} = 110\ 000$ units μg^{-1}

Recovery is determined by working out the total activity in the purified sample expressed as a percentage of the total activity in the crude serum.

Total activity in crude serum: $102.63 \times 38\ 000 \times 1000 = 3.89 \times 10^9$ units

Total activity in purified serum: $110\ 000 \times 40 \times 5 = 2.2 \times 10^7$ units

If you worked out these and divided the activity in the purified sample by that in the crude sample and multiplied by 100 to express as a percentage you would find the recovery to be 0.56%. This demonstrates to you the very high loss of cytokine during these tedious purification steps.

The purification factor is simply the specific activity of the purified fraction divided by that of the crude fraction:

$\dfrac{110\ 000}{102.63} = 1072$ times

7.7 This question describes the sort of results which were obtained in the early days when IL-2's from different species were characterised and you would have had to make some decisions as to what cytokine you had purified.

1) This could be correct since the presence of the 24 kD component could be explained by another cytokine with ability to support T cell growth. Although the T cell growth assay was thought to indicate only IL-2, more recently it has been found that IL-4 can under some experimental conditions give a positive result and is about that size (see Table 7.2).

2) The heterogeneity seen in isoelectric focusing could be due to multiple cytokines. It could also be due to post synthetic modifications which are common with cytokines. The addition of sialic acid moieties results in heterogeneity in the IEF spectrum.

3) This is highly unlikely since the presence of IL-2 is defined by the biological activity, ie supporting the growth of T cells.

7.8

1) Correct. As with B cell hybridomas, the unfused tumour cells do not survive in HAT and BW5147 cells do not have the HGPRT enzyme.

2) Incorrect. T cell clones are antigen specific but most produce more than one cytokine.

3) Incorrect. You cannot make this conclusion using biological assays alone. However, if you can calculate the specific activity during your purification procedure, say, in activity units per unit weight of recovered material then you obtain at least an indication of the degree of purity.

4) Correct. As we emphasised in the text, because you have to stimulate most cell lines to produce the cytokine, you are adding mitogens which could induce proliferative activity on the cells used in the bioassay and hence affect the bioassays of the cytokines.

5) Incorrect. T cell cloning is obviously not a good approach for production of IL-2 **in bulk** since you have to use this cytokine to support the growth of the cells. In most cases, cell lines produce more cytokine than T cell clones but development of a T cell clone may be easier to produce a particular cytokine than finding a suitable cell line. Additionally, the question does not emphasise bulk production of cytokines and the statement is therefore incorrect since we get a lot of information from studying production of cytokines (even small amounts) by T cell clones.

7.9

The list should have been in this order: you will notice that every step is not necessarily included.

9) stimulate cells.

12) oligo-dT chromatography.

4) gel electrophoresis of RNA.

11) micro-inject oocytes.

5) synthesise cDNAs.

6) ligate into vectors.

1) replicate on nitrocellulose filters.

7) hybridise.

8) prepare autoradiographs.

2) identify clones/plaques.

10) sequence cDNA.

3) express cDNA.

7.10

1) Hybridisation translation involves the production of the cytokine in *Xenopus* oocytes injected with mRNA. Thus this matches with d).

2) In order to perform PCR you need to know the nucleotide sequence of the borders of the gene b).

3) Once the sequence of cytokine cDNA has been determined for a cytokine, this cDNA can then be used as a radiolabelled probe to detect by hybridisation the cytokine cDNA in the library of another species. Hence this matches with a).

4) The amino acid sequences of tryptic peptides can be used to synthesise oligonucleotides which can be used as probes to search for the cytokine cDNA using hybridisation methods. This then matches with c).

7.11 First, you find out how many picomols of the cytokine are bound to the total cells by dividing the bound dpm by the specific activity of the radiolabel.

No of picomols attached to cells = $\frac{4500}{1.7 \times 10^6}$ picomols

which is: $\frac{4500 \times 10^{-12}}{1.7 \times 10^6}$ mol

Since Avagadros number is the number of cytokine molecules per mol then the number of cytokine molecules attached to the total cells is:

$$\frac{4500 \times 10^{-12} \times 6 \times 10^{23}}{1.7 \times 10^6}$$

If you now divide this by the number of cells, you will obtain the number of cytokine receptors per cell.

$$\frac{\frac{4500 \times 10^{-12} \times 6 \times 10^{23}}{1.7 \times 10^6}}{4.5 \times 10^5} = 3529 \text{ cytokine receptors per cell}$$

It is obviously a useful exercise because it is applicable to determining the number of receptors on any type of cell.

7.12

1) Incorrect. Since radiolabelled IL-2 is covalently bound to both receptor chains of the IL-2R an anti-IL-2 antibody will precipitate both receptor chains. However, antibodies directed against the p55 chain only will only precipitate the p55-IL-2 complex and not the p75-IL-2 complex. Antibodies directed against p55 cannot therefore substitute for the anti-IL-2 antibodies.

2) Incorrect. The molecular weight of the p75-IL-2 complex is 90 kD. Since this includes the 15 kD IL-2, the receptor chain has a molecular weight of about 75 kD.

3) Correct.

4) Correct.

5) Incorrect. The cell blot assay only detects the presence of the cytokine not its biological activity.

7.13 These were quite straightforward and you should have had no difficulty matching the correct pairs.

Northern analysis, 1) matches with electrophoresis, b). MAPPing, 2) matches with PCR, a). RIA, 3) matches with radiolabelled cytokines, d). ELISA, 4) matches with enzyme anti-cytokine antibodies, e). *In situ* hybridisation, 5) matches with tissue section, c).

7.14

1) This best describes mRNA assays since you can detect cytokine production from single cells using, for instance the MAPPing technique. There are some cases where the ELISA (a binding assay) can also detect single cell production such as in the Elispot assay.

2) This is an mRNA assay, specifically, the *in situ* hybridisation technique.

3) Cytokines in body fluids are best detected by binding assays, eg the ELISA, RIA or IRMA.

4) T cell growth is a biological function and demands a bioassay.

5) Since immunoreactive cytokines indicate a cytokine product these would be assayed using binding assays.

Responses to Chapter 8 SAQs

8.1

1) Correct. Anti-idiotype antibodies bind to antigen specific T cells or B cells and, therefore, these anti-idiotype antibodies may be used to target the destruction of specific T or B cells.

2) Incorrect. Although these antibodies are being used for this purpose, they are not ideal since all T cells are eliminated leaving the patient susceptible to infection and cancer.

3) Correct.

4) Incorrect. IgG antibodies can be measured this way but IgE antibodies are only present in very low amounts in serum and cannot be detected by this method.

5) Incorrect. Although most antibodies do reflect the past history of disease, some do not. For instance you have A, B blood group antibodies which are not produced in response to a previous infection and mothers may have anti-paternal HLA antibodies in their bloodstream due to previous pregnancies.

8.2

	Type of cell targeted to tumour		
Component/process involved	**T cell**	**Macrophage**	**NK**
1) Anti-CD3	+	-	-
2) ADCC	-	+	+
3) Anti-CD2	+	-	+
4) Anti-TCR	+	-	-
5) Anti-tumour cell	+	+	+
6) Anti-FcR	-	+	+
7) MHC restricted	-	-	-
8) Cytotoxicity	+	+	+
9) Phagocytosis	-	+	-
10) Cytokine production	+	+	+

We will give you two examples as to how you can work these out. If a bispecific antibody has anti-CD3 activity, it will bind with cells with CD3 surface components. These are T cells (item 1). Phagocytosis (item 9) is a feature only of macrophages, T cells and NK cells do not exert their cytotoxic effects by phagocytosis.

8.3

1) Incorrect. The fusion is of the order of 10^{-4}.

2) Incorrect. If you work out the number of possible different antibodies you will realise that the bispecific antibody of choice will be in a minority. In successfully fused cells you could possibly detect the following products: either the monospecific anti-tumour antibody or the monospecific anti-T cell antibody or any combination of the types of heavy chains and the two types of light chains.

3) Correct. Because it is in a minority and possesses both specificities, it would be very difficult to purify.

4) Correct. Some antibodies will result from pairing of heavy and light chains with specificity for the T cell only.

5) Correct. For a similar reason to 4).

8.4

The statement should read as follows:

Recombinant globulins are molecules composed of **constant domains** of antibody molecules attached to other proteins. **Immunoadhesins** are one form of recombinant globulins which target cells for **cytotoxicity** or phagocytosis. CD4-immunoadhesin extends the **half life** of CD4 and also promotes the **killing** of cells infected with virus which enter the cell through the CD4 molecule. ICAM-1 immunoadhesin prevents the binding of malaria parasite infected erythrocytes to venule endothelium through the ICAM-1 molecule and promotes destruction of the infected red cells and the parasite by **Fc receptor bearing** cells such as **macrophages.**

8.5 Ricin alone 1) would kill the tumour cells and any other cells present. The procedure described in 2) would also be effective, the idea being that the attachment of B chains to A-IM by the anti-A antibodies would enhance uptake of the immunotoxin. We would expect 3) to work since the bispecific antibodies would bind the A chains to the tumour cells. However, the A chain alone may not be very cytotoxic. The protocol described in 4) would not work since the A chains have no mechanism for attaching to the tumour cells since the antibodies are not linked. The protocol in 5) would result in killing of the tumour cells, the presence of the anti-B chain antibodies is unlikely to affect the outcome.

8.6

1) Incorrect. Although all B cells are killed, new B cells, which can replace the normal B cells killed, are constantly being made whereas the tumour cells would not be replaced.

2) Incorrect. The major side effect is uptake of the immunotoxin by non-target cells due to Fc receptor interactions.

3) Correct. The toxin must gain entry into the cell by receptor mediated endocytosis.

4) Correct. For instance, CD19 is not expressed on stem cells and so anti-CD19 can be used in an immunotoxin to eliminate B cell cancers. Although this will also kill normal B cells, these will be replaced by CD19 negative stem cells. The CD19 molecule is expressed once the stem cell has partly developed into a B cell.

5) Correct. This immunotoxin is more selective since it will only kill T cells which have become activated by the foreign transplant sparing nonactivated T cells.

8.7 This question was to help you remember the differences between recombinant globulins and immunotoxins and you should have succeeded with most of the answers. We have assumed that the majority of immunotoxins are composed of antibody fragments and hence do not possess the Fc end of antibodies. You will also notice that whereas recombinant globulins always have an antibody component, immunotoxins do not, for instance, in IL-2 toxin the cytokine is used to target the cell.

Characteristic	Recombinant globulin	Immunotoxin
1) Single chain constructs	+	+
2) Many are antigen specific	-	+
3) Immunoadhesins	+	-
4) Fc receptor involved	+	-
5) Protein A purification	+	-
6) Variable regions	-	+
7) Genetically engineered	+	+
8) Needs endocytosis	-	+
9) May be of non-antibody constriction	-	+

8.8

1) Incorrect. In adoptive immunotherapy, T cells are extracted from the tumour mass and expanded *in vitro* using IL-2.

2) Incorrect. Figure 8.9 does show that the mice are only immune to tumour A. This is not due to IL-2 being antigen specific but because the mouse was first treated with IL-2 cDNA transfected tumour A cells. This induced the rapid expansion of tumour A specific T cells under the influence of IL-2. These cells subsequently killed the normal tumour A cells injected 6 weeks later. They could not kill tumour B cells as they are only specific for tumour A cells.

3) Correct. Since the tumour cell constitutively produces the cytokine in large amounts it promotes its own destruction.

4) Incorrect. There are instances where IL-2 alone has been effective against some cancers.

5) Correct. Examination of Figures 8.8 and 8.9 would lead you to this conclusion.

8.9

Characteristic	TIL	LAK
1) T cells	+	-
2) Tumour specific	+	-
3) Expanded in IL-2	+	+
4) Natural killer cells	-	+
5) Kills more than one type of tumour cell	-	+
6) MHC restricted	+	-
7) Effective against all cancers	-	-

Suggestions for Further Reading

BIOTOL series, Cellular Interactions and Immumobiology, Butterworth-Heinemann, Oxford, ISBN 0 7506 0564 2.

Benjamini E. & Leskowitz S., Immunology - a short course, Wiley, Chichester, 1991, ISBN 0 471 567515.

Bryant, N.J., Laboratory Immunology and Serology (3rd Edition), Saunders & Baillére Tindall, London, 1992, ISBN 0 721 64212 8.

Clements, M.J., Cytokines, Blackwell Scientific Publications, Oxford, 1991, ISBN 1 872 74870 8.

Coligan, J., Kruisbeek, A.M., Margulies, D.H., Shevach, E.M. & Strober, W., Current Protocols in Immunology, Wiley, Chichester, (on going), ISBN 0 471 52276 7.

Coutinho A. & Kazatchkine, M., Autoimmunity, Physiology and Disease, Wiley, Chichester, 1994, ISBN 0 471 59227 7.

Ferencik, M., Handbook of Immunochemistry, Chapman Hall, 1993, ISBN 0 412 359 80.

Golub E.S. & Green D.R., Immunology: a synthesis, Freeman, Oxford, 1991, ISBN 0 878 93 263 1.

Herzenberg, L.A., Haughton, G., and Rajewsky, K., CD5 B Cells in Development and Disease, The New York Academy of Sciences, London, 1992, ISBN 0 897 66701 8.

Kerr M.A. & Thorpe R., Immunochemistry Labfax, Blackwell Scientific Publications, Oxford.

Kuby, J., Immunology, Freeman, Oxford, 1992, ISBN 0 878 92 2257 7.

Ollier, W. & Symmons, D.P.M., Autoimmunity, Blackwell Scientific Publications, Oxford, 1992, ISBN 1 872 74850 3.

Paul W.E., Immunology, Recognition and Response, Freeman, Oxford, 1991, ISBN 0 878 9223 2.

St. Georgiev, V & Yamaguchi, H., Immonomodulating Drugs, The New York Academy of Sciences, London, 1993, ISBN 0 897 66771 9.

Stuart, T.H.M., Cellular Immune Mechanisms and Tumour Dormancy, CRC, London, 1992, ISBN 0 8493 4520 0.

Up-to-date articles may also be found in a wide range of journals dedicated to immunology and to related topics including cancer, virology, microbiology, cell biology, biochemistry and biotechnology.

Index

A

D

E

I

J

K

L

M

N

O

P

S